VOLUME ONE HUNDRED AND TWENTY NINE

Advances in
COMPUTERS
Perspective of DNA Computing in Computer Science

VOLUME ONE HUNDRED AND TWENTY NINE

ADVANCES IN
COMPUTERS

Perspective of DNA Computing in Computer Science

Edited by

SUYEL NAMASUDRA

Department of Computer Science and Engineering,
National Institute of Technology Agartala,
Tripura, India

ACADEMIC PRESS

An imprint of Elsevier

ELSEVIER

Academic Press is an imprint of Elsevier
50 Hampshire Street, 5th Floor, Cambridge, MA 02139, United States
525 B Street, Suite 1650, San Diego, CA 92101, United States
The Boulevard, Langford Lane, Kidlington, Oxford OX5 1GB, United Kingdom
125 London Wall, London, EC2Y 5AS, United Kingdom

First edition 2023

ISBN: 978-0-323-85546-4
ISSN: 0065-2458

For information on all Academic Press publications
visit our website at https://www.elsevier.com/books-and-journals

Publisher: Zoe Kruze
Developmental Editor: Nadia Santos
Production Project Manager: Sudharshini Renganathan
Cover Designer: Mark Rogers
Typeset by STRAIVE, India

Working together
to grow libraries in
developing countries

www.elsevier.com • www.bookaid.org

Contents

4. A novel image encryption and decryption scheme by using DNA computing   129
Chiranjeev Bhaya, Arup Kumar Pal, and SK Hafizul Islam

5. Hiding information in an image using DNA cryptography   173
P. Bharathi Devi, P. Ravindra, and R. Kiran Kumar

6. The design of a S-box based on DNA computing and chaos theories   211
Jun Peng, Shangzhu Jin, Yingxu Wang, Xi Zheng, and Xiangren Wang

Contributors

Divyansh Agrawal
School of Computing Science and Engineering, Galgotias University, Greater Noida, India

Chiranjeev Bhaya
Department of Computer Science and Engineering, Indian Institute of Technology (Indian School of Mines), Dhanbad, Jharkhand, India

Francisco M. Couto
LASIGE, Departamento de Informática, Faculdade de Ciências, Universidade de Lisboa, Lisbon, Portugal

Jérémie Decouchant
Faculty of Electrical Engineering, Mathematics and Computer Science, Delft University of Technology, Delft, Netherlands

P. Bharathi Devi
Department of Computer Science, S.K.B.R. Government Degree College, Macherla, Andhra Pradesh, India

Maria Fernandes
SnT-University of Luxembourg, Esch-sur-Alzette, Luxembourg

Jiechao Gao
Department of Computer Science, University of Virginia, Charlottesville, VA, United States

Sathish Gunasekaran
Verena Haptic & VR Systems Pvt Ltd, Chennai, Tamil Nadu, India

SK Hafizul Islam
Department of Computer Science and Engineering, Indian Institute of Information Technology, Kalyani, West Bengal, India

Shangzhu Jin
School of Intelligent Technology and Engineering; Informatization Office, Chongqing University of Science and Technology, Chongqing, China

Manish Kumar
Department of Mathematics, Birla Institute of Technology and Science-Pilani, Hyderabad, Telangana, India

R. Kiran Kumar
Department of Computer Science, Krishna University, Machilipatnam, Andhra Pradesh, India

Tarun Kumar
Department of Computer Science and Engineering, National Institute of Technology Patna, Bihar, India

Sachin Minocha
School of Computing Science and Engineering, Galgotias University, Greater Noida, India

Suyel Namasudra
Department of Computer Science and Engineering, National Institute of Technology Agartala, Tripura, India

Arup Kumar Pal
Department of Computer Science and Engineering, Indian Institute of Technology (Indian School of Mines), Dhanbad, Jharkhand, India

Jun Peng
School of Mathematics, Physics and Data Science; School of Intelligent Technology and Engineering, Chongqing University of Science and Technology, Chongqing, China

Manojkumar Ramteke
Department of Chemical Engineering, Indian Institute of Technology Delhi, New Delhi, India

P. Ravindra
Department of Computer Science, K.B.N. College, Vijayawada, Andhra Pradesh, India

Deepak Sharma
Department of Chemical Engineering, Indian Institute of Technology Delhi, New Delhi, India

Xiangren Wang
School of Intelligent Technology and Engineering, Chongqing University of Science and Technology, Chongqing, China

Yingxu Wang
International Institute of Cognitive Informatics and Cognitive Computing (ICICC), Department of Electrical and Software Engineering, Schulich School of Engineering and Hotchkiss Brain Institute, University of Calgary, Calgary, AB, Canada

Tiange Xie
Institute of Information Engineering, Chinese Academy of Sciences, Beijing, China

Xi Zheng
School of Intelligent Technology and Engineering, Chongqing University of Science and Technology, Chongqing, China

Preface

With the advancement of many technologies, such as cloud computing, big data, and many more, users have started exchanging confidential data over the Internet. As there are malicious users and attackers all over the Internet, users' data face many security issues. Here, the main concern is that the data exchanging process must not breach data security. Sometimes, attackers procure confidential data by hacking secret keys, and they even replace sensitive data with fake data. Traditional cryptosystems involve large prime numbers and factorization operations in the key generation and data encryption algorithms. Moreover, the encryption and decryption algorithms of conventional cryptosystems use many complex mathematical operations that require enormous computing power.

To solve the above-mentioned issues, deoxyribonucleic acid (DNA) computing is one of the best solutions. DNA computing is an emerging branch of computing that uses DNA sequences, biochemistry, and hardware for encoding genetic information on computers. Here, information is represented using four DNA bases, namely A (adenine), G (guanine), C (cytosine), and T (thymine), instead of the binary representation (1 and 0) used in traditional computers. There are many applications of DNA computing in the field of computer science. Nowadays, DNA computing is widely used in cryptography and steganography for enhancing data security, so that unauthorized users are unable to retrieve the original data. As four DNA bases are used, DNA computing supports more randomness and makes it more complex for attackers or malicious users to hack data. DNA computing is also used for data storing because a large number of data items can be stored inside the condensed volume. One gram of DNA holds approximately 10^{21} DNA bases or approximately 700 TB. However, it takes approximately 233 hard disks to store the same data on 3 TB hard disks, and the weight of all these hard disks can be approximately 151 kg. DNA computing is also used in cloud environments and wireless sensor networks (WSNs) for enhancing security and improving performance.

This book provides perspectives on DNA computing in computer science. Here, Chapter 1 presents an introduction to DNA computing, whereas Chapter 2 discusses security, privacy, and trust management in DNA computing. Chapter 3 explains how DNA computing can be used in cryptography and steganography. In this chapter, many existing DNA computing-based schemes are briefly presented. Chapters 4 and 5 describe novel schemes based on DNA computing to improve data security. Chapter 6 proposes a novel S-box based on DNA computing and chaos theory. Chapters 7 and 8 discuss how DNA computing can be utilized in two emerging areas of computer science, namely big data and cloud computing. Chapter 9 provides an analytical study on secure data communication in WSNs using DNA computing. Chapter 10 concludes the book by discussing research challenges and future work directions in DNA computing. This chapter can be highly beneficial for researchers, academicians, and industry professionals conducting research on DNA computing.

SUYEL NAMASUDRA
Editor

> CHAPTER ONE

Introduction to DNA computing

Tarun Kumar[a] and Suyel Namasudra[b]

[a]Department of Computer Science and Engineering, National Institute of Technology Patna, Bihar, India
[b]Department of Computer Science and Engineering, National Institute of Technology Agartala, Tripura, India

Contents

Abstract

Currently, Deoxyribonucleic Acid (DNA) computing is considered as one of the advanced fields of Information Technology (IT) industries. DNA computing is a technique inspired from biological science that makes the use of DNA bases, namely Adenine (A), Guanine (G), Thymine (T), and Cytosine (C), for operations and as an information carrier. L. M. Adleman first brought forth this concept in 1994. DNA computing has many advantages, such as parallel computing, large storage capability, minimal power requirement, and molecular computation. Nowadays, it is used in different fields, including cryptography, steganography, big data storage, quantum computing, DNA chip, and medical application. In this chapter, the fundamentals of human DNA and DNA computing, including the structure of DNA, polymerase chain reaction, history, advantages, disadvantages, operations of DNA computing, etc., have been discussed in detail. In addition, many applications of DNA computing in different fields, namely cryptography, steganography, big data, cloud computing, DNA chip, medical research are also presented, which can be highly beneficial for the researchers, academicians, and other professionals doing their research in DNA computing.

Abbreviations

A	adenine
ABE	attribute-based encryption
AES	advanced encryption standard
ATP	adenosine triphosphate

Advances in Computers, Volume 129
ISSN 0065-2458
https://doi.org/10.1016/bs.adcom.2022.08.001

1

C	cytosine
DNA	deoxyribonucleic acid
dNTPs	deoxynucleoside triphosphates
ER	endoplasmic reticulum
G	guanine
GA	genetic algorithm
H	hydrogen
HPP	hamiltonian path problem
IBE	identity-based encryption
IT	Information Technology
Mg	magnesium
Mn	manganese
NP	nondeterministic polynomial
OH	hydroxyl group
PCR	polymerase chain reaction
RNA	ribonucleic acid
T	thymine
U	uracil

1. Introduction

DNA is a long molecule that contains the majority of the genetic information needed for the reproduction and development of the body. All cells in the human body have the same DNA, which is responsible for all aspects of the cell. The coding scheme of the components of the DNA molecule is mainly responsible for the complexity and organization of all human beings. A double helix structure is formed by two biopolymer strands curving in the opposite direction in a DNA molecule [1]. Here, a nucleotide can be defined as an organic molecule, which is the basic building block of DNA and RNA. Each nucleotide is made up of three distinct parts: (1) a 5-carbon sugar (2) a nitrogen base, and (3) a phosphate group. There are four nitrogen bases in DNA identifying A, G, T, and C. In DNA, A is always paired with T, and C is always paired with G as per the base-pairing law [2]. Combinations of random DNA bases can be used to produce a large number of DNA sequences. The structure of DNA is depicted in Fig. 1. In DNA computing, genetic information is encoded in many computer systems using the concept of DNA, molecular biology, hardware, and biochemistry. It is a form of parallel computing that takes the advantage of a large number of different DNA molecules. Experiment, theory, and implementation are the

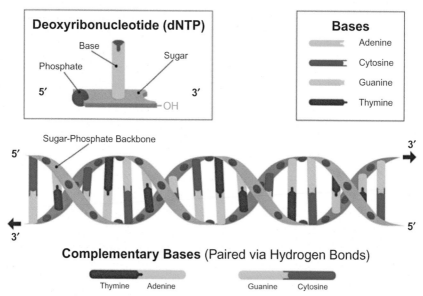

Fig. 1 DNA structure.

fields of DNA computing-based research and development. Unlike the conventional binary digits, i.e., 1 and 0, data or file in DNA computing is primarily expressed using four genetic alphabets, i.e., A, G, T, and C. It is possible because short DNA molecules of some random sequences can be easily synthesized. If an algorithm is considered, DNA molecules can be used to give an input of that algorithm with some specific sequences, and the instructions are given to the molecules by laboratory operations to produce the final collections of molecules.

In 1994, a scientist at the University of Southern California, L. M. Adleman, has first suggested utilizing the theory of DNA for computation [3]. Adleman has demonstrated a proof-of-concept by solving the seven-point Hamiltonian Path Problem (HPP) using DNA as a method of computation. This novel technique was used to resolve Nondeterministic Polynomial time (NP-hard) problems. It has been quickly realized that DNA computing may not be the best solution for this problem [4]. In 1999, computer scientists, Ogihara and Ray [5] have implemented DNA computing-based Boolean circuits. DNA computing is currently one of the trending fields of biology and computer science. Significant knowledge in both computer engineering and DNA molecule is needed to develop an efficient DNA computing-based algorithm. DNA computing can be used to solve and calculate scientific problems and equations that are difficult to solve using the current data sharing and storage techniques. There are two main reasons for its popularity. It requires: (1) a least amount of storage, and (2) minimal power. DNA can store memory space at a density of about 1 bit per cubic nanometer [6]. High computation power by combining the DNA bases and the complex structure of DNA attracts a large number of researchers to use DNA computing in a variety of fields. One of the most critical issues in DNA computing is how researchers can reduce the likelihood of errors occurring during execution. Several studies have been carried out to show that by using the appropriate encoding technique, the efficiency of DNA computing can be greatly improved [7].

DNA computing is now widely used in many fields, including cryptography [8], nanotechnology [9], combinatorial optimization [10], Boolean circuit construction [11], data storage [12], quantum computing [13], and many more [14]. Nowadays, around 2.5 quintillion bytes of data are produced every day by humans. As most of the data are transmitted over the internet, numerous attackers and malicious users always try to achieve unauthorized access to the data. Thus, the application of DNA computing in cryptography is most popular to encrypt any data. In DNA-based cryptography, data are encrypted by using the nitrogen bases, i.e., A, T, G, and C, instead of 0 and 1. Many researchers have proposed schemes to improve data security by using DNA computing [15]. Here, a complete DNA encoding character set must include alphabets, numbers, and special characters. The mapping table, which provides the complete characters of DNA must be automatically generated [16]. Otherwise, again data may face security issues. The term steganography refers to the practice of concealing the presence of data. In DNA-based steganography, data are hidden using the DNA bases. This technique is straightforward that only transforms the plaintext into a DNA sequence. Basically, it does not encrypt any data, and hides the plaintext, i.e., DNA sequence of the plaintext within other DNA strands. Anyone who knows the primers can effortlessly find the location of the original DNA sequence or plaintext.

DNA computing is also used in quantum computing. Quantum computing aims to create computers focused on quantum theory that is capable of dealing with the existence and behavior of all quantum aspects, while DNA computing can focus on storage. Therefore, a combination of both fields can change the world. The DNA molecules can be directly used in quantum computers either by using nuclear magnetic resonance or by dopping the DNA molecules for implementing the quantum gate. Companies, such as Microsoft, IBM, Google, etc., are investing in quantum computing.

DNA computing is also used in cloud computing. In the IT industry, cloud computing is widely used for storing users' confidential data. As the size of data is gradually increasing, it also creates issues for a user to store the big data. DNA computing can be very useful to store any big data in a cloud environment as 1 g of DNA bases can store around 700 TB of data. Thus, some gram of DNA bases can store all data of the world. Here, the data owners of a cloud environment first convert the data into its binary form, and then, the binary form of the data is converted into a DNA sequence using any DNA encoding rule followed by any complementary pair–rule [17]. DNACloud, a tool that is designed to store the data in the form of DNA sequences [18]. This tool mainly performs three tasks: (1) encoding of data into DNA sequences (2) estimating the requirements for the storage of data on DNA, and (3) decoding of DNA sequencing to the original data. Many researchers are exploring this field [19].

The main aim of this chapter is to discuss all the basic details of DNA computing, so that researchers and professionals can use the concept of DNA computing in related fields. The fundamentals of DNA are discussed on the molecular level, including Polymerase Chain Reaction (PCR), the history of DNA computing, and many more. Furthermore, this chapter discusses the advantages and disadvantages of DNA computing. Many applications of DNA computing are also discussed, which can be highly beneficial to the researchers doing their research in DNA computing. The key contributions of this chapter are as follows:

(1) Fundamentals of human DNA and the concept of polymerase chain reaction are discussed in this chapter.
(2) The basics of DNA computing and Adleman's experiment to solve the HPP are presented in detail.
(3) Finally, many applications of DNA computing in several fields are also explained.

The rest of the chapter consists of several parts. In Section 2, all the fundamentals of DNA computing, such as the background of DNA, PCR, DNA computing, history, advantages, and disadvantages of DNA computing, are discussed in detail. Applications of DNA computing are discussed in Section 3. Finally, the entire chapter is concluded in Section 4 along with some future works.

2. Fundamentals of DNA computing

In this section, background and structure of human DNA, polymerase chain reaction, advantages, disadvantages, and operations of DNA computing, are discussed in detail.

2.1 Background of human DNA

Biologically, deoxyribonucleic acid is a lengthy molecule that stores the instructions of all the workings of human bodies, and it is transferred from parents to the child. It contains information that supports the cells to make proteins, and proteins are responsible for many essential functions of the human body, such as digestion, cell building, muscle movements, and many more [20]. A cell can be defined as the building block of the human body. All the cells of the human body hold the same DNA, and there are approximately trillions of cells in the human body for providing the main structure of the human body. Human cells have many parts as mentioned below, which are responsible for different functions:

(1) **Cytoplasm:** It is made up of cytosol, which is a jelly-like fluid. Cytoplasm mainly covers the nucleus along with other structures.

(2) **Cytoskeleton:** It is a long network of fiber, which builds the cell's framework. The cytoskeleton is responsible for the shape of the cell, cell division, movement of cells, and many more.

(3) **Endoplasmic Reticulum (ER):** ER is responsible for transporting the molecules, i.e., inside or outside the cell towards their destination.

(4) **Golgi Apparatus:** Golgi apparatus bundles the processed molecules, and the endoplasmic reticule transfers Golgi apparatus to the other cells as per requirement.

(5) **Lysosomes and Peroxisomes:** These organelles are mainly the centre of recycling in the cell, which digest bacteria for invading the cell, rid the cell, and recycle the cell.

(6) **Mitochondria:** Mitochondria is one of the complex organelles, which converts the energy into the form for use by the cell. To separate the mitochondria from the DNA in the nucleus, mitochondria has the genetic material.

(7) **Plasma Membrane:** It is the exterior outer layer of each cell. The plasma membrane separates each cell from its own environment and allows the materials to exit and join the cell.

(8) **Ribosomes:** The organelles that process the cell's genetic instructions for the formation of protein are known as ribosomes. In the cytoplasm, ribosomes can move freely.

Genes are basically the functional and physical unit of heredity. These are made up of DNA. In a human body, genes usually vary in size from a small number of DNA bases to more than 2 million DNA bases. Many genes act as the instructions for making the molecules called proteins. But, some genes do not code for proteins. There are two copies of every gene in the human body in which a single copy is inherited from the mother's genes, and another copy is inherited from the father's genes. Most of the genes in all human bodies are almost the same. However, there are differences in the number of genes among people. This difference is less than 1% of the total genes. Here, Alleles are the forms of the identical gene with minor changes in the sequence of DNA bases, and these minor changes are responsible for the unique physical features of every person.

2.1.1 Structure of human DNA

DNA molecule plays an important role in DNA computing. The biochemical field consists of small and large molecules, monomers, and polymers. DNA is a kind of polymer made up of monomers known as deoxyribonucleotides. DNA is a vital molecule in a cell, and it has an interesting structure that supports two primary functions of DNA: (1) self-replication, and (2) protein synthesis coding, which ensures that the same copy is passed to offspring cells. As mentioned earlier, deoxyribonucleotide is made up of three components: (1) sugar (2) nitrogenous base, and (3) phosphate group. Deoxyribose, the sugar of DNA explains the prefix term, i.e., deoxy-ribo. The term nucleotide is used instead of deoxyribonucleotides to simplify the terminology. Deoxyribose is made up of five carbon atoms, which are numbered for ease of reference. As the base also consists of carbons, the sugar's carbons are numbered from $1'$ to $5'$ to avoid confusion. The $1'$ carbon is attached to the base, while the $5'$ carbon is attached to the phosphate group. A Hydroxyl Group (OH) is bound with the $3'$ carbon in the sugar structure [11].

The only difference among nucleotides is their bases, which are divided into two categories: (1) purines, and (2) pyrimidines. In nucleotides, there are two purines: (1) adenine, and (2) guanine, and two pyrimidines: (1) cytosine, and (2) thymine. As nucleotides are only distinguished by their bases based on the type of the base, they are referred to as A, G, C, and T.

Fig. 2 depicts the composition of nucleotide, where one of the possible DNA bases is at node B, a phosphate group is P, and the sugar base (carbons $1'$ to $5'$). Fig. 3 shows the standard and simplified chemical structure of a nucleotide.

Ribonucleic Acid (RNA) is another important polymer for human cells. It has an almost similar structure to DNA. It is made up of ribonucleotides, which are monomers. Ribonucleotide with Adenine base and a triple phosphate group is referred to as an Adenosine Triphosphate (ATP)

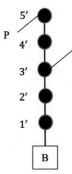

Fig. 2 A simple representation of a nucleotide.

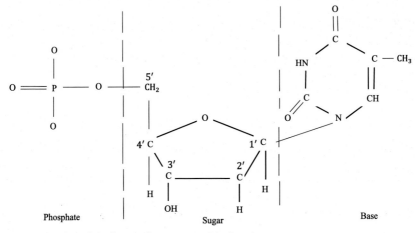

Fig. 3 A nucleotide's chemical structure with thymine.

molecule. It is the primary source of energy in human cells. There are two ways to distinguish ribonucleotide and nucleotide:

(i) It contains a ribose sugar, which is distinguished from deoxyribose sugar by the presence of the hydroxyl group on the $2'$ carbon rather than the Hydrogen (H).

(ii) In RNA, the Uracil (U) base replaces the thymine base. Thus, there are four bases in RNA, namely A, U, C, and G.

Nucleotides can be bound to each other in two ways:

(i) As shown in Fig. 4, 3'-OH of one nucleotide is bound to the 5'-phosphate group of another nucleotide. Thus, it forms a phosphodiester bond, which is a tight covalent bond. It gives the direction

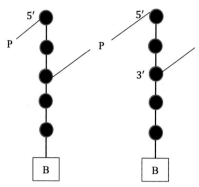

Fig. 4 Phosphodiester bond.

of molecules referred to as the 5′-3′ or 3′-5′ direction. These directions are critical for understanding the functionality and the entire processing of DNA.

(ii) A weak hydrogen bond is formed, when one nucleotide's base interacts with the base of another nucleotide. The base pairing constraints are A is always paired with T and C is always paired with G. No other base pairing is possible.

The pairing principle is known as the Watson-Crick complementary pairing rule. In 1953, James D. Watson and Francis H. C. Crick have introduced this famous double helix structure of DNA [21]. It is crucial to comprehend the function and structure of DNA. A thin wiggly line between the bases shown in Fig. 5 reflects the fact that the hydrogen bond is much weaker than the phosphodiester bond. Most importantly, due to the pairing of A and T, there are two hydrogen bonds between the two nucleotides, and three hydrogen bonds occur because of the pairing between C and G. Therefore, the pairing between C and G is more powerful than the pairing between A and T, and more energy is required to separate the pairing between C and G. For reflecting the differences, it uses two wiggly lines for the pairing between A and T and three wiggly lines for the pairing between C and G pairing.

The single-stranded DNA can be formed by using phosphodiester bonds. As shown in Fig. 6, the nucleotides with the free 5′ phosphate and free 3′ hydroxyl are given on the left end and right end, respectively. As nucleotides can be identified by their bases, when naming them by using A, G, C, and T, a single strand can also be represented as a sequence of letters. For example, 5' ACG is the single strand form in Fig. 6.

In practice, the hydrogen bond between single nucleotides is too weak to hold the two nucleotides bound together, thus, longer stretches are required

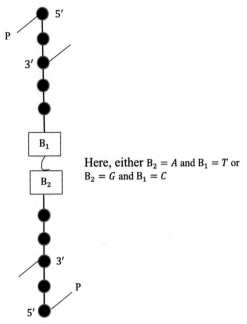

Here, either $B_2 = A$ and $B_1 = T$ or
$B_2 = G$ and $B_1 = C$

Fig. 5 Hydrogen bond.

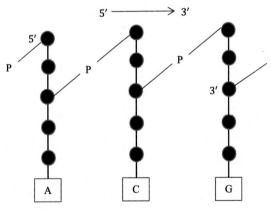

Fig. 6 Single-stranded DNA.

to bond them together. A stable bond is created by the cumulative effect of
hydrogen bonds between complementary pair bases in DNA.

2.1.2 Structure of double helix

The double-stranded DNA molecule can be formed from the single-
stranded DNA molecule by using the Watson–Crick complementary rule.

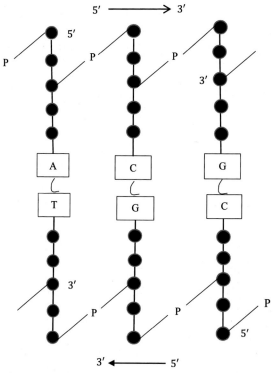

Fig. 7 Forming double strands.

Fig. 7 depicts how to join two single-stranded DNA molecules by using a hydrogen bond. In a double-stranded molecule, two single strands are placed in opposite directions. Here, one nucleotide's 5′ end is bonded 3′ end of another nucleotide. It is a standard practice to draw the upper strand of a double-stranded molecule from left to right in the 5′ to 3′ direction, and the lower strand must be drawn from left to right in the 3′ to 5′ direction [11]. It is shown in Fig. 7 that the upper strand is 5' ACG and the lower strand is 3' TGC.

The double-stranded representation of a DNA molecule in terms of two linear strands bound to each other by the Watson-Crick complementary pair-rule has a significant simplification because two strands in a DNA molecule are wrapped around each other to form the structure of the double helix as depicted in Fig. 8.

In vivo DNA replication, the condition is very complicated as a massive DNA molecule must fit into a small cell. This process is very complex, and it is performed hierarchically in many stages in more complex cells (eukaryotes). When considering processes in human cells, the shape of a DNA

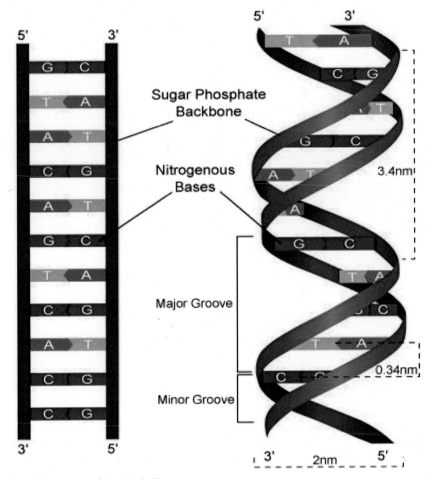

Fig. 8 Structure of double helix.

molecule is extremely important. It also must be noted that the human DNA molecules are linearly structured. However, bacterial DNA is often circular. Circular molecules can be simply constructed by forming a phosphodiester bond between the last and first nucleotide. The ability for processing DNA is very important in genetic engineering, as well as in DNA computing [21].

2.1.3 Polymerase chain reaction (PCR)

PCR is a widely used technique for making millions to billions of copies of a particular DNA sample. It is used for the amplification of a small DNA strand to a large amount of DNA bases. The copies can be partial copies or complete copies. Karry Mullis, an American biochemist, the recipient of the 1993 Nobel Prize, has invented it in 1983 [22]. This method is based on the natural processes that a cell goes through as it replicates a new strand of DNA. For PCR, only a few biological components are required. Thermal cycling is used in the majority of PCR processes, which is the process of heating and cooling reactants repeatedly in order to enable temperature-dependent reactions, including DNA melting and enzyme-driven DNA replication. There are two key reagents in PCR: (1) primers, and (2) DNA polymerase. These are short single strands of DNA with a complementary sequence to the target DNA region, typically about 20 nucleotides long. In PCR, at first, two strands of the DNA, i.e., double helix are separated physically at a very high temperature, which is known as nucleic acid denaturation. Then, it is heated at comparatively less temperature and the primers are bound to the complementary DNA sequences. Thus, two separate DNA strands become templates for DNA polymerase to enzymatically assemble a new DNA strand from the free nucleotides. When PCR progresses, the newly generated DNA is used as a template to replicate. Therefore, a chain reaction is generated by which the original or main DNA template or sequence is exponentially amplified.

Maximum PCR application processes support a heat-constant DNA polymerase like Taq polymerase [23]. Taq polymerase is suitable for PCR because of its heat stability. Otherwise, DNA polymerase must be added in every cycle, which is a costly and tedious process. Several components and reagents are needed as mentioned below for a simple PCR setup:

(1) DNA template containing the DNA target region for amplifying.

(2) DNA polymerase, which is an enzyme to polymerase new strands of DNA. Here, Taq polymerase is commonly used.

(3) Two DNA primers, which are complementary to 3′ ends of the DNA target's sense and anti-sense strands.

(4) Deoxynucleoside Triphosphates (dNTPs), the molecules from which a new DNA strand is generated by DNA polymerase.

(5) DNA polymerase uses deoxynucleotide triphosphate nucleotides with triphosphate groups as building blocks to create a new DNA strand.

(6) Buffer solution that gives an environment for optimal DNA polymerase operation and stability

(7) Bivalent cations, classically Manganese (Mn) or Magnesium (Mg) ions are commonly used.

As depicted in Fig. 9, there are six main steps in PCR, namely initialization, denaturation, annealing, extension, final elongation, and final hold, that are

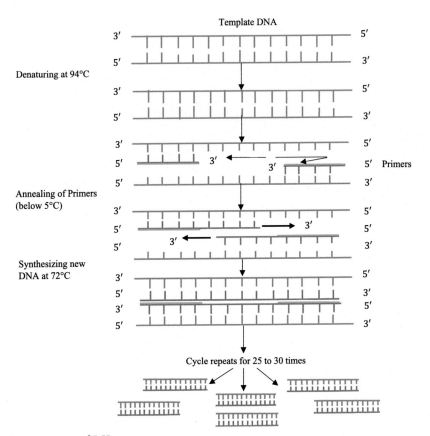

Fig. 9 Process of PCR.

repeated 20 to 40 times in which temperatures are changed, known as thermal cycles. In each cycle, there are 2 to 3 discrete temperature steps [24].

(1) **Initialization:** In the initialization step, heat is activated for DNA polymerase. Here, the reaction chamber is heated at the temperature of 94 °C to 96 °C. Here, 98 °C temperature is used, if tremendously thermostable polymerase is used. This process lasts for 1 to 10 min.

(2) **Denaturation:** The reaction chamber is heated at 94 °C to 98 °C for 20 to 30 s in this stage, which is the first routine cycling event. As the hydrogen bonds between the complementary pair bases are broken, this denaturation process results in the double-stranded DNA template to melt and two single-stranded DNA molecules are generated.

(3) **Annealing:** This step enables primers of both the single-stranded DNA templates to be annealed by lowering the reaction temperature

(i.e., 50 °C to 65 °C) for 20 to 40 s. In most cases, the two distinct primers are used in the reaction each for both single-stranded complements comprising the mark or target region. Since the annealing temperature has a significant impact on efficiency and specificity, determining the correct temperature for this step is very difficult. This determined temperature must be low, so that the primer must be only bound to the flawlessly complementary portion of the single strand. The primer cannot bind properly, when the temperature is much less. There is also a possibility that the primer may not bind at all, if the temperature is much high. When the primer closely matches the main or template sequence, stable hydrogen bonds are formed between the complementary pair bases. In this step, DNA formation starts by binding the polymerase with the primer-template hybrid.

(4) **Extension:** In this step, the temperature is determined based on the used DNA polymerase. The optimal temperature for Taq polymerase is approx. 75 °C to 80 °C. However, is widely used for this enzyme. By inserting free dNTPs from the reaction, which is complementary in the direction 5′ to 3′ of the template, the DNA polymerase produces a new strand complementary to the main or template DNA strand. The size of the target DNA region for the amplification, as well as the used DNA polymerase, determine the exact time needed for elongation. At the optimal temperature, a thousand DNA bases are polymerized per minute by most of the DNA polymerases. The amount of target DNA sequences gets double in each elongation phase under ideal conditions. The original main DNA strands along with all the newly formed DNA strands become the template strands for the next elongation cycle, which results in an exponential amplification of the same target DNA sequences in each subsequent cycle.

All of the above steps are part of the same cycle. Numerous cycles are needed for amplifying the target DNA sequences millions of times. The total amount of DNA copies generated after a specified number of cycles is calculated by using the formula 2^n, where n is denoted as the number of rounds or cycles. Therefore, a reaction of 30 cycles produces 2^{30} or 1,073,741,824 copies of the original target DNA sequence.

(5) **Final Elongation:** The final elongation step is optional. However, it is executed for 5 to 15 min after the last PCR cycle at 70 °C to 74 °C temperature. This temperature is optimal for mostly used polymerases in PCR for ensuring that any leftover single-stranded DNA must be completely elongated.

(6) **Final Hold:** This step is mainly used for cooling the reaction chamber from 4 °C to 15 °C for an unspecified period of time.

2.1.4 Applications of polymerase chain reaction

There are many applications of PCR in interdisciplinary research areas by making it possible to retrieve DNA sequences. Some are discussed in this subsection.

(1) **Human Health and Genome Project:** For improving human health and quality of life, PCR is an essential technique. PCR is mainly used in biological science to identify infectious organisms and to diagnose disorders in genes. It can also be used to analyze virus DNA to diagnose deadly diseases like AIDS, and it is more accurate than ELISA's standardized identification. Furthermore, PCR may be utilized to diagnose diseases, including Lyme by recognizing the function of bacterial DNA strands. In addition, PCR is used to inhibit sexual transfer diseases. The research on all human genes is generally referred to as the human genome project. The identification of unique genes, mutations in these genes, and rates of mutation are greatly aided by PCR.

(2) **Genetic Fingerprinting:** In forensics, genetic fingerprinting is used to match a person's DNA with a blood sample, parent's DNA, etc. Here, a very small amount of DNA is used in most cases. So, the main role of PCR is to amplify the small amount of DNA from blood, hair, semen, and other tissues. Then, the gel electrophoresis technique is further used to examine the amplified DNA fragments.

(3) **Detection of Hereditary Disease:** As hereditary diseases are directly connected to genes, it is difficult to identify them. However, PCR is one of the best solutions to analyze this complex process. PCR can help to amplify the number of DNA, and an appropriate marker can be used to identify the disease-causing gene.

(4) **Cloning:** In industry and research laboratories, gene cloning has many applications and PCR can be utilized for gene cloning. PCR is used for amplifying the gene, if a gene needs to be cloned. Then, it is loaded into a vector followed by transporting the gene by a vector. Thus, PCR is essential to study gene cloning [25].

(5) **Ancient DNA Analysis:** In ancient DNA analysis, PCR is used. Here, the DNA separated from Egyptian mummies is amplified by using PCR to identify a person and further studies. The gene may also be associated with the most current gene and used to do evolutionary research.

(6) Gel Electrophoresis: This technique separates DNA strands in different sizes by pulling fragments of DNA through a gel matrix with an electric current. Here, DNA ladder, a unit of size of DNA sequence is usually included to determine the size of the fragments in PCR. By analyzing the size of the DNA sequences, many medicines are developed for critical diseases.

2.2 DNA computing

DNA computing is a branch of natural computing that uses the molecular properties of human DNA to perform logical and arithmetic operations instead of binary bits 0 and 1. This enables massively parallel computation, making it possible to solve complex mathematical problems in a fraction of time. Therefore, in DNA computing, computation is much more effective with a large amount of self-replicating DNA than with a conventional computer. A computation can be considered as an execution of an algorithm that has some well-defined set of instructions, which accepts inputs, processes them, and generates some outputs as the result. Developing an algorithm for DNA computing requires a lot of experience in molecular biology and computer science. Instead of using binary digits (0 and 1) in an electronic computer, information or data is stored in DNA computing-based computer by using the four genetic alphabets, namely A, T, G, and C. The ability to synthesize short DNA sequences artificially allows these sequences to be used as inputs in algorithms. DNA computing-based simulations use one billion times less energy than a traditional machine. DNA computing also supports storing data in less than one trillion time volume than traditional compact discs. Furthermore, DNA computing is extremely parallel as chemical reactions by using billions and trillions of DNA molecules can be executed at the same time.

In 1994, L. M. Adleman has first used DNA computing to solve the HPP, a renowned NP-complete problem [3]. An NP-complete problem cannot be solved by a deterministic algorithm in polynomial time. A directed graph $G(V, E)$ having V and E as a set of vertices and edges, respectively with V_i and V_o as designated vertices such that V_i, $V_o \in V$ is said to have a Hamiltonian path, if and only if there exists a path between V_i and V_o that traverse every vertex only once [4]. Adleman has solved the HPP for a Graph G having 7 vertices with $V_i = V_0$ and $V_o = V_6$ and uses 50 pmol of each nucleotide for the DNA molecule encoding. The non-deterministic algorithm using DNA computing suggested by Adleman to solve the HPP is as follows.

Step 1: The first step encodes the vertices and edges of the graph into DNA bases and generates a random path through the graph. For every vertex $u \in V$, a random strand consisting of 20 nucleotides is synthesized.

$$S_u = \underbrace{AATGCCTAGT}_{a_u}\underbrace{TCCTAATGGC}_{b_u}$$

$$S_v = \underbrace{TAGGACTAGG}_{a_v}\underbrace{CTTCAAGTAT}_{b_v}$$

$$S_w = \underbrace{CCGTATGATC}_{a_w}\underbrace{CGTACGGCTT}_{b_w}$$

Here, S_u, S_v and S_w represent the strand for vertex u, v and w, respectively. For every edge $u \rightarrow v \in E$, an opportune strand is synthesized by concatenating the $3'$ end of S_u and $5'$ end of S_v, i.e., b_u and a_v, respectively. In the similar way, the edge $v \rightarrow w$ can be synthesized by concatenating the $3'$ end of S_v and $5'$ end of S_w, i.e., b_v and a_w, respectively as given below:

$$S_{u \rightarrow v} = \underbrace{TCCTAATGGC}_{b_u}\underbrace{TAGGACTAGG}_{a_v}$$

$$S_{v \rightarrow w} = \underbrace{CTTCAAGTAT}_{b_v}\underbrace{CGTACGGCTT}_{a_w}$$

The complement of each vertex $v \in V$ is generated by exchanging the base pairs to get the edges and it is calculated as:

$$\overline{S_v} = ATCCTGATCCGAAGTTCATA$$

where $\overline{S_v}$ is the complementary edge of S_v. The ligation reaction takes place on encoded vertex and edges to generate DNA strands that give a random path through the graph.

Step 2: The DNA sequences generated in step 1 are expanded by PCR with S_0 and $\overline{S_6}$ as primers, a short sequence with complementary bases. This step gives the routes from the starting to the final specified node of the graph.

Step 3: In this step, the length is calculated on the result of step 2 using the agarose gel to select the route with exact 7 nodes, which implies $20*7 = 140$ basepair. This step selects the routes that visit the required number of nodes.

Step 4: The fourth step is used to verify the traversing of each node of the graph. The strand $S_{u \to v}$ corresponding to the vertex v is selected as it belongs to the route incoming for the vertex v.

Step 5: Finally, the Hamilton path is calculated in this step. If a path is generated in step 1, all the steps are skipped. Otherwise, steps 2 to 4 are executed.

The aforementioned steps use DNA computing to get the Hamiltonian path. Adleman also observes that the time is increased linearly with the size of the graph. Thus, Adleman has proved high parallelism capabilities of DNA computing by solving NP-complete problem, i.e., HPP, in polynomial time, which needs exponential time by the traditional system [3].

Although the HPP is a NP-complete problem, but, Adleman's experiment doesn't ensure that DNA computing can be used to solve any other NP-complete problem. Adleman's DNA computing-based design is trivial as when the number of nodes is increased, the number of DNA computing-based components is also increased. So, biochemists and computer scientists have started to explore this new field. Lipton has extended Adleman's model in such a way that biological computers can change the way of computation [26]. At first, Lipton has generalized Adleman's solution and tried to solve the satisfiability problem, i.e., a basic NP-complete problem. The main contribution of Lipton's solution is the way to encode a binary string into DNA strands. However, there are some major concerns in DNA computing:

(1) Its computational model is mostly based on molecular techniques to solve any particular issue. The variety of problems leads to discrepancies in computing systems and there is also no coding model and standard in DNA computing.

(2) DNA computing is prone to errors. These are mainly randomly generated DNA sequences and the chances to occur error increase in all the experimental stages.

(3) There are no mathematical models to prove the concept of DNA computing, which is very critical in many sensitive computational problems.

2.2.1 History of DNA computing

The concept of DNA gets popularity, when Watson and Crick have discovered the double-helix structure of DNA in 1953. Then, scientists and researchers have gained a better understanding of DNA replication and genetic regulation of cellular activities. During that time, researchers were much focused on the computing capability of living systems. The primary objective was to implement the concept of living systems in computing devices. For example, wireless automatic system, genetic algorithm, neural network, etc.

In the early 1960s, Richard Feynman has invented the first molecular computer [27]. The concept of DNA was first used for computing in 1994 by L. M. Adleman. Adleman has demonstrated that DNA can be used for executing computations in massively parallel manners. By using the four DNA bases, Adleman has encoded a classic "hard" problem called Traveling Salesman Problem into DNA strands and used the property of molecular biology for finding the answer. Then, Adleman has also demonstrated the uses of DNA in computing to solve various problems like the 3-SAT problem, the 0/1 Knapsack Problem, and many more. After that, many Turing machines were proved by using the concept of original Adleman's experiments.

In 1995, Baum [28] has introduced the idea of DNA computing-based storage. He has shown that a tiny volume of DNA strands can be used to store a large amount of data due to the ultra-high density of DNA. In 1997, Ouyang et al. [29] have proposed an experimental explanation for the "maximum clique" dilemma based on molecular biology. In 1997, researchers from the University of Rochester uses DNA computing to create logic gates.

Liu et al. [30] have developed a DNA computing-based model in 2000, where a multi-based encoding strategy is used in a surface-based DNA measurement approach [31]. In 2002, researchers from the Weizmann Institute of Science have demonstrated a programmable computing system based on enzymes and DNA. Reif et al. [32] have first introduced the idea to use DNA computing in robotics. They have utilized molecular biology to generate energy for the walker. As it was the first demonstration, an extensive variety of DNA computing-based walkers are demonstrated by many researchers. This machine is capable of diagnosing cancer in a cell.

In 2013, the first biological transistor was created based on DNA computing [33]. Nowadays, DNA computing is used in many fields, including cryptography, steganography, authentication, and many more [34–35]. As four DNA bases are used in DNA computing, it supports much randomness in key generation and data encryption phases in cryptography.

2.2.2 Advantages of DNA computing

DNA computing has many advantages over traditional computing. As DNA strands can hold a huge amount of data, it supports solving decomposable problems in less time. In this subsection, some advantages of DNA computing are discussed in detail [36].

(1) **Performance:** The performance rate of DNA strands can be exponentially improved by running millions of operations simultaneously. Adleman's experiment was executed with 1014 operations/s by using DNA computing, which is equal to 100 Teraflops or 100 trillion floating-point operations/s. At the same time, the world's fastest supercomputer has the capability of 35.8 teraflops. It is one of the best solutions for a problem that involves a large number of calculations, such as clustering optimization, scheduling problem, etc.

(2) **Parallel Processing:** The ability to execute a great number of tasks in a parallel manner is a major benefit of DNA computing over classical computing. This countless degree of parallelism benefits a wide range of applications like cryptography, steganography, etc. DNA computers' massively parallel computing capability can enhance the speed of the operation. A group of 1018 DNA strands can enhance the speed of the advanced supercomputers by 10,000 times of it [36].

(3) **Storage Capability:** Today, knowledge is not only in the form of words or documents, but also in the form of photographs, video, and other formats, and all of these require a large amount of storage space. DNA computing can be one of the best solutions for addressing today's data storage issue. As compared to digital data storage, DNA can store a lot of information in a limited amount of space. Standard storage media, including videotapes require 10^{12} cubic nanometer of space for storing one bit of information. However, DNA computing requires 1 cubic nanometer for one bit. 1 cm^3 of DNA strand can hold more data than a trillion of compact discs. This is due to the DNA molecule's data density that is 18 Mbits per inch. A few gram of DNA can store all the data of the world.

(4) **Minimal Power Consumption:** The computers based on DNA computing are small and light. As DNA computers are mainly based on basic biological operations and chemical bond does not require any electricity, these need low power or electricity. Here, power is only required to prevent DNA denaturation.

2.2.3 Disadvantages of DNA computing

Along with the advantages of DNA computing, it also has some disadvantages [36]. As the size of a DNA strand is long, there can be errors in DNA computing, when nucleotides are paired with each other. Here, all the processes require human intervention. There are some other major disadvantages of DNA computing as mentioned below:

(1) **Requires High Memory:** In DNA computing, a large amount of memory is required to generate a set of solutions for relatively simple problems. Even though DNA can store or save a trillion times more data, if a significantly large problem needs to be resolved, the technique by which the data is handled requires a large number of DNA strands.

(2) **Accuracy:** The synthesis of DNA is prone to errors like mismatching pairs, and it is mainly dependent on the accuracy or the correctness of the involved enzymes. The likelihood of error increases exponentially and must limit the number of operations until the risk is higher than producing the correct outcome. Due to this, DNA computing cannot be applied in sensitive applications.

(3) **Resource-Intensive:** All the phases of parallel operations must require time in days or hours determined by mechanical or human intervention in many steps. As DNA strands are made for a particular problem, for each and every new problem, a new DNA strand should be created that is more critical and time-consuming. Because of the vast parallelism of DNA computing, algorithms can be performed in polynomial time. However, they are very limited to applying for small instances of the problem as they need the creation of an unobstructed space of solution.

(4) **Non-Replaceable:** Traditional computers cannot be replaced by DNA computers at this time due to their high cost. In addition, DNA computers are not flexible, as well as are not easily programmable. Here, users cannot just sit with a familiar keyboard and start typing for programming [37]. They must need to understand all the basic details of molecular biology, DNA computing, hardware, and computer programming language.

2.3 Operations of DNA computing

As discussed earlier, DNA computing utilizes four DNA bases, i.e., A, C, G, and T for computation. Here, the Watson-Crick complementary pair-rule is also used in which A is paired with T and C is paired with G. However, anyone can design any other complementary pair-rule [15]. In this subsection, many operations of DNA computing are discussed in detail.

(1) **DNA Synthesis:** In reality, it is one of the most fundamental biological operations of DNA computing. The solid-state DNA synthesis method

is based on the binding of the first nucleotide to a solid support followed by the step-by-step addition of subsequent nucleotides in a reactant solution from 3′ to 5′ direction. One drawback of this operation is that a laboratory method can produce only 20–25 strands [38].

(2) **DNA Replication:** It is accomplished by the polymerase chain reaction, which includes the enzyme DNA polymerase. PCR replication process involves a template, which is a single-stranded guiding DNA, and a primer, which is a template annealed oligonucleotide. The primer is needed to start the polymerase enzyme's synthesis reaction. By sequentially adding the nucleotides at primer's one end, the primer is extended to 3′ end until the required strand is acquired which begins with a primer and complements the template. Here, the primer is extended only in the 5′ to 3′ direction.

(3) **Short by Length:** A technique known as gel electrophoresis is used to shorten the length of a DNA sequence [38]. Here, negatively charged DNA molecules are positioned in "wells", i.e., on one side of the poly-acrylamide gel. The gel is then passed into an electric current with the negative pole on one side of the wells and the positive pole on another side. The DNA molecule is attracted to the positive pole, and the larger molecules move through the gel more slowly. After an interval, mol-ecules are separated into discrete bands based on their size.

(4) **Separating DNA Sequence:** This operation supports researchers to remove a DNA sequence from the solution that contains the desired sequence or DNA strand. This operation is executed by generating the DNA sequence, whose complement is the desired DNA sequence. Then, a magnetic substance is used to attach the newly generated DNA strand with it. This magnetic substance is used for extracting the strands after the annealing process.

(5) **ASCII Rule:** Although this operation is not directly associated with DNA computing, it is considered one of the important operations. The main purpose of this operation is to confuse the attackers and malicious users for unauthorized access to any confidential data. In this operation, an 8-bit binary value of the plaintext is converted into corresponding ASCII values ranging from 0 to 255. Then, this decimal value is again converted into their corresponding binary values for fur-ther processes of DNA computing. As the plaintext of data is converted into another binary form before some other operations of DNA com-puting, it supports massive randomness. Table 1 shows an example of the ASCII rule.

Table 1 ASCII rule.

Binary value	ASCII	Binary value
00000000	12	00001100
00000001	220	11,011,100
.	.	.
.	.	.
.	.	.
11,111,111	10	00001010

Table 2 2-Bit binary encoding rule.

00	01	10	11
A	T	G	C
A	T	C	G
.	.	.	.
.	.	.	.
.	.	.	.
G	T	C	A

(6) **DNA Encoding:** In DNA encoding, the binary value of any data or plaintext is converted into DNA bases by using DNA computing. It is one of the popular operations of DNA computing. For example, a DNA encoding rule can be defined as 00-A, 01-T, 10-G, and 11-C. This implies if there are binary values as 00, 01, 10, and 11 in the plaintext, they can be converted into A, T, G, and C, respectively. It does not matter in which manner or sequence the binary values are represented in the plaintext. Anyone can assign any rule to convert the binary values into DNA bases. Thus, a DNA encoding rule can be applied to the binary values of any plaintext. Table 2 shows an example of the DNA encoding rule.

(7) **DNA Complementary Rule:** The complementary pair-rule assigns a DNA base to each nucleotide base. To make an algorithm more complex, a researcher can create his or her own complementary rule by using the DNA bases. As there are four DNA bases, i.e., A, C, G, and T, the complementary pair-rule can be assigned as A-T and

Table 3 DNA XOR rule.

XOR	A	T	G	C
A	A	G	C	T
T	C	T	G	A
G	G	C	T	A
C	T	G	C	A

C-G, which means if there is A, it can be replaced with T and vice versa. As there are 4 DNA bases, there can be 4!, i.e., 24 encoding rules to transform binary values of any data into DNA bases. After converting the binary values into DNA bases, DNA bases can be converted into another DNA bases by using the complementary pair-rule [15].

For example, if the binary value of any data is 00011011, it can be converted into DNA bases as ATGC by using any DNA encoding rule. Here, A, T, G, and C are 00, 01, 10, and 11, respectively. Then, ATGC can be converted into TACG by using the complementary pair rule. This rule is popular in DNA cryptography.

(8) **DNA XOR Rule:** Because of the recent advancement of DNA computing, many researchers are nowadays using the DNA XOR rule in related fields. In DNA computing, the DNA XOR operation is identical to the XOR operation. Table 3 shows a DNA computing-based XOR rule.

For example, if there is any DNA sequence ATGCCGTA and a key or DNA strand CTAGTAGC, it can be converted into another DNA sequence TTGCGGGT by using the DNA XOR rule of Table 3. Then, the original DNA sequence can be retrieved by using the newly generated DNA sequence, i.e., TTGCGGGT and original key CTAGTAGC. To retrieve the original DNA sequence, again the same DNA XOR rule must be used.

3. Applications of DNA computing

Nowadays, DNA computing is used in many sectors like DNA chips, cryptography, data storage, computation, and many more. In this section, many applications of DNA computing are discussed in detail.

(1) **DNA Chip:** The recent technology has been drastically changed due to the use of DNA chips and microarrays [39]. DNA chip technology uses microarrays of molecules immobilized on any solid surface for

biochemical analysis. Microarrays are used by scientists and researchers for expression analysis, genotyping of the genome, DNA resequencing, and polymorphism detection. DNA computing-based chips are like silicon chips that are mainly used for data storage in the form of DNA sequences [40]. These chips are mainly made up of a huge amount of embedded spots on the solid surface in which spot keeps probes. Here, probes can be referred to as a small DNA sequence. In each spot, when a DNA sequence is tied with these probes, data or files are electronically calculated on the basis of the ratio of combining a probe with the DNA sequence. DNA computing is attractive to store data because a large number of data items or files can be stored within the condensed volume. 10^{21} DNA bases are there in 1 g DNA that implies approx. 700 TB data can be stored in 1 g of DNA [41]. Therefore, all the files of the universe can be stored in a few grams of DNA. The speeds of reading and writing in DNA computing are high that also motivates many researchers or other professionals to use DNA computing.

Nowadays, much emphasis is given by the manufacturers for developing small biochips with high data handling capacity. These types of biochips can be highly beneficial in many areas, including cryptography. This cryptographic technique consists of the following steps:

Step 1: At first, a group of specific probes is considered as the encryption key, and a collection of corresponding probes containing complementary DNA sequences is considered as the decryption key. Both the keys are generated by a key generator. The encryption key and decryption key are sent to the sender and receiver, respectively, securely.

Step 2: In the second step, the plaintext of a file or data is transformed into its binary form. Then, this binary form is embedded into a DNA chip in terms of DNA sequence. It must be noted that no one is able to decrypt the ciphertext without the decryption key.

Step 3: The receiver decrypts the ciphertext, i.e., cipher DNA using the complementary DNA sequence or decryption key. Then, the receiver can use an appropriate software to retrieve the original plaintext.

(2) **Genetic Programming:** A Genetic Algorithm (GA) can be defined as a soft computing technique based on natural evolution. GA replicates the natural selection process in which the fittest of individuals are selected to reproduce offspring of the subsequent generation. It is mainly used to search for optimal values in each successive generation. The major advantage of DNA-GA approaches is the combination of the

high storage capacity and massive parallelism of DNA with the search capability of GAs. In DNA computing, GA is one of the best solutions to break the limit of the brute-force method. As 1 g of DNA has 10^{21} DNA bases, the results of each generation of GA can be easily encoded in DNA bases by using the binary form. A large population can carry a large range of genetic diversity. Thus, in a few generations, it can generate high-fitness chromosomes, and the size of the search space is effectively reduced. In addition, if there is an experiment in vitro operations on DNA, it integrally includes errors. These errors can be ignored by executing GAs than by implementing deterministic algorithms. Here, errors can be referred to as contributing factors in GA.

In 1997, Deaton et al. [42] have proposed a GA based on DNA computing for efficient encoding. Yoshikawa et al. [43] have combined pseudo-bacterial GA with the DNA encoding method. In 1999, Chen et al. [44] have implemented DNA computing-based GA in a laboratory to solve some problems, such as the royal road, Max 1 s, and cold war problems. Wood and Chen [45] have proposed a DNA strand design that is well-matched for the royal road problem by using a GA. They have used the vitro evolution started with a haphazard population-based on DNA bases. In 2004, Yuan et al. [46] have designed a DNA computing-based GA to solve the maximal clique problem that is accomplished to produce an accurate solution within a few rounds. The simulation of Yuan and Chen indicates that the time required for their scheme is linear with the number of vertices of the particular network.

(3) **Cryptography:** Information security plays a major role in several areas like military relations, confidential business, financial institutions, and so on. In conventional systems, customers or users are restricted to their own domain. In today's advanced technologies, users are much interested to store and access data from outside of their domain. As a result, data security has become extremely important due to the involvement of numerous attackers and hackers, who are constantly attempting to hack users' personal and sensitive data or file for their own gain or to generate revenue. They often use their fake data to replace the original data. Thus, users' data may again face data security problems.

Several cryptographic techniques are already proposed by researchers for data transfer. Identity-Based Encryption (IBE) was first introduced by Shamir [47]. Here, the data sender defines an identity that should be matched by the recipient to decrypt the data. After some years, a novel model, namely fuzzy identity-based encryption has been suggested in which a user's identity is represented with several descriptive attributes [48]. Sahai and Waters [49] have proposed Attribute-Based Encryption (ABE) for providing access to complex data. ABE is divided into two types: ciphertext policy-based ABE and key policy-based ABE [50]. Most of the conventional approaches are mainly based on complex mathematical equations in which researchers mainly focus on increasing the complexity through the modification of the equations [51]. To improve data security, some conventional approaches encrypt data, and then, insert the encrypted data into various multimedia data or files as the cover media, such as image, video, audio, etc. [52]. Hence, malicious users or hackers are not able to find the required data to breach the cryptosystems.

DNA cryptography is currently one of the fastest-growing technologies based on the concept of DNA computing. In DNA cryptography, DNA computing is used to encrypt the data to achieve a strong security mechanism that prevents unauthorized, attackers and malicious users from reading the original data content. Here, DNA bases, namely A, T, G, and C are used for data encryption. Many DNA computing-based cryptographic algorithms have been proposed, including asymmetric and symmetric key cryptosystems, triple stage DNA cryptography, DNA-based chaotic computation, and many more. For a strong DNA-based encryption scheme, there are a few conditions as mentioned below:

In [53], a novel access control model is introduced in which decimal encoding rule, complementary pair-rule, ASCII values, and DNA encoding rule are used for data encryption. In the coupled map lattice-based scheme [54], authors have used coupled map lattice and DNA strands for improving data security. Along with DNA cryptography, this scheme uses the SHA-256 algorithm. In 2019, Wang et al. [55] have proposed a recombinant DNA technique for data security. A bio-inspired cryptosystem was proposed by Reddy et al. [56] based on DNA computing, and central dogma molecular biology is used to encrypt and decrypt any data. This scheme consists of three main phases: (1) key generation (2) encryption, and (3) decryption. However, this scheme does not support much randomness and takes much time for data encryption.

(4) Polymerase Chain Reaction: As discussed earlier, PCR is much effective in producing a large number of copies from a small amount of DNA. Here, primer is required to design an amplification process. Nowadays, PCR is also used for cryptography by making the encryption key or secret key compound. This means that the encryption key contains both the PCR primer pairs and public key, while the decryption key contains both the complementary primer pairs and private key [57]. To communicate any data, two primers are shared between the sender and receiver in this technique through a safe channel before starting the data encryption processes.

In the process of data encryption, algorithms like RSA, Advanced Encryption Standard (AES), etc., may be used as the preprocessing stage. Then, a coding rule converts the ciphertext into the corresponding DNA strand. Thus, a new ciphertext is generated. As shown in Fig. 10, DNA cipher refers to the ciphertext of the data in the form of a DNA sequence and plaintext refers to the binary form of the data. The DNA cipher is surrounded by secret primers, which are then combined with other unknown DNA strands. PCR technique is used to generate these unknown DNA strands. The sender then sends the mixed DNA to the sender. The receiver obtains the block DNA cipher by performing PCR with the aid of the secret primer, and then, reverses the entire data encryption operation. No one can retrieve the DNA cipher without any prior knowledge of two primers.

The above-mentioned PCR approach has several implications, such as Cui et al. [58] have suggested a data encryption approach based on PCR amplification and DNA coding. Tanaka et al. [59] have introduced a public key scheme based on a DNA computing-based one-way function. Yamamoto et al. [60] have proposed a novel encryption scheme that uses both modern algorithms and molecular techniques to provide two-level security by using the PCR-based large-scale DNA space. If one security level is breached, the system is kept secure by using another level. However, one of the major issues of this PCR-based security is the exchanging of the secret key between the sender and recipient.

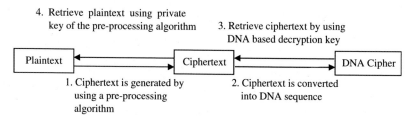

Fig. 10 PCR based cryptography.

(5) **DNA-based Steganography:** In DNA-based steganography, one or more DNA sequences are used as inputs in terms of the plaintext of the data. Randomly generated DNA-based secret key strands are attached to the DNA input sequence. The resulting plaintext DNA sequence is then hidden by mixing it with a large number of randomly generated DNA strands. In the data decryption phase, the secret key is used to decode the DNA strands by using a collection of known DNA separation methods. Such plaintext messages or DNA sequences can be separated by using the hybridization process with the help of the secret key strand's complements. As the DNA-based steganography approach minimizes the total cost of the encryption phase, it is efficient. But, it is not secured for the applications, which are based on statistical analysis.

Risca [61] has used standard biological protocols to propose a DNA-based steganographic technique. This approach encrypts the information by encoding it as a DNA sequence flanked by two separate private primers. Gehani et al. [62] have suggested an improvement technique by distinguishing the distracted DNA stands from the probability distribution of the plaintext of the file or data or message. Roy et al. [63] have proposed a DNA computing-based steganographic scheme in which confidential data are embedded into a document, i.e., a Word document. Here, at first, the plaintext is embedded into a DNA strand, and then, this DNA strand is again encrypted by using DNA computing. In 2015, Gupta and Singh [64] have proposed a DNA computing-based data hiding technique by using two main processes: (1) substitution process, and (2) central dogma. They have converted data into a DNA sequence. Then, it is transformed into a ribonucleic acid followed by transforming into a protein sequence. Tuncer and Avci [65] have proposed a novel probabilistic secret sharing scheme by utilizing DNA XOR operation. In this scheme, data is embedded into green, red, and blue channels of a random cover image. However, this scheme is vulnerable to man-in-the-middle attacks. Wang et al. [66] have proposed a data hiding scheme in 2017 by using a randomly generated DNA XOR rule. This scheme consumes much time, and it is mainly based on two algorithms: (1) embedding watermarking, and (2) extracting watermarking.

(6) **DNA Fingerprint:** DNA computing-based fingerprint involves detecting variations of DNA bases in a particular region of the DNA sequence as a small change of DNA bases is replicated several times in the actual sequences. During density gradient centrifugation,

these repeated DNA peaks are isolated from bulk genomic DNA as distinct peaks. It is a lab procedure, which is mainly used to solve a criminal case. Here, the DNA sample of the crime scene is matched with a suspect's DNA sample. If both the DNA samples are the same, then in most of the cases, the suspect is considered guilty for the crime. Otherwise, an investigation is conducted in some other aspects. DNA fingerprinting is also used for paternity determination [67].

(7) **Boolean Circuit:** The Boolean circuit is considered an important Turing-based correspondent model of parallel computation. Here, a Boolean circuit that is n input bounded fan in can be visualized as an acyclic directed graph, S. There are mainly two types of node in the graph, namely input node with zero in–degree and gate node with at most two in–degree. Each input node is connected with a Boolean variable from the input set and each gate node is connected with a Boolean function. The collection of all such functions is known as circuit basis.

Researchers have developed DNA computing-based Boolean circuits. The first DNA computing-based Boolean circuit was proposed by Ogihara and Ray [5]. They have implemented a real-time simulation of the depth of the Boolean circuit. In [68], it has been estimated that the time complexity of the Boolean circuit must be proportional to the circuit's size. DNA computing-based Boolean circuit is based on the polymerase chain reaction. Here, the logical gates consist of one DNA strand only, which significantly decreases leakage reactions and restoration steps of signal such that the performance of the circuit is improved. A large logical circuit can be built from the gates by simple cascades. In particular, it has been shown a short and compact logic circuit that calculates the four-bit input square root function [69]. Here, R1, R2, R3, and R4, are the input bits and the total number of output bits is 8. The first four output bits represent one position of the square root, while the next four bits represent the first decimal position.

(8) **Clustering:** Clustering deals with the construction of meaningful relationships in a high-dimensional complex dataset. DNA computing can be used to develop clustering techniques. This technique is suitable, when dealing with huge datasets and there are heterogeneous characters in the dataset. Nowadays, clustering is used in many application areas, such as facility allocation [70], signal processing [71], astronomy [72], medical data analysis [73], and many more. The main

challenge of clustering is the combinatorial explosion of alternatives, which must be carefully considered. This clustering's combinatorial efforts are enormous in essence. There must be a computation and process of all combinations of data points based on all possible numbers of clusters for constructing an optimal resolution for the clustering object. For a given large number of calculations, the silicon-based computer may not be very efficient, when dealing with huge problems with high dimensionality.

DNA computing-based clustering technique helps to reduce the complexity of time by using the massively parallel computation of DNA computing. Here, all the possible solutions can be easily explored all at once in a parallel manner. A silicon-based computer solves a problem by assigning tasks to numerous processes. Thus, the allocation technique results in sequential or linear processing. On the other hand, DNA computing-based technique can solve such problems by using a single parallel process. Interestingly, this novel technology also offers to manage energy efficiently. Each DNA computing-based clustering algorithm is based on the certain underlying rationale and supports a positive understanding of the data. In addition, such an algorithm also derives from its fundamental optimization technique, computational enhancements, and validation tools. Many researchers also have proposed clustering techniques by using fuzzy clustering [74] and curve techniques [75].

(9) **DNA Computer:** DNA computer uses DNA bases, namely A, T, G, and C, as the memory units. Here, recombinant DNA techniques are used for executing fundamental operations. Test tubes are used in DNA computers for computations. Sometimes, a glass-coated in 24 K gold is also used for computations. In a DNA computer, both input and output are DNA strands in which information is encoded, and a program is executed in terms of many biochemical operations that have many effects, such as synthesizing, modifying, extracting, and cloning the strands of DNA. The main difference between DNA computers and traditional computers is the storage capacity. Due to the storage capacity of DNA, DNA computers support a large memory by using DNA bases, namely G, A, C, and T. On the other hand, electronic computers support comparatively less storage capacity by using 0 and 1.

(10) **Scheduling:** Nowadays, DNA computing is used for job scheduling problem algorithms due to its unique properties, such as vast parallel,

tiny parts, and organic edges. The job scheduling problem can be easily managed by a standard computer or by a human. In [76], Ibrahim et al. have proposed a DNA computing-based algorithm to solve the job scheduling problem. They have added some operations to the evolutionary operations to achieve better performance. Tian et al. [77] have proposed another DNA computing-based algorithm in which an encoding scheme for the job shop scheduling problem is developed along with some novel DNA computing-based operations. Here, all possible solutions are made after an initial result or solution is constructed. Then, DNA computing-based operations are utilized for finding the optimal schedule. The complexity of DNA computing-based algorithm of Tian et al. is $O(n^2)$. In addition, the final strand's length of the optimal schedule is always within the appropriate range.

4. Conclusions and future works

In this digital era, almost all transactions are executed over the internet. As there are numerous attackers and hackers, users' confidential data face security issues. DNA computing can be used to improve data security. DNA computing is a form of molecular computing in which DNA bases (A, T, G, and C) are used for encoding and processing. However, DNA computing is still in its early phase. In this chapter, the fundamentals of human DNA, such as structure, polymerase chain reaction, history, etc., were presented. The basics of DNA computing and its importance, including its advantages, disadvantages, and operations of DNA computing were discussed. Moreover, many applications of DNA computing in different fields, such as DNA chips, genetic programming, data storage, cryptography, and many more, are presented in detail.

As DNA computing is very emerging, it can be used in many advanced technologies, such as cloud computing, big data, quantum computing, and many more, to improve efficiency and performance. In a cloud computing environment, data owners store their data on the cloud server. DNA computing-based cryptography can be used to develop a novel model for the cloud computing environment to improve data security. In the future, DNA computing can also be used to store big data by using DNA bases. Moreover, there is a huge scope to develop DNA computers for fast processing and to use the concept of DNA computing in quantum computing to achieve better performance.

References

[1] L.M. Adleman, Computing with DNA, Sci. Am. 279 (2) (1998) 54–61.

[2] J.D. Watson, T.A. Baker, S.P. Bell, A. Gann, M. Levine, R. Losick, Molecular Biology of the Gene (International Ed.), Pearson Education, 2004.

[3] L.M. Adleman, Molecular computation of solutions to combinatorial problems, Science 266 (5187) (1994) 1021–1024.

[4] D.I. Lewin, DNA computing, Comput. Sci. Eng. 4 (3) (2002) 5–8.

[5] M. Ogihara, A. Ray, Simulating Boolean circuits on a DNA computer, Algorithmica 25 (2) (1999) 239–250.

[6] S. Tagore, S. Bhattacharya, M. Islam, M.L. Islam, DNA computation: application and perspectives, J. Proteom. Bioinform. 3 (7) (2010).

[7] D. Boneh, C. Dunworth, R.J. Lipton, J. Sgall, Making DNA computers error resistant, DNA Based Comput. II 44 (1996) 163–170.

[8] S. Namasudra, G.C. Deka, Introduction of DNA computing in cryptography, in: S. Namasudra, G.C. Deka (Eds.), Advances of DNA computing in cryptography, Taylor & Francis, 2018, pp. 17–34.

[9] R. Deaton, J. Chen, J.-W. Kim, M.H. Garzon, D.H. Wood, Test tube selection of large independent sets of DNA oligonucleotides, in: Nanotechnology: Science and Computation, Springer, 2006, pp. 147–161.

[10] G.G. Owenson, M. Amos, D.A. Hodgson, A. Gibbons, DNA-based logic, Soft Comput. 5 (2) (2001) 102–105.

[11] G. Paun, G. Rozenberg, A. Salomaa, DNA Computing: New Computing Paradigms, Springer, 2005.

[12] C. Bancroft, T. Bowler, B. Bloom, C.T. Clelland, Long-term storage of information in DNA, Science 293 (5536) (2001) 1763.

[13] S. Namasudra, G.C. Deka, R. Bali, Applications and future trends of DNA computing, in: S. Namasudra, G.C. Deka (Eds.), Advances of DNA Computing in Cryptography, Taylor & Francis, 2018, pp. 181–192.

[14] P. Pavithran, S. Mathew, S. Namasudra, G. Srivastava, A novel cryptosystem based on DNA cryptography, hyperchaotic systems and a randomly generated Moore machine for cyber physical systems, Comput. Commun. 188 (2022) 1–12.

[15] S. Namasudra, Fast and secure data accessing by using DNA computing for the cloud environment, IEEE Trans. Serv. Comput. (2020), https://doi.org/10.1109/TSC.2020.3046471.

[16] H. Dongming, D.-F. Zhou, Y. Sheng, Y. Shaoliang, Z. Luozhi, Z. Xin, Image encryption using exclusive-OR with DNA complementary rules and double random phase encoding, Phys. Lett. A 383 (9) (2019) 915–922.

[17] L. Kari, R. Kitto, G. Thierrin, Codes, involutions, and DNA encodings, in: Formal and Natural Computing, Springer, 2002, pp. 376–393.

[18] S. Shah,, D. Limbachiya, and M. K. Gupta, "DNACloud: A potential tool for storing big data on DNA," arXiv preprint arXiv:1310.6992, 2013.

[19] A.C. Patel, C.G. Joshi, Deoxyribonucleic acid as a tool for digital information storage: an overview, Indian J. Vet. Sci. Biotechnol. 15 (1) (2019) 1–8.

[20] D.L. Nelson, A.L. Lehninger, M.M. Cox, Lehninger Principles of Biochemistry, Macmillan, 2008.

[21] J.D. Watson, F.H.C. Crick, A structure of deoxyribose nucleic acid, Nature 171 (1953) 737–738, https://doi.org/10.1038/171737a0.

[22] L. Garibyan, N. Avashia, Research techniques made simple: polymerase chain reaction (PCR), J. Invest. Dermatol. 133 (3) (2013), https://doi.org/10.1038/jid.2013.1.

[23] W.B. Coleman, G.J. Tsongalis, Laboratory approaches in molecular pathology—the polymerase chain reaction, in: Diagnostic Molecular Pathology, Academic Press, 2017, pp. 15–23.

[24] L. Gonick, M. Wheelis, The Cartoon Guide to Genetics, Harper Perennial, 1991.

[25] K. Drlica, Understanding DNA and gene cloning: a guide for the curious, in: Understanding DNA and gene cloning: a guide for the curious, ed. 2, John Wiley & Sons, 1992.

[26] R.J. Lipton, Using DNA to solve NP-complete problems, Science 268 (4) (1995) 542–545.

[27] R.P. Feynman, D. Gilbert, Miniaturization, Reinhold, New York, 1961, pp. 282–296.

[28] E.B. Baum, Building an associative memory vastly larger than the brain, Science 268 (5210) (1995) 583–585.

[29] Q. Ouyang, P.D. Kaplan, S. Liu, A. Libchaber, DNA solution of the maximal clique problem, Science 278 (5337) (1997) 446–449.

[30] Q. Liu, L. Wang, A.G. Frutos, A.E. Condon, R.M. Corn, L.M. Smith, DNA computing on surfaces, Nature 403 (6766) (2000) 175–179.

[31] L.M. Smith, R.M. Corn, A.E. Condon, M.G. Lagally, A.G. Frutos, Q. Liu, A.J. Thiel, A surface-based approach to DNA computation, J. Comput. Biol. 5 (2) (1998) 255–267.

[32] J.H. Reif, T.H. LaBean, M. Pirrung, V.S. Rana, B. Guo, C. Kingsford, G.S. Wickham, Experimental construction of very large scale DNA databases with associative search capability, in: International Workshop on DNA-Based Computers, Springer, Berlin, Heidelberg, 2001, pp. 231–247.

[33] M. Sarkar, P. Ghosal, S.P. Mohanty, Exploring the feasibility of a DNA computer: design of an ALU using sticker-based DNA model, IEEE Trans. Nanobioscience 16 (6) (2017) 383–399.

[34] S. Namasudra, R. Chakraborty, A. Majumder and N. R. Moparthi, "Securing multimedia by using DNA based encryption in the cloud computing environment", in ACM Trans. Multimedia Comput. Commun. Appl., vol. 16, no. 3s, pp. 1–19, 2020, DOI: https://doi.org/10.1145/3392665.

[35] S. Namasudra, P. Roy, Time saving protocol for data accessing in cloud computing, IET Commun. 11 (10) (2017) 1558–1565.

[36] https://cs.stanford.edu/people/eroberts/courses/soco/projects/2003-04/dna-computing/evaluation.htm, 2003 [Accessed on 12 May, 2021].

[37] J. Watada, R.B.A. Bakar, DNA computing and its applications, in: 2008 Eighth international conference on intelligent systems design and applications, vol. 2, 2008, pp. 288–294.

[38] L. Kari, S. Seki, P. Sosík, DNA Computing: Foundations and Implications, Handbook of Natural Computing, Springer, 2012, pp. 1073–1127.

[39] G. Ventimiglia, S. Petralia, Recent advances in DNA microarray technology: an overview on production strategies and detection methods, BioNano Sci. 3 (4) (2013) 428–450.

[40] E. Czeizler, E. Czeizler, A short survey on Watson-Crick automata, Bull. EATCS 88 (3) (2006) 104–119.

[41] G.Z. Cui, Y. Liu, X. Zhang, New direction of data storage: DNA molecular storage technology, Comput. Eng. Appl. 42 (26) (2006) 29–32.

[42] R. Deaton, R.C. Murphy, J.A. Rose, M. Garzon, D.R. Franceschetti, S.E. Stevens, A DNA based implementation of an evolutionary search for good encodings for DNA computation, in: Proceedings of 1997 IEEE International Conference on Evolutionary Computation (ICEC'97), 1997, pp. 267–271.

[43] T. Yoshikawa, T. Furuhashi, Y. Uchikawa, The effects of combination of DNA coding method with pseudo-bacterial GA, in: Proceedings of 1997 IEEE International Conference on Evolutionary Computation (ICEC'97), 1997, pp. 285–290.

[44] J. Chen, E. Antipov, B. Lemieux, W. Cedeño, D.H. Wood, DNA computing implementing genetic algorithms, in: Evolution as Computation, 1999, pp. 39–49.

[45] D.H. Wood, J. Chen, Physical separation of DNA according to royal road fitness, in: Proceedings of the 1999 Congress on evolutionary computation-CEC99 (cat. No. 99TH8406), vol. 2, 1999, pp. 1011–1016.

[46] Y. Li, C. Fang, Q. Ouyang, Genetic algorithm in DNA computing: a solution to the maximal clique problem, Chin. Sci. Bull. 49 (9) (2004) 967–971.

[47] A. Shamir, Identity-based cryptosystems and signature schemes, in: Workshop on the theory and application of cryptographic techniques, Springer, Berlin, Heidelberg, 1984, pp. 47–53.

[48] R. Yanil, G. Dawu, W. Shuozhong, Z. Xinpeng, New fuzzy identity-based encryption in the standard model, Informatica 21 (3) (2010) 393–407.

[49] A. Sahai, B. Waters, Fuzzy identity-based encryption, in: Annual international conference on the theory and applications of cryptographic techniques, Springer, Berlin, Heidelberg, 2005, pp. 457–473.

[50] J. Bethencourt, A. Sahai, B. Waters, Ciphertext-policy attribute-based encryption, in: 2007 IEEE symposium on security and privacy (SP'07), IEEE, 2007, pp. 321–334.

[51] C.C. Lin, N.-L. Hsueh, A lossless data hiding scheme based on three-pixel block differences, Pattern Recognit. 41 (4) (2008) 1415–1425.

[52] S. Voloshynovskiy, T. Pun, J. Fridrich, F.P. González, N. Memon, Security of data hiding technologies, Signal Process. 83 (10) (2003) 2065–2067.

[53] S. Namasudra, P. Roy, P. Vijayakumar, S. Audithan, B. Balamurugan, Time efficient secure DNA based access control model for cloud computing environment, Future Gener. Comput. Syst. 73 (2017) 90–105.

[54] W. Xingyuan, Y. Hou, S. Wang, R. Li, A new image encryption algorithm based on CML and DNA sequence, IEEE Access 6 (2018) 62272–62285.

[55] Y. Wang, Q. Han, G. Cui, J. Sun, Hiding message based on DNA sequence and recombinant DNA technique, IEEE Trans. Nanotechnol. 18 (2019) 299–307.

[56] M.I. Reddy, A.P.S. Kumar, K.S. Reddy, A secured cryptographic system based on DNA and a hybrid key generation approach, Biosystems 197 (2020).

[57] M. Roy, S. Chakraborty, K. Mali, S. Mitra, I. Mondal, R. Dawn, D. Das, S. Chatterjee, A dual layer image encryption using polymerase chain reaction amplification and dna encryption, in: 2019 International Conference on Opto-Electronics and Applied Optics (Optronix), IEEE, 2019, pp. 1–4.

[58] G. Cui, L. Qin, Y. Wang, X. Zhang, An encryption scheme using DNA technology, in: 2008 3rd International Conference on Bio-Inspired Computing: Theories and Applications, IEEE, 2008, pp. 37–42.

[59] K. Tanaka, A. Okamoto, I. Saito, Public-key system using DNA as a one-way function for key distribution, Biosystems 81 (1) (2005) 25–29.

[60] M. Yamamoto, S. Kashiwamura, A. Ohuchi, M. Furukawa, Large-scale DNA memory based on the nested PCR, Nat. Comput. 7 (3) (2008) 335–346.

[61] V.I. Risca, DNA-based steganography, Cryptologia 25 (1) (2001) 37–49.

[62] A. Gehani, T. LaBean, J. Reif, DNA-based cryptography, in: Aspects of Molecular Computing, Springer, 2003, pp. 167–188.

[63] S. Roy, S. Sadhukhan, S. Sadhu, S.K. Bandyopadhyay, A novel approach towards development of hybrid image steganography using DNA sequences, Indian J. Sci. Technol. 8 (22) (2015) 1–7.

[64] R. Gupta, R.K. Singh, An improved substitution method for data encryption using DNA sequence and CDMB, in: International Symposium on Security in Computing and Communication, Springer, Cham, 2015, pp. 197–206.

[65] T. Tuncer, E. Avci, A reversible data hiding algorithm based on probabilistic DNA-XOR secret sharing scheme for color images, Displays 41 (2016) 1–8.

[66] B. Wang, Y. Xie, S. Zhou, C. Zhou, X. Zheng, Reversible data hiding based on DNA computing, Comput. Intell. Neurosci. 2017 (2017), https://doi.org/10.1155/2017/7276084.

[67] P. Helminen, M-L. Lokki, C. Ehnholm, A. Jeffreys, and L. Peltonen, "Application of DNA fingerprints to paternity determinations," Lancet, vol. 331, no. 8585, pp. 574–576, 1998.

[68] M. Amos, P.E. Dunne, DNA simulation of Boolean circuits, in: Proceeding of 3rd Annual Genetic Programming Conference, 1997, pp. 679–683.

[69] Z. Ekmekc, B. Ulu, E. Ekmekci, A new molecular logic circuit with 4 bit input, Sens. Actuators B 231 (2016) 655–658.

[70] R.B.A. Bakar, J. Watada, W. Pedrycz, A DNA computing approach to data clustering based on mutual distance order, in: Proceedings 9th Czech-Japan Seminar, 2006, pp. 139–145.

[71] K.L. Oehler, R.M. Gray, Combining image compression and classification using vector quantization, in: IEEE transactions on pattern analysis and machine intelligence, 1995, pp. 461–473. vol. 17, no. 5.

[72] K.L. Wagstaff, V.G. Laidler, Making the most of missing values: object clustering with partial data in astronomy, in: P.L. Shopbell, et al. (Eds.), Astronomical data analysis software and systems XIV, vol. 347, Astronomical Society of the Pacific Conference Series, 2005, pp. 172–176.

[73] J. Lin, D. Karakos, D. Demner-Fushman, S. Khudanpur, Generative content models for structural analysis of medical abstracts, in: Proceedings of the BioNLP workshop on Linking Natural Language Processing and Biology at HLT-NAACL, 2006, pp. 65–72.

[74] W. Pedrycz, Knowledge-Based Clustering: From Data to Information Granules, John Wiley & Sons, 2005.

[75] L. Cleju, P. Fränti, X. Wu, Clustering based on principal curve, in: Scandinavian Conference on Image Analysis, Springer, Berlin, Heidelberg, 2005, pp. 872–881.

[76] G.J. Ibrahim, T.A. Rashid, A.T. Sadiq, Evolutionary DNA computing algorithm for job scheduling problem, IETE J. Res. 64 (4) (2018) 514–527.

[77] X. Tian, X. Liu, H. Zhang, M. Sun, Y. Zhao, A DNA algorithm for the job shop scheduling problem based on the Adleman-Lipton model, Plos one 15 (12) (2020).

About the authors

Mr. Tarun Kumar is a research scholar in the Department of Computer Science and Engineering at the National Institute of Technology Patna, Bihar, India. He is also an Assistant Professor in the School of Computing Science and Engineering, Galgotias University, Greater Noida, India. He has 16 years of experience in academics. His research interests are Cloud Computing and DNA Computing. He has published several papers in peer reviewed journals and international conferences. He also organized and attended several workshops.

Suyel Namasudra is an assistant professor in the Department of Computer Science and Engineering at the National Institute of Technology Agartala, Tripura, India. Before joining the National Institute of Technology Agartala, Dr. Namasudra was an assistant professor in the Department of Computer Science and Engineering at the National Institute of Technology Patna, Bihar, India, and a post-doctorate fellow at the International University of La Rioja (UNIR), Spain. He has received Ph.D. degree in Computer Science and Engineering from the National Institute of Technology Silchar, Assam, India. His research interests include DNA computing, blockchain technology, cloud computing, and IoT. Dr. Namasudra has edited 4 books, 5 patents, and 60 publications in conference proceedings, book chapters, and refereed journals like IEEE TII, IEEE T-ITS, IEEE TSC, IEEE TCSS, ACM TOMM, ACM TALLIP, FGCS, CAEE, and many more. He has served as a Lead Guest Editor/Guest Editor in many reputed journals like ACM TOMM (ACM, IF: 3.144), CAEE (Elsevier, IF: 3.818), CAIS (Springer, IF: 4.927), CMC (Tech Science Press, IF: 3.772), Sensors (MDPI, IF: 3.576), and many more. Dr. Namasudra has participated in many international conferences as an Organizer and Session Chair. He is a member of IEEE, ACM, and IEI. Dr. Namasudra has been featured in the list of the top 2% scientists in the world in 2021 and 2022, and his h-index is 25.

CHAPTER TWO

Security, privacy, and trust management in DNA computing

Maria Fernandes[a],*, Jérémie Decouchant[b], and Francisco M. Couto[c]
[a]SnT–University of Luxembourg, Esch-sur-Alzette, Luxembourg
[b]Faculty of Electrical Engineering, Mathematics and Computer Science, Delft University of Technology, Delft, Netherlands
[c]LASIGE, Departamento de Informática, Faculdade de Ciências, Universidade de Lisboa, Lisbon, Portugal

Contents

Abstract

DNA computing is an emerging field that aims at enabling more efficient data storage and processing. One principle of DNA computing is to encode some information

---

*Current affiliation: Big Data Institute, University of Oxford, Oxford, United Kingdom; Wellcome Centre for Human Genetics, University of Oxford, Oxford, United Kingdom.

Advances in Computers, Volume 129
ISSN 0065-2458
https://doi.org/10.1016/bs.adcom.2022.08.009

(e.g., image, video, programming scripts) into a digital DNA-like sequence and then synthesize the corresponding DNA molecule. Synthesizing this molecule using digital or real human genomic fragments theoretically opens the possibility for privacy attacks, which have been demonstrated on a large array of human genomic data. These privacy attacks aim at breaching the privacy of DNA samples, allowing an attacker to discover privacy-critical information from the partial or complete DNA information of an individual. In the context of DNA computing, novel privacy attacks will certainly emerge and could consist in discovering a part of a particular script or video that is privacy-critical. It is therefore important to consider whether privacy attacks and defense mechanisms can be used when manipulating genomic data. First, this chapter provides the background about genomic data, and its modern generation and processing. It then provides a survey on known genomic privacy attacks, and presents the privacy-enhancing technologies that have been designed to protect genomic data. Later, this chapter also introduces the current trust management methods one can rely on to further secure DNA storage and processing methods, before discussing how DNA computing currently relates to those attacks and privacy-preserving technologies. Finally, this chapter presents future research avenues.

Abbreviations

1000 GP	1000 Genomes Project
AES	Advanced Encryption Standard
CSP	Cloud Service Provider
dbGAP	database of Genotypes and Phenotypes
DNA	Desoxyribonucleic Acid
GA4GH	Global Alliance for Human Genetics
GWASs	Genome-Wide Association Studies
IBS	Identical-By-State
LD	Linkage Disequilibrium
NGS	Next Generation Sequencing
SGX	Software Guard Extensions
SMPC	Secure Multiparty Computations
SNPs	Single Nucleotide Polymorphisms
TEEs	Trusted Execution Environments
WGS	Whole Genome Sequencing

1. Introduction

Human Deoxyribonucleic Acid (DNA) is the genetic material contained in human cells that encode all the information necessary for an organism's functioning and reproduction. Therefore, understanding how DNA modulates all those processes became a hot topic due to its potential contributions on fields such as healthcare and forensics.

The DNA molecule is composed of four nucleotides, i.e., adenine (A), thymine (T), guanine (G), and cytosine (C), that bind together to form a double helix. The two strands that compose the double helix are antiparallel, and, for each individual, contain genomic variations at certain positions in the genome. Genomic variations are alternative genomic subsequences where individuals might differ, the most common among people are Single Nucleotide Polymorphisms (SNPs). SNPs are genomic variation where there is a single nucleotide difference, either because a nucleotide has been replaced by another one, or because a nucleotide has been inserted or deleted. SNPs are also the most studied genomic variations because of their low complexity. Genomic variations among and between individuals are interconnected by complex statistical relations, such as Linkage Disequilibrium (LD) and kinship, which respectively describe dependencies between different regions in the genome, and between genomes from relatives. LD describes the nonrandom associations of a group of genomic variations, which comes from their simultaneous transmission during cell divisions. The resulting statistical associations can be exploited to infer a genomic variation when others are known. *Kinship* or kin relationships describe hereditary connections and marriage ties between individuals, including direct bonds (e.g., children, parents, grandparents) and collateral bonds (e.g., siblings, cousins, aunts, uncles). Kin relationships result in genomic similarities between members of the same family because of the biological inheritance process that dictates the transmission of genetic information from parents to their children.

The notions of genotypes and phenotypes are important to understand the genomic data field, and are defined and correlated as follows. An individual's genotype corresponds to its set of genomic variations. The phenotype of an individual is the set of its observable physical characteristics, such as its appearance (e.g., skin color/type, hair color/type, body silhouette), development (e.g., blood cells, hormones production), and behavior. The phenotype is the result of the expression of the information in the genotype in combination with the environment interactions (i.e., epigenetics). Since the environment has a great impact on the phenotype, even individuals with similar genotypes, such as twins, can present different phenotypes.

The inclusion of genomic data in multiple scientific areas was promoted by the advances of Next Generation Sequencing (NGS) technologies, which decreased the data generation cost and, consequently, increased the availability of genomic data. The first step to reveal the DNA information encoded by a sample is to sequence it using a NGS technology. NGS technologies are machines that perform a chain of chemical reactions on a biological sample (e.g., a blood sample) to translate it into its digital equivalent, in the form of sequences of nucleotide called reads. After sequencing, the information retrieved is treated in a processing pipeline to identify its special features, i.e., its genomic variations.

Let us briefly introduce the main steps of this processing pipeline: read alignment and variant calling. Read alignment is the first step required to determine the biological information contained on the reads produced by NGS technologies. In this step the reads are mapped to a reference genome to determine their original position in the genome. The reference genome is a synthetic genome sequence containing the most common genomic variations in the global population. The human reference genome was assembled based on the genome of several individuals from all around the world. Therefore, it represents a global synthetic sequence and not a single individual's genome. Variant calling is the step where the aligned reads are compared to the reference genome to identify the positions at which they differ. In this step, a quality score is used to distinguish from real genomic variations and sequencing errors.

DNA data have been increasingly used in healthcare data processing pipelines, such as personalized medicine and disease predisposition testing, research, direct-to-consumer services, and forensics [1]. The high throughput of sequencing machines, which encourages huge DNA data production, and the intensive computations required to process genomic data, often leads these processing pipelines to be outsourced to cloud environments that provide powerful computational resources at an affordable cost. Cloud-based environments for biomedical data have been described in Refs. [2, 3]. However, although public clouds provide powerful computational resources at an affordable cost, they are managed by a third party, i.e., a Cloud Service Provider (CSP). Using public cloud therefore raises new challenges in order to keep genomic data secure [4]. Furthermore, the increasing availability of genomic data and its large array of potential applications encouraged data sharing to accelerate the understanding of DNA functioning, and allow the largest number to benefit from the information it encodes. Due to the important expected applications of DNA data, and because of its greater availability, several genomic data repositories were created to support and speed up knowledge acquisition and creation. Some examples of those repositories include the 1000 Genomes Project (1000 GP), the database of Genotypes and Phenotypes (dbGaP), and the 100,000 Genomes projects. The 1000 GP is a publicly available repository of human genomes, launched in 2008 with the goal of creating a resource on human genetic variations. dbGaP is a repository containing genotype and phenotype data, which is an important collection for Genome-Wide Association Studies (GWASs), and genome-diseases correlation studies. The 100,000 Genomes Project is an England effort to provide a repository of cancer and rare diseases genomic data to boost research in these fields. GWASs are a particular DNA study whose goal is to link observed genomic variations with particular diseases. GWAS consists of a massive scan over multiple individuals' genomes to search for particular patterns that help to predict occurrences of a disease. Once those patterns are identified, they can be used to study the contribution of genes to the disease, and improve its diagnostic and treatment. Along with the increase of genomic data processing, new requirements have emerged such as the needs for high performance and privacy. The high-performance demand pushed for the use of scalable and cost-efficient environments, such as public clouds, that provide powerful computational resources and large storage capacity. However, as DNA data is directly linked to its owner identity, privacy breaches can occur when data is not protected before it is sent to the cloud.

The emphasis put on genomic data security has been growing with the application of genomic data in developing fields. Enforcing data security in an information system requires providing both privacy and trust. Privacy challenges may arise when genomic data, which is sensitive information, is stored and processed in a cloud environment, or shared. The privacy risks that appear when outsourcing biomedical data to public clouds without adequate protections are discussed in Refs. [5, 6]. Genomic data carry sensitive information such as predisposition to genetic diseases, physical traits and familial relations. As have seen described, members of a family share genomic traits, and genome correlations. Humbert et al. [7] demonstrate that the genomic privacy of a target individual decreases when genomic information from its family members is shared. This work highlights the increasing of privacy risks, since human genomes sequencing is constantly growing. In addition, the nonrevocable nature of DNA makes any potential data leakage result in privacy loss that can never be attenuated. Therefore, such data should be kept secret to prevent any harm to the owner, such as genetic discrimination, which could result in denial of health insurance, education, and employment, or blackmail [8]. From a security perspective, biological correlations between human genomes should also be considered when designing DNA data processing algorithms, since they can be exploited by an adversary to infer further information based on a partial genomic sequence and known statistics and/or to relate family members. Therefore, it is important to protect family relationship information to ideally prevent privacy attacks. A deeper discussion on privacy attacks on genomic data and existing privacy-preserving techniques can be found in Ref. [9].

DNA computing is an emerging field that aims at using DNA to store information to perform computations through chemical reactions. For this purpose, synthesized DNA is produced, by first composing DNA sequences in silico and then producing the corresponding DNA molecules in laboratory. Since each molecule is synthesized to store digital information, i.e., video, photo, code, its sequence is obtained by using a binary correspondence for each nucleotide. For example, a basic encoding of nucleotides over 2 bits could be: A = [00], T = [01], C = [10], and G = [11]. DNA computing promises the development of massive parallel computing technologies, which would allow complex problems to be solved in a short amount of time, instead of requiring weeks using conventional computers. These promises rely on the fact that millions of DNA molecules can interact simultaneously. However, this also increases the complexity of the output that a DNA computer would provide. Human DNA and synthesized DNA (generated for DNA computing) share the same structure; however, since they encode different information, they can have different properties. Yet, an interesting practical application of DNA computing, in which the synthesized DNA has similar properties as the DNA found in the human body, is the synthesis of DNA molecules to detect cancerous or damaged cells with the goal of triggering the repairing response on them. This process allows the prevention of the rapid multiplication of such cells and therefore slows the effects of the resulting illnesses. This process has been described by Shapiro *et al.* [10]. In this context, the synthesized DNA generated, which uses as template the human DNA, presents the statistical correlations it possesses such as LD and kinship. Therefore, similar to the sequenced human DNA, the synthesized DNA is vulnerable to security attacks performed on human DNA. Such attacks aim at disclosing sensitive information about the owner based on his/her DNA sequence or on the DNA sequence of its relatives. This chapter focuses on the security, privacy, and trust management aspect of human DNA. The goal of this chapter is to show the privacy risks that synthesized DNA and human DNA share. We therefore describe security attacks that have been performed on the latter and provide an overview of the state-of-the-art techniques one can use to prevent such attacks. In addition, this chapter also provides guidelines to maintain trust when designing privacy-preserving solutions for human–like DNA. Overall, this chapter makes the following contributions.

1. First, this chapter describes the security and privacy challenges in the context of human DNA and DNA computing, and put the emphasis on existing privacy attacks on genomic data.
2. Second, it surveys the scientific community's efforts to develop privacy-preserving techniques for genomic data.

3. Third, it highlights how trust can be maintained while processing genomic data.

4. Finally, this chapter discusses the relations and impact of privacy-preserving techniques and genomic data privacy attacks on DNA computing.

The remainder of this chapter is organized as follows. Section 2 provides an overview on the existing security attacks on genomic data. Section 3 presents the community effort to develop privacy-preserving solutions to prevent the reported security attacks. Section 4 describes the important trust management aspects for the design of secure genomic data processing and storage solutions. Section 5 discusses the previous sections, and the evolution of the scientific community best practices. Section 6 concludes this chapter and provides some insights for future work.

2. Security attacks

Over the past decades, many privacy attacks on genomic data have been reported. These attacks explored genomic data features obtained from a single individual, from a family, or from a target group. These attacks alerted the research community of the need to develop privacy-preserving genomic data processing and storage methods to benefit from cloud environments and keep private information secret.

Security attacks occur when an adversary has access or is able to modify data to which authorized access was not granted. Nowadays, DNA computing is used in biomedical sciences, with applications in healthcare and personalized medicine, where human DNA is used as a template, such as DNA molecules synthesis for abnormal cells detection. In this context, the synthesized DNA need to present properties similar to those of the DNA naturally found in the human cells. However, these similarities make the synthesized DNA susceptible to the privacy attacks reported on human DNA. Therefore, this section describes the security attacks on human DNA, which one should keep in mind when applying DNA computing for biomedical applications. When launching privacy attacks against human DNA, the adversary aims at discovering sensitive and not released information about a target individual or group. Privacy attacks were reported since 2006 and further exploitation of nonprotected genomic data was also described, with possibly severe consequences to the data owner, in particular, possible insurance denial and employment refusal [11, 12]. Due the demonstrated misuse of genomic data, researchers aimed at developing privacy-preserving techniques to adequately protect genomic data and prevent future harm for the data owners.

Fig. 1 presents a summary of all the different reported attacks on human DNA over the years for all the attacks categories. The genomic privacy attacks described in the literature can be classified into four categories, which are further discuss in this section:

1. Inference attacks
2. Re-identification attacks
3. Membership attacks
4. Recovery attacks

The main differences between the attack categories are the background knowledge (i.e., the information) the adversary has initially access to, and the kind of information the adversary tries to learn. Table 1 summarizes the adversary's background knowledge and target information for each category. The amount and quality of the background information directly impacts the outcome information of the attack, and consequently its success.

At the end, this section also discusses system exploits, which may target information systems that manipulate human DNA. Such exploits should be prevented so that privacy-preserving methods remain robust to attacks.

2.1 Inference attacks

Inference attacks aim at discovering additional sensitive information based on a partial or full genomic sequence. This kind of attack was also commonly used to infer the health status of target individuals based on their genomic sequence and on background knowledge about disease-related genes. The information retrieved from this kind of attack can be used by the other attacks categories. The background knowledge used for these attacks is some genomic information about a target individual (e.g., SNPs information, whole or partial genomic sequence) and population statistics such as allele frequencies or LD.

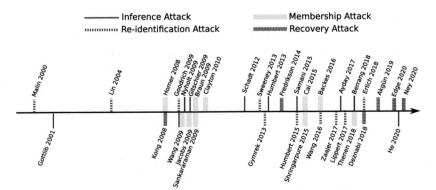

Fig. 1 Overview of privacy attacks.

Table 1 Security attacks: Background and discovered information.

Attack category	Possible background info.	Desired info.
Inference	Partial or full DNA sequence Population statistics	Hidden/nonobserved genomic variations Disease predisposition
Re-identification	Partial or full DNA sequence Available medical data Population statistics Demographic information Familial relationships Phenotypic traits GWAS statistics	Individual's identity
Membership	Partial or full DNA sequence Reference population statistics GWAS statistics Genotype frequencies Gene expression profiles	Participation of an individual in a group of interest
Recovery	Partial or full DNA sequence of relatives Familial relationships Genomic variation statistics	Reconstruction of partial or complete DNA sequence of a target individual

Gottlib [8] reported the use of genetic testing by US employers to infer the genetic disease susceptibilities of employees who did not provide their consent. Performing these tests without consent represents a privacy violation that can lead to employment discrimination. Nyholt et al. [13] demonstrated that even after hiding a particular gene, it might be possible to infer it from its neighboring regions. Despite Professor Watson's DNA sequence not containing information about the APOE gene—one of the main genes known to be related to Alzheimer's disease—the authors were able to discover its value using the neighboring regions of the hidden gene. This was made possible by the biological relations that link close regions in the human DNA. Wang et al. [14] proposed an inference attack using integer programming to infer hundreds of SNPs based on known pair-wise correlations between SNPs in the human genome. In this paper the authors also propose a membership attack that is discussed later in this chapter (see Section 2.3). Gitschier [15] described a method to infer haplotypes from the Y chromosome by exploring genealogical relations between men. The proposed method was used to infer surnames of individual from the Utah Residents with Northern and Western European Ancestry population present in the 1000 Genomes Project. Therefore, it may also enable re-identification. Schadt *et al.* [16] performed an inference attack based on gene expression data, which allows the prediction of the genomic sequence that leads to an observed expression data. They then proposed the use of the predicted genomic sequence to perform a re-identification attack and discover individuals in large populations. Humbert *et al.* [7] demonstrated how to infer the genomic sequence of family members, related to an individual whose genomic sequence is observed. This attack uses reproduction statistics and known relationships between genomic variations in the human genome, and uses belief propagation, which in the context of genomic data is used to compute unobserved genomic variations that have a correlation with observed ones. Samani *et al.* [17] proposed an attack that takes advantage of publicly available high-order correlations among single nucleotide variations existing in the human genome, in particular recombination rate and diploid genotypes, to discover hidden or nondisclosed genomic variations. This attack has a higher inference power than previous work that considered only lower-order correlations. Ayday *et al.* [18] inferred hidden or nonobserved DNA information based on the partial DNA sequence of a target individual or a DNA sequence from a member of his family and publicly available phenotypic information. Berrang *et al.* [19] proposed a method that uses Bayesian networks, which leverage the combination of different types of information, for inference risk evaluation. The proposed method is used to perform inference of mother–child relationships based on DNA methylation profiles and genomic information, which the authors named a linking attack. He *et al.* [20] developed an inference attack based on publicly available genomic information and personal traits revealed by target individuals or their relatives. The attack allows an adversary to predict nonobserved genotypes and traits.

Table 2 summarizes the techniques used to perform inference attacks, the key findings, and the possible harm caused to the genomic data owner.

2.2 Re-identification attacks

Re-identification attacks aim at associating a given DNA sequence with its owner. The background knowledge that is required for performing such an attack might include publicly available DNA statistics, the genomic sequence of target individuals or publicly available genomic sequences, medical records with additional personal details (e.g., name, age, gender, geography), and genealogical databases that contain familiar relationships.

Table 2 Inference attacks.

Technique	Key finding	Outcome
Genetic testing	Genetic diseases susceptibility	Employment denial
Statistics-based inference	Hidden genomic regions inference	Disease susceptibility disclosure
Integer programming	Hidden genomic variations	Genomic information and disease susceptibility disclosure
Genealogy-based inference	Genomic profiles inference	Allow re-identification
Gene expression based inference	Genomic sequence inference	Allow re-identification and membership attacks
Belief propagation with genomic statistics	Hidden genomic variations	Genomic information and disease susceptibility disclosure
Genomic correlations-based inference	Hidden genomic variations	Genomic information and disease susceptibility disclosure
Phenotype-based inference	Hidden genomic variations	Genomic information and disease susceptibility disclosure
Bayesian networks	Mother–child relations	Familial relationships disclosure
Traits-based inference	Hidden genomic variations	Genomic information and personal traits disclosure

Malin and Sweeney [21] developed CleanGene, a software that assesses the identification risk associated to DNA sequences. This software performs re-identification based on available health-related data, i.e., from pharmacy records and hospital records, and disease knowledge from publicly available repositories. Later, the same authors proposed REIDIT [22], an algorithm that performs data re-identification. The proposed algorithm uses deterministic methods to link genomic data to named individuals, whose information is available in published records. The authors define a trail attack as a variant of a re-identification attack where the identity of a target individual is discovered using information that is collected from different independent sources (e.g., from different studies in which the individual participated). In the end, the adversary collects sufficient information to perform re-identification, even if the information of each source alone is not sufficient. Li *et al.* [23] demonstrated that 75 statistically independent SNPs are sufficient to unequivocally identify an individual. The study was based on common random SNPs. However, if rare SNPs are observed the number of SNPs required for a successful re-identification would be even lower. This work provided an important baseline for further design of privacy-preserving approaches. Goodrich *et al.* [24] proposed the Mastermind attack, which demonstrates that even after applying cryptographic techniques to protect genomic sequences' privacy, it is still possible to learn some information at each guess attempt which goal is to learn hidden information (e.g., genomic variations). Like in the mastermind game, at each guess attempt, correctly guessed DNA positions are reported and if so, an adversary already learns part of the sequence. This attack allows the discovery of the identity of a genomic sequence owner with a limited number of guess attempts. Sweeney [25] proposed a re-identification method using demographic information applied to de-identified health data. This attack demonstrates that based on few attributes (i.e., place, gender, and birth data), which are combined with voter registration lists, it is possible to identify 53% of the American population. The information used in this attack was easy to obtain, since the voter registration lists was obtained for 20 dollars, and the health data was publicly or semipublicly available. Gymrek *et al.* [26] combined DNA information from the Y chromosome with publicly available genealogical information, in particular surnames. Then, taking advantage of the father to son surname heritage, the authors were able to de-anonymize 131 genomes from the 1000 Genomes Project. However, since this attack is based on the Y chromosome, it only applies to male individuals. Sweeney *et al.* [27] proposed an attack that links anonymized genomic data from the Personal Genome Project to their owner's name by combining publicly available records (i.e., voting lists) and demographic information (i.e., birth date, postal code, gender). Humbert *et al.* [28] performed a re-identification attack on anonymized publicly available genomic data that takes advantage of genotype–phenotype relationships, either obtained from public repositories that report SNPs and phenotypic trait relations or computed from genotype–phenotype databases. Wang *et al.* [29] demonstrated that it is possible to infer the identity of target individuals from aggregate statistics, such as GWASs statistics, even if they are differentially private, using Bayesian networks. Although the success of identity inference attacks decreases when rigorous privacy protection is applied to the GWAS statistics, the authors showed that the success probability of such attacks increases with the background knowledge and that it is higher than that of random guesses. Lippert *et al.* [30] proposed a method that allows re-identification of individuals based on the prediction of traits applying phenotyping, and statistical modeling on Whole-Genome Sequencing (WGS) data. The authors relied on maximum entropy algorithm to use trait predictions for determining the genomic sample and phenotype profile that originated from the same individual. The authors show that phenotypic prediction from WGS data can enable re-identification without further data required. Zaaijer *et al.* [31] proposed a new method to perform re-identification of human DNA samples in a fast and inexpensive way, called MinION sketching. They demonstrated that analyzing 60–300 randomly selected SNPs and relying on Bayesian inference it is possible to link anonymized genomic samples with their owners. Erlich *et al.* [32] proposed a re-identification attack based on long-range familial searches. The proposed attack was performed on a dataset with 1.28 million individuals collected from direct-to-consumer services, such as 23andMe and Ancestry. Considering US individuals with a European ancestry, which represented 85% of the individuals from the dataset, it was possible to reach a third cousin or closer relationship in 60% of the searches. After finding a relative in a long-range familial search, the authors demonstrated that it is possible to perform re-identification using common demographic identifiers (e.g., age, gender, geography).

Table 3 summarizes the main techniques and background information used for re-identification attacks, and their key findings. In this kind of attacks the goal and harm caused is always identity disclosure by linking de-identified genomic data with the owner's identity, independently of the information and method used.

2.3 Membership attacks

Membership attacks focus on inferring the participation of a given individual in a study group. The attacks in this category compare the statistics of the target DNA statistics with the statistics of the target group and of a reference population. With this method, the closer the statistics of an individual are to a certain group, the higher is the probability that she belongs to that group.

Homer *et al.* [33] were the first to propose a membership attack that inferred the participation of an individual in a given study group. Using genomic variations expression or allele frequency information, the authors propose some statistics that compare the value obtained for a target

Table 3 Re-identification attacks.

Technique	Knowledge	Key finding
Medical data crossing	Pharmacy, hospital records, and diseases information	CleanGene software for re-identification risk assessment of DNA sequences
Deterministic methods to link DNA data to individuals	Public genetic and nominative information from different sources	REIDIT software performs re-identification; definition of trails attacks
Common allele-based statistics	Common SNPs information	75 Independent SNPs allow unequivocally identification of individuals
Data crossing	Demographic and de-identified health information	The identity of 53% of the American population was disclosed
Genealogical correlations	Y chromosome and genealogical information	Surnames disclosure for 131 genomes from the 1000 GP
Genotype–phenotype correlations	Genotype and phenotype information	Physical traits based re-identification
Bayesian networks	Aggregated statistics	Re-identification of a target individual participating in a GWAS
Phenotypic-based correlations	Whole-genome sequencing (WGS) data and physical traits	Re-identification without requiring further information
Long-range familial searches	Familial relations and demographic information	Familial relations reveal close relatives, and demographic identifiers allow re-identification

individual's DNA sequence with the ones obtained for a reference population and for a particular study group. Then, one can conclude that the target individual belongs to the study group, if its statistic are closer to the value of the study group. Interestingly, this attack showed that studying 50,000 independent SNPs is sufficient to infer the participation of an individual in a group of 100 people, or 10,000 SNPs if the group is composed of 10 people. Wang et al. [14] proposed an extension of the Homer's attack by adding into consideration existing statistical correlations among SNPs on the DNA, i.e., LD. Such correlations allow the inference of further nonobserved SNPs that when included in the attacker knowledge empower the attack statistics. This work shows that the optimized attack is able to determine the participation of some individuals in a particular Genome-Wide Association Study (GWAS) even using a low-quality reference population. In addition, the attack used a couple hundred SNPs, which represent around 30× less SNPs than in Homer's attack to achieve the same attack power. Braun et al. [34] proposed the use of empirical tests to assess the participation of a target individual in a study group. The empirical test is performed using the target individual genotype information and the marginal allele frequencies of the group one is studying. Jacobs et al. [35] demonstrated how likelihood-based statistics can be used to infer the participation of a target individual or his close relatives in a GWAS. The statistics are computed using genotype frequencies and individual genotypes. In addition, this paper evaluates the membership attack power for different sample size and considering different sets of SNPs for computing the statistics. Sankararaman et al. [36] developed the SecureGenome tool, which enables the detection of a target individual on a study group based on the summary statistics from a GWAS. The membership attack compares the target individual alleles profile with the allele frequencies of the study group and the allele frequencies of a reference population, similar to Homer's attack [33]. Clayton et al. [37] designed a membership attack using a Bayesian approach. This attack considers prior probability knowledge about the participation of an individual in a certain sample. Shringarpure and Bustamante [38] demonstrated an attack on the Beacons Project. This project designed a platform for secure querying of genomic information where only Boolean answers are returned to the user, with the purpose of limiting the amount of private information disclosed. Although the information obtained per query is limited, the proposed attack shows that by querying 250 genomic variations and combining their results, it is still possible to discover the participation of an individual in a beacon with 64 European individuals. Re-identification was also deemed possible; however, it required a much higher number of queries (1000 genomic variation queries). This attack showed that beacons are susceptible to membership attacks and may also leak phenotypic information about the participants they study. Cai et al. [39] proposed an attack that takes as input a dataset and the GWAS statistics from 25 randomly selected genomic variation sites to infer whether an individual participated in the case group. More precisely, this attack requires the genotype of a target individual and compares it with the genotypes inferred from the case group using the GWAS statistics, and if a match is found, the target individual is considered to have participated in the case group. From then on, it is also possible to perform re-identification of case individuals. Backes *et al.* [40] proved that genomic data is not the only type of omics data that can be used to perform membership attacks. They used expression data, in particular from microRNA and showed how to infer the participation of a target individual in a particular group. Since microRNA data is more affected by the health status than genomic data, it is therefore more informative about the group the individual belongs to, be it the control or the case group. Thenen *et al.* [41] improved the membership attacks proposed by Shringarpure and Bustamante. With only 5 queries (50× less queries than the ones required by the former attack), they were able to infer the participation of individuals in a beacon with 95% of confidence, using the same beacon configuration used in previous works. The attack improvements are due to the use of high-order Markov chains to infer high-order relations on the genomic data. Another important finding of this work is that current privacy protection measures, which include particular genomic regions hiding and the implementation of a query budget, are not efficient against the proposed membership attacks.

Table 4 summarizes the techniques and background information used in the membership attacks, and their key findings. All the attacks in this category require some genomic information of the target individual(s).

2.4 Recovery attacks

Recovery attacks focus on inferring the DNA sequence of a target individual aided by publicly available DNA statistics, such as allele frequency in the reference population, and/or kin relations if the genomic sequences of relatives are available. The inferred DNA sequence can then be used to perform attacks from the previous categories, which assume the availability of the DNA sequence of a target individual.

Kong *et al.* [42] exploited kin relationships to infer haplotypes of target individuals based on observable genomic information of their relatives. The proposed inference method incorporates information about recurrent mutations transmitted from parents to their children and fine-scale recombinations. Wang *et al.* [14] also performed recovery of nonobserved sequences, overall 100 sequences containing a total of 174 SNPs were recovered, based on single and pair-wise allele frequencies. Those recovered sequences were then used in the proposed inference and membership attacks. Fredrikson *et al.* [43] showed that personalized medicine models

Table 4 Membership attacks.

Technique	Information	Key finding
Queries knowledge combination	Genomic variations expression data or statistics	50,000 SNPs are sufficient to disclose membership in a group of 100 people and 10,000 SNPs for a 10 people group
Statistics comparison	Genomic variations expression data or statistics and statistical correlations	Including statistical correlation in the genome requires 30× less SNPs to achieve the same attack power
Empirical tests	Allele frequencies of the study group	Membership disclosure
Likelihood-based statistics	GWAS statistics	Membership inference of a target individual and close relatives
Bayesian methods	Target group statistics	Membership disclosure
Queries knowledge combination	Genomic variations queries	For a beacon with 64 European individuals 250 queries allow membership disclosure
Higher-order Markov chains	Genomic variations information	Membership prediction using 50× less queries for the same result of similar attacks
Disease and gene expression correlations	Expression data	Membership disclosure in disease-related studies

can leak information about an individual's DNA sequence. By combining the information from a pharmacogenetic model, which was used to design particular medicines for a patient, and demographic information from the same patient, the authors proved that is possible to discover some hidden regions of the DNA sequence of a patient. Deznabi et al. [44] described how to discover parts of the genomic sequence of a target individual based on familial relations, public phenotype information, and other available data from online repositories (e.g., social networks). Akgün [45] proposed an active recovery attack which allows the adversary to discover genomic data from an individual using SNP statistics. This attack consists of the manipulation of the weights attributed to the SNPs used in a test so that it is easier to infer the SNPs of a target individual from the test results. This attack was one of the few assuming a dishonest party. Edge and Coop [46] proved that using publicly available genomic data an adversary is able to learn the genomic sequence of a target individual using Identical-By-State (IBS) tiling. This technique consists in matching known genomic sequences against an unknown one to obtain information about it. The authors showed that applying IBS tiling for 900 genomes from the 1000 Genomes Project reveals at least one allele from 82% of the SNP sites of an individual with European ancestry. In addition, the authors proposed a variant of IBS tiling, called IBS probing, that allows the adversary to learn if the target individual' genome contains a specific disease-related allele, whose neighboring sequence is known. A related attack was described by Ney et al. [47]. In this attack, the authors demonstrated that an adversary could almost learn the entire genomic sequence of a target individual from GEDmatch—a US direct-to-consumer online service that compares DNA data files. Two possible ways are described to learn the individual's genomic sequence: (i) by uploading artificial nearly-all-heterozygote genome and examining the resulting IBS segments (similar to Ref. [46]), and (ii) by uploading an all-heterozygote genome and examining the resulting images.

Table 5 summarizes the different techniques and background information used on recovery attacks, and their respective key findings.

2.5 System exploits

System exploits can affect all information systems, and are therefore not specific to genomic data processing systems. They explore system vulnerabilities and they must be taken into account when designing privacy-preserving systems in order to ensure their long-term security. System exploits and intrusions can lead to user's data exposure and consequently to information leakage. In order to be secure, systems must ensure confidentiality, integrity, and authentication. Confidentiality focuses on preventing unauthorized access to the data. Integrity ensures that the data is not modified by unauthorized users and is also in charge of reporting those changes in case they happen.

Table 5 Recovery attacks.

Technique	Information	Key finding
Kin-based inference	Kin relationships and relatives partial genomic data	Target individual sequence reconstruction
Information integration	Pharmacogenetic models and demographic information	Target individual genomic sequence reconstruction
Identical-by-state tiling	Known and unknown genomic sequences	Multiple SNPs inference that allow genomic sequence reconstruction

Last but not the least, authentication is the property that allows the verification of users' identity and then grant or deny them access to the system.

Malicious attacks are also part of system exploits that can lead to privacy breaches and data theft. A successful attack can result in data loss if backup data copies are not maintained.

Finally, in the context of DNA data, maintaining secure systems is of paramount importance, due to the previously discussed sensitive information encoded in the DNA and the reported privacy breaches. There are two main scenarios for genomic data systems: (i) the data is stored locally or in a private server, or (ii) the data is stored in a public cloud, e.g., because of its large size.

In the first scenario, the system designer is responsible for placing protection methods to prevent system exploits. To strengthen the protection, protection techniques can be also applied at the data level, for example, data obfuscation and data encryption. While, for the second scenario, the system protection is of the responsibility of the CSP. Therefore, the user should protect his data before sending it to the public cloud. Commonly, this process is made through data encryption since it prevents the cloud or any other entity that obtains an access to the data to learn its real value.

3. Privacy-preserving techniques

Traditional privacy-preserving techniques need to be adapted to be used on genomic data, since, as discussed previously, genomic data itself contains re-identifiable information. Several approaches were proposed to perform some computations on genomic data in a privacy-preserving way, such as in GWASs statistics computation, DNA sequences alignment, and genomic database queries. GWASs consist in the analysis of several genomes over multiple genomic variation positions to find the relation between genotypes and diseases. Such solutions use different techniques, such as data obfuscation, cryptography, and trusted hardware. DNA computing itself is a paradigm that can be used to design novel privacy-preserving techniques; however, this field is in its infancy. Briefly, work on this field described DNA cryptography, which is described later in this section.

As the previous section detailed, several privacy attacks on genomic data have been described in the literature. Following these findings, the potential impact of privacy attacks on data owners made the research community focus its effort on the development of methods to prevent successful attacks. These methods can be categorized as follows:

1. De-identification methods
2. Data augmentation methods

3. Cryptography-based approaches
4. Secure Multiparty Computations (SMPC)

3.1 De-identification methods

De-identification consists in removing all the personal identifiers from the data in order to keep secret the identity of the data owner. The main goal of de-identification is to prevent the direct association of genomic data with their owners, and consequently, protect the owners identity. In other words, de-identification aims at providing data anonymity. This is a technique widely implemented for biomedical data. However, as demonstrated by re-identification attacks, applied alone this method is often not enough to protect the data owner. Indeed, these methods are particularly inefficient in protecting genomic data since they do not remove the identifying information contained in the genomic data itself (i.e., rare genomic variations). In addition, providing anonymity has become more difficult with greater availability of information in online platforms, which can be related with genomic data and contribute to individuals' identification.

K-anonymity is a widely applied paradigm to enforce a stronger data de-identification, which consist in modulating the data attributes in such a way that based on those attributes an adversary is not able to distinguish an individual from k-1 other individuals [48, 49]. The two methods mainly used to achieve k-anonymity are suppression and generalization. Suppression consists in removing the attributes that can lead to direct identification of an individual, such as those that are not shared with other individuals in the dataset and are not generalizable. Generalization consists in translating an attribute value in a broader class. For example, it is common to replace a numerical value by an interval that contains it. Emam *et al.* [50] developed a de-identification algorithm that ensures k-anonymity on health datasets. Other commonly used methods to achieve anonymity are l-diversity and t-closeness. l-Diversity [51] was proposed as an improvement of k-anonymity that preserves privacy when the diversity in the attribute values is low, and assumes that the adversary has access to background knowledge. t-closeness [52] is a refinement of l-diversity where the difference between the distribution of the sensitive attributes for a given class and the distribution of the same attributes in the full table is at most t. Although this method improves privacy, it also implies some utility loss at the data management and data mining levels.

DNA Lattice Anonymization (DNALA) [53] was proposed to anonymize pairs of genomic sequences, which are represented by the sequence that represents the minimal distance to both. This is a generalization process that resembles k-anonymity. The proposed method was tested on human genomic sequences publicly available. The main limitations of the proposed generalization are the following: (i) it is dependent of the pair of sequences considered; and (ii) it is limited to two sequences.

Lin *et al.* [54] proposed a generalization method based on data binning. The proposed approach ensures that no unique record is present in the database released to the users. The bin size works as an anonymity level indicator, since a larger bin size makes data less specific and detailed, and consequently, provides a higher level of anonymity.

In conclusion, although de-identification methods were reported to be insufficient to prevent re-identification [38, 55, 56], they do complicate them. It is also important to consider that anonymity in the context of genomic data is different from other data types, since genomic data contains personal identifiable information itself. Furthermore, its combination with other metadata, such as name, gender, age, and geographic details, power the privacy attacks described in the previous section. DNA computing, as an emerging paradigm, could also contribute for the development of de-identification methods, allowing the data to be de-identified as soon as it is produced by the sequencing machines. This could be done, for example, by removing the need of metadata, such as identifiers, by also encoding them in a DNA sequence format.

Fig. 2 summarizes common de-identification techniques used for medical records. For the name and diagnosis columns, it is used pseudo-anonymization where the real names and diagnosed disease are replaced

Name	Age	Num. exams	Diagnosis
Alice	34	3	Diabetes
Bob	25	10	Cancer
Claire	29	7	Diabetes
David	31	4	Diabetes
Eva	28	2	Cancer
Frank	28	9	Cancer

Name	Age	Num. exams	Diagnosis
P1	30-34	1-5	D1
P2	25-29	6-10	D2
P3	25-29	6-10	D1
P4	30-34	1-5	D1
P5	25-29	1-5	D2
P6	25-29	6-10	D2

Fig. 2 De-identification.

by a unique identifier. For the age we have two classes (25–29 and 30–34 years) that grant 4-anonymity for the 25–29 class and 2-anonymity for the 30–34 class. In other works, this means that at least four records have the same information in a given data column. For the number of exams, there are two classes (1–5 and 6–10 exams) and this generalization grants 3-anonymity.

The referred techniques are applied to metadata that usually is together with the DNA data; however, DNA data contains identifiable information itself. Therefore, other techniques are required to completely anonymize the DNA data.

3.2 Data augmentation methods

Data augmentation consists in applying generalization or obfuscation in order to protect data. In the context of genomic data, it consists in making the data of different individuals indistinguishable by generalizing or obfuscating the information it contains, so as to prevent their unequivocal identification.

Generalization, at the genomic sequence level, consists in the representation of two or more sequences with the most common sequence among them. In other words, a set of genomic reads are represented by the most common genomic variations they contain [53, 57].

Data masking consists in hiding sections of the data in order to make them unobservable and unpredictable for an adversary. The data that is masked corresponds to the sensitive information one wants to protect from unauthorized access. Cogo et al. [58] proposed the first automated sensitive DNA short sequences detection, which relies on Bloom filters. This approach improves privacy protection by allowing the user to store and process the sensitive and insensitive DNA sequences differently. Later, Decouchant et al. [59] proposed an automated sensitive information detection for long DNA sequences. This approach allows the efficient privacy-preserving processing of DNA sequences with a lower performance overhead and higher precision than Cogo's approach. Extending this approach, Fernandes et al. [60] designed a sensitivity levels classification method-based DNA properties, such as allele frequency and LD.

Differential privacy is another data augmentation technique, which is used to make an aggregate result indistinguishable whether a single individual participates or not in that result through the addition of noise [1, 61, 62]. With differential privacy, the greater the noise added, the higher the privacy protection is. However, the addition of noise reduces the data utility. Consequently, applying this technique requires studying the trade-off between data utility and privacy protection to ensure that subsequent data analysis will not be compromised.

The techniques in this category make re-identification attacks harder to perform, since their main purpose is to hide sensitive information or genomic regions, possibly by adding noise.

3.3 Cryptography-based approaches

Cryptography-based approaches are characterized by the protection of the input and output using an encryption scheme. These approaches can be mainly used for two purposes: storage or processing. They differ by the encryption schemes used since the processing scenario requires operations to be allowed on the encrypted data. Cryptography-based approaches are interesting because of their guaranteed privacy-protection; however, their application is limited due to their longer computational time. Garbled circuits [63] allow two parties to perform secure computations. For example, a user can send his encrypted data to a server where some computations are performed and, then, the encrypted results is sent back to the user. Garbled circuits protect the input and intermediate results, which are never revealed, since they are always manipulated encrypted on the server side.

Homomorphic encryption is a particular subject of cryptography which allows mathematical operations on the encrypted data. This allows some computations to be outsourced to untrusted environments, such as public clouds, without having to decrypt the data. Atallah *et al.* [64] designed a privacy-preserving strings comparison algorithm using homomorphic encryption, which computes the edit distance between two DNA sequences. Kantarcioglu *et al.* [65] proposed a cryptography-based approach that allows genomic sequences sharing and querying.

Despite the performance limitations of cryptography-based approaches, there are still some applications where encryption can be practical, such as determining disease susceptibility through genetic testing [66] and secure datasets querying [67]. He *et al.* [68] proposed a cryptography-based approach to identify relatives. In this approach a pair of individuals only share the necessary encrypted genomic information to determine if they are relatives or not. The results obtained showed that this approach is able to find relationships up to third cousins level while preserving the privacy of the individuals.

Cryptography-based approaches provides a high level of protection; however, current encryption schemes are not designed to protect genomic data for its full lifetime [7]. This is a real challenge since an individual's genome is partially inherited from previous generations and transmitted to the future ones.

In the context of DNA computing, DNA cryptography is emerging. This technology uses DNA algorithms that convert the information to be protected into DNA, first converted to the digital DNA sequence and then to the DNA molecule. This process is equivalent to the operations performed by standard cryptographic schemes, where plaintext information is converted into encrypted data. In this field two types of DNA cryptography methods exist: DNA-based data hiding schemes and DNA-based encryption schemes. DNA-based data hiding schemes consist in converting the message to DNA nucleotides and then mix the real message DNA with fake DNA sequences and send the mix to a receiver. DNA-based data hiding schemes consist in encoding the information in DNA nucleotides instead of using binary form. Then, the double helix is created following the complementarity property to increase the complexity [69, 70].

Fig. 3 summarizes the process of encryption of DNA sequences. Usually, the DNA sequence is translated into a 2-bit sequence and, then, each 2-bit is encrypted individually as represented in the figure. Following, computations can be performed on the encrypted data and in the end the encrypted result is decrypted to reveal the result in clear text.

3.4 Secure multiparty computations

SMPC allow the secure collaboration of several institutions, e.g., hospitals, biocenters, researcher centers, and universities. SMPC enable computations

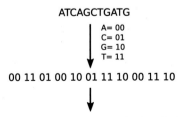

Fig. 3 DNA encryption.

over a dataset that is distributed among the participating institutions while keeping the data private [71]. This technique allows third parties to perform computations on encrypted data, without learning any information about the data, the computed results, or the contribution of any party participating in the computation. Therefore, SMPC does not require to trust a third party. In addition, since data is transferred encrypted, data usability is not compromised while providing privacy protection. On the other hand, the computational overhead and communication costs between participants are high, which may make SMPC not practical for some applications. SMPC can use standard cryptographic algorithms, such as AES, which provide strong privacy and confidentiality properties. Some examples of SMPC-based approaches include the work described in Refs. [72–74]. Aziz et al. [72] proposed a SMPC-based approach that used the Paillier encryption scheme to enable queries on genomic data. The results were obtained considering five geodistributed parties and show that this method requires a computational time comprised between 7 min and 4 h. Cho et al. [73] proposed secure GWAS analysis through SMPC. Another closely related contribution is METIS [74], which uses SMPC to make the clients, the server, and data owners cooperate to enable computations on genomic data provided by the data owner without disclosing it to the other parties. This approach intents to give the genomic data owner the full control regarding his/her genomic data analysis permissions. At the same time, the data owner has no computations overhead.

Table 6 summarizes the advantages and disadvantages of the privacy-preserving techniques discussed in this section.

Table 6 Overview of privacy-preserving techniques.

Technique	Advantages	Disadvantages
De-identification	Personal identifiers removal Replace identifiers by pseudo information.	Not efficient. Genomic data is identifying information itself Permit to link genomic data to owner's identity
Data augmentation	Hidden information Real value replaced by more abstract information	Data utility can decrease Information precision decreases

Continued

Table 6 Overview of privacy-preserving techniques.—cont'd

Technique	Advantages	Disadvantages
Cryptography	Highest protection Prevent unauthorized access to the data Computations secure computations running on untrusted environments	Low performance Data size increases Computational overhead increases Limited operations allowed
SMPC[a]	Allow secure collaboration Computations run on encrypted data Data utility preserved No computational overhead on the client side	Large data analysis require nonnegligible data transfers High computational overhead

[a]Secure multiparty computations (SMPC).

4. Trust management

Trust management considers other possible sources of exploits that can compromise the genomic data privacy. The data producer, the users, and the storage service providers play an important role on the genomic data cycle. Privacy protection for genomic data must consider the vulnerabilities to potential attacks on the data producer side, on the user side, and on the storage provider. This topics are discussed in the following sections.

4.1 Genomic data ownership

An important question regarding human DNA privacy and trust management is to determine to whom the sequenced genomic data belongs to, and who should keep it. For several researchers the sequenced genomic data belong to its donor and he/she should own the full rights regarding that data. However, nowadays, DNA sequencing and processing involves different institutions (i.e., sequencing center, biocenter, hospital) which make ownership complex with several entities requesting and having access to the DNA sequence. As discussed in the next section, the data provider is a commonly trusted entity, whose main role is to produce the DNA sequence. However, assuming that the DNA sequence should be kept only by its donor, the traditional processing pipeline needs to be modified so that the data provider or any other entity involved in the data processing is not able to keep a copy of the data.

In the processing phase, genomic data owner should be able to define for which applications he/she allows the data to be used, in other words, the data owner must define the data access and usage policies. There are two main kinds of data use, primary and secondary. The primary use of the data grants the permission to use the data for a particular application (e.g., research, GWAS, genetic testing), which is often the original reason behind its collection. The secondary use allows the data to be used for other applications than the initial use case. In both cases, the data is controlled by the service provider. Therefore, the data owner needs to define the access and usage policies before the access to the data is granted. Furthermore, at any moment the data owner must be able to update the consent accordingly to his/her preferences, which should be immediately implemented. Although, the data owner might allow the use of his/her data and later change it, once the data usage has been granted for some application, it is impossible to ensure that the involved entities destroy their data copy.

4.2 Trusting the data provider

One way to get a genome sequenced is to participate in genetic studies by donating a biological sample, which is handled by the data provider. The data provider, commonly biocenters with sequencing facilities, is the entity responsible to generate the data. The data providers for human sequencing data is the sequencing center. In the DNA computing context, the data provider is the laboratory where the DNA sample is synthesized.

Other widely known data providers are direct-to-consumer sequencing services, such as 23andMe, Helix, MyHeritage, and Ancestry, that provide genome sequencing service and allow the user to explore its own genomic information as a recreational process. The cost of those services has been dropping in the last decade (23andMe: from $99; MyHeritage: 79€) and their use has increased. However, when using such a service the user must inform himself about the policies of data ownership. It must be clarified by the service provider if the spare biological sample is discarded and whether the genomic data is completely deleted upon request. Nevertheless, most of these direct-to-consumer sequencing services store the genomic data in servers they rent from public clouds, where it can be accessed by the user. Therefore, the user must also get informed about the data protection policies while choosing a service provider.

One other point of discussion is whether or not the data provider should keep a copy of the data. From the privacy point of view, it would open a window of vulnerability if a copy is kept, since the data is owned and managed by another entity. On the other hand, having a copy of the data allows better availability and reduces the risk of data loss. This decision must take into account the specific data storage and sharing policies of each data provider.

4.3 Access control

Access control consists in regulating who has access to the data following stipulated access policies and agreements. In addition, access control is also in charge of ensuring the traceability of the access to a particular file or data. In other words, it can be described as registering all the accesses made to the data, which in case of data modification or corruption would allow the identification of its author [62, 75]. Although this technique alone is not privacy-preserving, since it does not itself provide data protection, it is widely used in combination with other techniques.

Nowadays, the most commonly used access control model is the role based access control (RBAC) method, in which access is granted to the users that have a role that justifies data access for specific tasks they are in charge of. In the context of DNA data, the access to the data would be granted to the specialists the owner accepts to share data with, e.g., doctors and researchers.

Several genomic sequence repositories require a data access request that is evaluated by a specialized committee. Although this process is essential to protect the data from being misused, it can take several months until it concludes. This process can even be longer in the case of collaborative repositories since all the involved institutions need to approve the access. Another limitation of this methodology is that the access to the data is commonly granted for a short period of time, and in case the access period needs to be extended, the decision process needs to be repeated [76]. This process can sometimes limit accessibility to the datasets in exchange for more control over data accesses and, consequently, reduce the possibility of privacy breaches.

Erlich *et al.* [77] described an alternative to the classic access control, bilateral consent framework (BCF), to facilitate genomic data sharing while protecting data privacy. In the BCF, the participants can directly decide who can access their data instead of requiring a decision from the access control committee. This streamlines the process of obtaining access to data and also to apply changes on the access rights. However, although such a model seems to be promising, the privacy sensitive nature of genomic data makes many institutions keep utilizing more classic access control schemes. This demonstrates that dynamic access control solutions require further development.

Blockchain is an emerging technology that allows the collaboration between multiple parties removing the need of an intermediary trusted centralized party to authenticate the interactions. In the context of healthcare, blockchain applications have been increasing, with particular interest for patients identity's validation, to facilitate the management of permissions, and for access control to biomedical data [78, 79].

In the DNA computing field, Namasudra proposed a secure and fast access control model to address the system overhead and the long accessing time required when searching for data stored in the cloud [80]. This scheme keeps a fast data access list on the cloud side, and uses a 1024-bit DNA computing based key for data encryption, which increases data security.

4.4 Cloud environment: Storage and processing

Genomic data processing requires extensive resources due to the huge amount of data to be analyzed and the intensive computational processing. Therefore, this processing is often outsourced to public clouds, which provide performance improvement and larger storage memory. However, clouds are managed by a CSP that provides limited control to the user over its own data. Furthermore, the data can be manipulated (i.e., copied, transferred) by the CSP without users awareness and stored in multiple locations for data availability. These properties make the data more prone to possible unauthorized access by the CSP or an intruder when it is stored in a public cloud [81]. In order to prevent the unauthorized access to genomic data on clouds, the user can apply some protective techniques and must agree with the CSP on the conditions to handle the data [82].

In particular for genomic data, it is also important to know the data properties to limit the potential information leakage in case the data outsourced to the cloud might suffer some privacy breach. In Ref. [83] the authors define some data aggregation properties, in particular the ratio between number of genomes and number of genomic variations that users should follow when releasing genomic data to prevent information leakage. Cryptography-based schemes have been proposed to allow the outsourcing of intensive computations, such as sequence comparison, to public clouds while protecting data privacy [84]. Chen *et al.* [85] proposed a privacy-preserving reads alignment approach that combines processing on public and private clouds. The main concept of this approach is to assign hash values to sensitive information and, then, process the hashes in a public cloud. Further processing on the sensitive information is performed in the private cloud. Although this process outsources part of the computations, significant computations still run in the private cloud. Balaur [86] performs alignment on hybrid clouds based on MinHash and k-mer voting. This approach was developed to allow the user to transfer part of the processing to the public cloud while guaranteeing privacy protection and ensuring high accuracy. The candidate positions step uses privacy sensitive information, i.e., the genomic sequence and the reference genome in plaintext, and is performed in the trusted environment (private cloud), while the secure voting that selects the best position a read aligns to is performed using hash values in the public cloud. DepSky [87] focused on providing secure storage in a cloud-of-clouds relying on encryption, encoding, and data replication. Charon [88] was designed to provide secure storing and sharing of genomic data on cloud-of-cloud systems. GenoShare [89] is a tool that was developed to support conscious genomic data sharing. This tool takes as input the genomic information to be shared, the adversary's background knowledge, and other publicly available information, for example, previous data releases. It then simulates three privacy attacks (membership, phenotype inference, and kinship inference) to determine which data can safely be shared and which data should not be shared. More recently, Cogo *et al.* [90] proposed a privacy-preserving efficient and dependable cloud-based storage approach for human genomes. This approach combines sensitive information detection and deduplication. The sensitive information detection method is applied to human genomic reads for privacy protection improvement, while the deduplication method allows the optimization of the storage space. This work represents an important effort toward privacy-preserving outsourcing of genomic data to the cloud environment.

Trusted Execution Environments (TEEs) are secure and isolated memory and code partitions located on a processor, which are not accessible to the remaining system. They were developed to provide data confidentiality while ensuring the integrity of the code they execute. Consequently, if communications with the TEEs are secured, the data and intermediary results are protected. Some examples of TEEs include ARM TrustZone and Intel SGX. However, some side channel and cache attacks were reported and showed vulnerabilities of TEEs. In reaction, mitigation methods have also been proposed [91, 92]. Some secure genomic data processing algorithms have been developed using TEEs, including secure genetic testing and privacy-preserving decentralized processing of genomic data [93–96]. However, to assume that the processing done inside a TEE is secure, it requires to trust the processor manufacturer, e.g., Intel for the SGX technology.

Privacy risks associated to the processing of biomedical data in cloud environments were studied in Ref. [97]. This study highlights the main benefits of cloud computing as the following: (i) vast resources availability, which is useful for parallelization; (ii) affordable cost; and (iii) the user is free from the maintenance duties. However, important security challenges also appear when outsourcing sensitive information to the cloud, which include: (i) control, management, and security of the data is the user responsibility and the CSP; therefore, they should agree on the policies; (ii) access rights need to be well defined and updated, since cloud environments can be accessed remotely; (iii) the user must prevent data loss by holding a backup copy, since cloud environments are shared resources and are susceptible to failures, as all computational systems.

Table 7 summarizes the discussed trust management fronts in this section, highlighting the main challenges and techniques used.

Table 7 Trust management.

Topic	Challenges	Techniques
Data ownership	Define who owns the data	Access and usage agreement
Data provider	Ensure the data is used for the only purpose it was created Guarantee that no copy is used without permission.	Informed consent when sequencing
Access control	Prevent unauthorized access to the data Provide access traceability	Blockchain Data access lists Bilateral consent framework
Cloud environment	Protect data from being accessed by intruders and the Cloud Service Provider (CSP) Agreement between CSP and user regarding data storage and access	Data encryption Trusted execution environments Hash functions

5. Discussion

Genomic data presents a wide range of applications, which boosted its production and availability, creating massive amounts of genomic data to be processed. This led the research community to develop high-throughput algorithms and search how to leverage more efficient environments for genomic data processing, such as public clouds. However, several privacy attacks on genomic data have been reported in the literature, demonstrating that data was not being protected adequately. Such privacy attacks are classified into four main classes: inference attacks, re-identification attacks, membership attacks, and recovery attacks. The main difference between the different types of attacks is the information the adversary intents to discover, ranging from obtaining additional genomic information about a target individual to discovering the data owner identity. This applies to human DNA data, which is characterized by natural correlations that occur between regions in the genome and between the genomes of members of the same family. Privacy attacks on human DNA use those correlations to improve the power of the attack. For synthesized DNA such correlations exist if it was created based on the human genome. If it is created independently, the properties can be different; however, since they share the same format, similar attacks might be devised to disclose information contained in synthesized DNA. The potential harm caused by privacy attacks alerted the researchers about the urgent need for efficient privacy-preserving approaches for storing, sharing, and analyzing it. Although several privacy-preserving approaches were proposed in the last decades, the field is still at an early development stage and reaching a good balance between privacy and utility is still a challenge in many applications. The main difficulty to address is that privacy-serving techniques commonly used are not sufficient for genomic sequences since they contain personal and identifiable information themselves.

Privacy-preserving techniques applied to genomic data include de-identification methods, data augmentation methods, cryptography-based approaches, and SMPC. In the de-identification methods, anonymization techniques are widely used for genomic data; however, they have been shown to be insufficient to prevent re-identification attacks [26]. For genomic data, anonymization is not sufficient since the data itself contains personal identifiable information. Nonetheless, those techniques are still applied to genomic data jointly with other protection methods, since they make the identity inference task harder. Data augmentation methods are more efficient than anonymization since they aim at hiding or generalizing properties that are inherent to genomic data. Data masking and differential privacy are commonly applied methods. Such methods target the prevention of re-identification attacks, even though they might not completely prevent them. A drawback data augmentation methods is reaching an acceptable trade-off between privacy and utility loss. In particular for differential privacy, in order to achieve good privacy protection, considerable noise is added to the data; however, its utility might become compromised. Cryptography–based approaches provide the highest protection to genomic data and they are very efficient against all kinds of privacy attacks, since any data or metadata is not revealed to any unauthorized party. However, the required computational resources and time are not negligible and even unpractical for some processing tasks. Cryptography–based approaches are efficient to prevent all the described privacy attacks if the data is only handled in plaintext in trusted environment and by the authorized users. SMPC are often applied to allow collaboration between different entities, e.g., biocenters, that want to perform computations over all data parties without revealing each entity data share. Regarding privacy protection, SMPC is efficient since data is only transmitted encrypted and the computations are only decrypted by the legit users, which have the decryption key.

Giving a practical example, to prevent recovery attacks as described in Refs. [46, 47] that apply IBS tilling and combine publicly available genomic data and genealogical databases, two mitigation approaches could be implemented. First, often the IBS tiling reveals the location of the queried segments, which contribute to the inference of the target individual's genome. Therefore, hiding this information from the results would complicate the sequence inference since the adversary would only learn if the segment is present or not in the genome. Second, for the case of the use of artificial or manipulated genomic sequences, the prohibition of such sequences would prevent the techniques applied in Ref. [47]. However, authentication techniques for genomic sequences are still an open challenge.

For synthesized DNA, natural correlations occurring in the human DNA are not considered, which introduces some independence between synthesized DNA sequences. Thus, techniques such as data augmentation and cryptography–based techniques can be applied and might provide improved protection to synthesized DNA. To determine the information encoded on the synthetic DNA molecules, traditional sequencing technologies that have been designed for human DNA sequencing can be used. This process eventually returns the plaintext sequence of nucleotides corresponding to the synthetic DNA information, which in turn reveals information about the original data that was used to create it. Therefore, the sequenced data must be stored securely to prevent information leakage.

The described privacy-preserving techniques are yet insufficient to address the performance and utility that would fully take advantage of genomic data processing while ensuring the adequate privacy protection. Interoperability among genomic data from different institutions potentially geodistributed is still not fully addressed. In addition, in this scenario standardization and common laws are required to replace the existing per country genomic data privacy guidelines.

In the context of genomic data, trust managements need to consider several points. First, the data owner must have the right to manage the data, including data access and processing. However, there are open discussions regarding who should keep the data, with the following possible scenarios: (i) only the data owner has a copy of the data; (ii) the data provider has a copy of the data; and (iii) the entities involved in the processing have a copy of the data. In scenario (i), the main problems are that data availability and data protections are solely of the responsibility of the data owner. In this case, data access would need to be requested to the owner each time the data is requested. In addition, there is a higher risk of data loss and in case it happens the genome needs to be re-sequenced which incurs extra expenses and extra time. In scenario (ii), the data owner needs to allow the data provider to keep a copy of his/her data. This scenario might leave the genomic data more exposed than in scenario (i), since the spare copy is managed by the data provider and the data owner needs to trust it. However, in case the data owner loses his copy, it can ask the data provider a new copy. In scenario (iii), allowing the processing entities to also have a copy of the data, means even less control on the data accesses. Although currently access control is strictly managed, involving other entities increases the complexity of the process. On the other side, this scenario provides better data availability.

Although access control have been defined as essential on systems storing and processing genomic data, its implementation is not yet optimized. Nowadays, data access requests are reviewed by a specialized committee and last several months.

The storage and processing of genomic data are often outsourced to more cost-efficient and powerful environments, such as public clouds. However, processing biomedical data on public clouds can represent an increased information leakage risk when the data is not efficiently protected.

Those risks have been described in Refs. [5, 6]. Since the public is an environment accessible to multiple entities, the data stored in it is more prone to possible access attempts by unauthorized entities. In order to take advantage of the computational resources provided by cloud environments, several solutions have been proposed to perform privacy-preserving computations on genomic data. However, such solutions are limited to few possible computations and they require nonnegligible computation on a trusted environment. Therefore, further studies are required on this subject in order to allow the maximization of the resources without sacrificing performance or privacy.

To conclude, privacy-preserving processing of genomic data has been evolving in the last decades. Significant advances were made; however, the journey is still not over. In addition to the improvement points discussed in this section, some other open challenges include: (i) considering a more realistic threat model, which would include malicious adversaries; (ii) considering dynamic systems, which tolerate datasets modifications (Refs. [98, 99] are early examples for genome wide association studies and genomic beacons, respectively); (iii) ensuring interoperability between data from different sources and define genomic data privacy standards. First, the most common threat model considered for the privacy-preserving approaches design is the honest-but-curious adversary. This model describes an adversary that follows the protocol but might however try to learn further information about the data. The honest-but-curious adversary model is a moderate model, since the behavior of the adversary is somehow controlled. In practice, malicious adversaries, which are adversaries that intentionally try to deviate from the protocol to extract information they do not have the right to access, or perturb the system, form a more realistic threat model since no assumption is made about their behavior. Second, most of the current genomic data systems assume static databases, where the datasets are published and are rarely or never updated. In practice, databases can be updated by the addition or removal of information. However, in practice, and considering that the data owner should be able to revoke the access permissions to his/her data at any time, the databases must allow not only the addition of information, but also the removal of some data without privacy breaches. This is currently an open challenge. Last, ensuring the interoperability between data from different sources requires standardization of the data formats along with the definition of standards for genomic data privacy. However, currently, the legislation for genomic data protection is not globalized. So far, each country defines the laws related to genomic data privacy. This complicates data sharing and collaborations between geodistributed institutions. However, the research community is aware of the need for improving this field. Although some entities have been working in the field, such as Global Alliance (GA4GH), trying to establish some common guidelines and define standard methods for genomic data protection.

6. Conclusion and future work

In the context of DNA computing, some applications make use of synthesized DNA that is similar to the human DNA; thus, this chapter discussed the properties of human DNA and privacy attacks that consider it. Genomic data is used in multiple fields due to its informative nature; however, it encodes highly sensitive and personal information that is unique for each individual. Therefore, privacy is essential on the genomic data life cycle, i.e., storage, sharing, and processing. Furthermore, DNA information leakage can result in unwanted and irreversible harm. Once it is leaked, genomic privacy cannot be recovered since DNA is nonrevocable. Moreover, intra- and intergenomic data correlations, respectively, among different regions in a DNA sequence and among family members can leak additional information, including hidden information if not adequately handled. The multiple privacy attacks reported on genomic data demonstrated the main vulnerabilities of current processing algorithms which would not provide enough privacy protection to the data. This urged the development of privacy-preserving approaches. The main challenges raised in this field are mainly the protection of data privacy and the practical performance, since often privacy protection requires performance sacrifices. Privacy-preserving techniques have been used to allow multiple genomic data applications, although some applications still remain unprotected. One of the main reasons for these limitations is the performance and/or data utility sacrifice most of the privacy-preserving techniques imply in order to improve privacy protection.

Genomic data privacy research has to address several open challenges. The development of privacy-preserving systems that are able to tolerate malicious adversaries, that intent to explore vulnerabilities to gain unauthorized access to genomic data, is important since this threat model is more realistic than the often assumed honest-but-curious adversary. In addition, the improvement of current privacy-protection techniques to enable updates on the data while guaranteeing the adequate privacy protection would also be a valuable advance in the field. More dynamic datasets would allow the production of more accurate statistics to speed up research. Finally, standardization would benefit the genomic data privacy and allow better interoperability between data from different sources, for example, from different studies that collect data in a geodistributed fashion. Currently per-country laws limit the large potential of genomic data sharing and research collaborations.

References

[1] M. Naveed, E. Ayday, E.W. Clayton, J. Fellay, C.A. Gunter, J.P. Hubaux, B.A. Malin, X. Wang, Privacy in the genomic Era, ACM Comput. Surv. 48 (1) (2015) 1–44.

[2] P.E. Verissimo, A. Bessani, E-biobanking: what have you done to my cell samples? Secur. Priv. 11 (6) (2013) 62–65.

[3] A. Bessani, J. Brandt, M. Bux, V. Cogo, L. Dimitrova, J. Dowling, A. Gholami, K. Hakimzadeh, M. Hummel, M. Ismail, E. Laure, U. Leser, J.E. Litton, R. Martinez, S. Niazi, J. Reichel, K. Zimmermann, BiobankCloud: a platform for the secure storage, sharing, and processing of large biomedical data sets, in: Proceedings of the Big-O(Q)/DMAH@VLDB 2015, 2015, pp. 86–105.

[4] M. Fernandes, J. Decouchant, F.M. Couto, P. Esteves-Veríssimo, Cloud-assisted read alignment and privacy, in: Proceedings of the 11th International Conference on Practical Applications of Computational Biology & Bioinformatics, 2017.

[5] A. Michalas, N. Paladi, C. Gehrmann, Security aspects of e-health systems migration to the cloud, in: Proceedings of the IEEE 16th International Conference on e-Health Networking, Applications and Services, 2014, pp. 212–218.

[6] B. Fabian, T. Ermakova, P. Junghanns, Collaborative and secure sharing of healthcare data in multi-clouds, Inf. Syst. 48 (2015) 132–150.

[7] M. Humbert, E. Ayday, J.P. Hubaux, A. Telenti, Addressing the concerns of the lacks family: quantification of kin genomic privacy, in: Proceedings of the ACM SIGSAC Conference on Computer & Communications Security, 2013, pp. 1141–1152.

[8] S. Gottlieb, US employer agrees to stop genetic testing, Br. Med. J. 322 (7284) (2001) 449.

[9] M. Fernandes, Reconciling data privacy with sharing in next-generation genomic workflows, (PhD thesis), University of Luxembourg 2020.

[10] E. Shapiro, T. Ran, Molecules reach consensus, Nat. Nanotechnol. 8 (2013) 703–705.

[11] R. Klitzman, P.S. Appelbaum, W. Chung, Should life insurers have access to genetic test results? JAMA 312 (18) (2014) 1855–1856.

[12] A.M.Y. Goh, E. Chiu, O. Yastrubetskaya, C. Erwin, J.K. Williams, A.R. Juhl, J.S. Paulsen, Perception, experience, and response to genetic discrimination in Huntington's disease: the Australian results of The International RESPOND-HD study, Genet. Test. Mol. Biomarkers 17 (2) (2013) 115–121.

[13] D.R. Nyholt, C.-E. Yu, P.M. Visscher, On Jim Watson's APOE status: genetic information is hard to hide, Eur. J. Hum. Genet. 17 (2009) 147–149.

[14] R. Wang, Y.F. Li, X. Wang, H. Tang, X. Zhou, Learning your identity and disease from research papers: information leaks in genome wide association study, in: Proceedings of the ACM Conference on Computer and Communications Security, 2009, pp. 534–544.

[15] J. Gitschier, Inferential genotyping of Y chromosomes in Latter-Day Saints founders and comparison to Utah samples in the HapMap project, Am. J. Hum. Genet. 84 (2) (2009) 251–258.

[16] E.E. Schadt, S. Woo, K. Hao, Bayesian method to predict individual SNP genotypes from gene expression data, Nat. Genet. 44 (5) (2012) 603–608.

[17] S.S. Samani, Z. Huang, E. Ayday, M. Elliot, J. Fellay, J.P. Hubaux, Z. Kutalik, Quantifying genomic privacy via inference attack with high-order SNV correlations, in: Proceedings of the IEEE Security and Privacy Workshops, 2015, pp. 32–40.

[18] E. Ayday, M. Humbert, Inference attacks against kin genomic privacy, IEEE Secur. Priv. 15 (5) (2017) 29–37.

[19] P. Berrang, M. Humbert, Y. Zhang, I. Lehmann, R. Eils, M. Backes, Dissecting privacy risks in biomedical data, in: Proceedings of the IEEE European Symposium on Security and Privacy, 2018, pp. 62–76.

[20] Z. He, J. Yu, J. Li, Q. Han, G. Luo, Y. Li, Inference attacks and controls on genotypes and phenotypes for individual genomic data, in: Proceedings of the IEEE/ACM Transactions on Computational Biology and Bioinformatics, 2020, pp. 930–937.

[21] B. Malin, L. Sweeney, Determining the identifiability of DNA database entries, in: AMIA Symposium, 2000, pp. 537–541.

[22] B. Malin, L. Sweeney, How (not) to protect genomic data privacy in a distributed network: using trail re-identification to evaluate and design anonymity protection systems, J. Biomed. Inform. 37 (3) (2004) 179–192.

[23] Z. Lin, A.B. Owen, R.B. Altman, Genomic research and human subject privacy, Science 305 (5681) (2004) 183–183.

[24] M.T. Goodrich, The mastermind attack on genomic data, in: Proceedings of the 30th IEEE Symposium on Security and Privacy, 2009, pp. 204–218.

[25] L. Sweeney, Simple demographics often identify people uniquely, Health 671 (2000) 1–34.

[26] M. Gymrek, A.L. McGuire, D. Golan, E. Halperin, Y. Erlich, Identifying personal genomes by surname inference, Science 339 (6117) (2013) 321–324.

[27] L. Sweeney, A. Abu, J. Winn, Identifying Participants in the Personal Genome Project by Name, Data Privacy Lab, IQSS, Harvard University, 2013.

[28] M. Humbert, K. Huguenin, J. Hugonot, E. Ayday, J.P. Hubaux, De-anonymizing genomic databases using phenotypic traits, Privacy Enhanc. Technol. 2015 (2) (2015) 99–114.

[29] Y. Wang, X. Wu, X. Shi, Infringement of Individual Privacy Via Mining Differentially Private GWAS Statistics, in: Proceedings of the International Conference on Big Data Computing and Communications, 2016, pp. 355–366.

[30] C. Lippert, R. Sabatini, M.C. Maher, E.Y. Kang, S. Lee, O. Arikan, A. Harley, A. Bernal, P. Garst, V. Lavrenko, K. Yocum, T. Wong, M. Zhu, W.Y. Yang, C. Chang, T. Lu, C.W.H. Lee, B. Hicks, S. Ramakrishnan, H. Tang, C. Xie, J. Piper, S. Brewerton, Y. Turpaz, A. Telenti, R.K. Roby, F.J. Och, J.C. Venter, Identification of individuals by trait prediction using whole-genome sequencing data, Natl. Acad. Sci. 114 (38) (2017) 1–6.

[31] S. Zaaijer, A. Gordon, D. Speyer, R. Piccone, S.C. Groen, Y. Erlich, Rapid re-identification of human samples using portable DNA sequencing, eLife 6 (e27798) (2017) 1–17.

[32] Y. Erlich, T. Shor, I. Pe'er, S. Carmi, Identity inference of genomic data using long-range familial searches, Science 362 (6415) (2018) 690–694.

[33] N. Homer, S. Szelinger, M. Redman, D. Duggan, W. Tembe, J. Muehling, J.V. Pearson, D.A. Stephan, S.F. Nelson, D.W. Craig, Resolving individuals contributing trace amounts of DNA to highly complex mixtures using high-density SNP genotyping microarrays, PLoS Genet. 4 (8) (2008) 1–9.

[34] R. Braun, W. Rowe, C. Schaefer, J. Zhang, K. Buetow, Needles in the haystack: identifying individuals present in pooled genomic data, PLoS Genet. 5 (10) (2009) 1–8.

[35] K.B. Jacobs, M. Yeager, S. Wacholder, D. Craig, P. Kraft, D.J. Hunter, J. Paschal, T.A. Manolio, M. Tucker, R.N. Hoover, G.D. Thomas, S.J. Chanock, N. Chatterjee, A new statistic and its power to infer membership in a genome-wide association study using genotype frequencies, Nat. Genet. 41 (11) (2009) 1253–1257.

[36] S. Sankararaman, G. Obozinski, M.I. Jordan, E. Halperin, Genomic privacy and limits of individual detection in a pool, Nat. Genet. 41 (9) (2009) 965–967.

[37] D. Clayton, On inferring presence of an individual in a mixture: a Bayesian approach, Biostatistics 11 (4) (2010) 661–673.

[38] S. Shringarpure, C. Bustamante, Privacy risks from genomic data-sharing beacons, Am. J. Hum. Genet. 97 (5) (2015) 631–646.

[39] R. Cai, Z. Hao, M. Winslett, X. Xiao, Y. Yang, Z. Zhang, S. Zhou, Deterministic identification of specific individuals from GWAS results, Bioinformatics 31 (11) (2015) 1701–1707.

[40] M. Backes, P. Berrang, M. Humbert, P. Manoharan, Membership privacy in MicroRNA-based studies, in: Proceedings of the ACM SIGSAC Conference on Computer and Communications Security, 2016, pp. 319–330.

[41] N. von Thenen, E. Ayday, A.E. Cicek, Re-identification of individuals in genomic data-sharing beacons via allele inference, Bioinformatics 35 (3) (2018) 365–371.

[42] A. Kong, G. Masson, M.L. Frigge, A. Gylfason, P. Zusmanovich, G. Thorleifsson, P.I. Olason, A. Ingason, S. Steinberg, T. Rafnar, P. Sulem, M. Mouy, F. Jonsson, U. Thorsteinsdottir, D.F. Gudbjartsson, H. Stefansson, K. Stefansson, Detection of sharing by descent, long-range phasing and haplotype imputation, Nat. Genet. 40 (9) (2008) 1068–1075.

[43] M. Fredrikson, E. Lantz, S. Jha, S. Lin, D. Page, T. Ristenpart, Privacy in pharmacogenetics: an end-to-end case study of personalized Warfarin dosing, in: Proceedings of the USENIX Security Symposium, 2014, pp. 17–32.

[44] I. Deznabi, M. Mobayen, N. Jafari, O. Tastan, E. Ayday, An inference attack on genomic data using kinship, complex correlations, and phenotype information, IEEE/ACM Trans. Comput. Biol. Bioinform. 15 (4) (2018) 1333–1343.

[45] M. Akgün, An active genomic data recovery attack, Balkan J. Elect. Comput. Eng. 7 (2019) 417–423.

[46] M.D. Edge, G. Coop, Attacks on genetic privacy via uploads to genealogical databases, eLife 9 (2020) e51810.

[47] P. Ney, L. Ceze, T. Kohno, Genotype extraction and false relative attacks: security risks to third-party genetic genealogy services beyond identity inference, in: Proceedings of the Network and Distributed System Security Symposium, 2020.

[48] L. Sweeney, k-anonymity: a model for protecting privacy, Int. J. Uncertainty Fuzziness Knowl.-Based Syst. 10 (05) (2002) 557–570.

[49] K. El Emam, F.K. Dankar, Protecting privacy using k-anonymity, J. Am. Med. Inform. Assoc. 15 (5) (2008) 627–637.

[50] E. Jonker, T. Roffey, R. Vaillancourt, J. Bottomley, E. Cogo, F.K. Dankar, K. El Emam, S. Chowdhury, D. Amyot, R. Issa, J.P. Corriveau, M. Walker, A globally optimal k-anonymity method for the de-identification of health data, J. Am. Med. Inform. Assoc. 16 (5) (2009) 670–682.

[51] A. Machanavajjhala, D. Kifer, J. Gehrke, M. Venkitasubramaniam, L-diversity: privacy beyond k-anonymity, ACM Trans. Knowl. Discov. Data 1 (1) (2007) 3–54.

[52] N. Li, T. Li, S. Venkatasubramanian, t-Closeness: privacy beyond k-anonymity and l-diversity, in: Proceedings of the IEEE International Conference on Data Engineering, 2007, pp. 106–115.

[53] B.A. Malin, Protecting DNA sequence anonymity with generalization lattices, Methods Inf. Med. 44 (2005) 687–692.

[54] Z. Lin, M. Hewett, R.B. Altman, Using binning to maintain confidentiality of medical data, in: Proceedings of the AMIA Symposium, 2002, pp. 454–458.

[55] B.A. Malin, An evaluation of the current state of genomic data privacy protection technology and a roadmap for the future, J. Am. Med. Inform. Assoc. 12 (1) (2005) 28–34.

[56] E.C. Hayden, Privacy protections: the genome hacker. Yaniv Erlich shows how research participants can be identified from 'anonymous' DNA, Nature 497 (7448) (2013) 172–174.

[57] G. Li, Y. Wang, X. Su, Improvements on a privacy-protection algorithm for DNA sequences with generalization lattices, Comput. Methods Programs Biomed. 108 (1) (2012) 1–9.

[58] V.V. Cogo, A. Bessani, F.M. Couto, P. Verissimo, A high-throughput method to detect privacy-sensitive human genomic data, in: Proceedings of the ACM Workshop on Privacy in the Electronic Society, 2015, pp. 101–110.

[59] J. Decouchant, M. Fernandes, M. Völp, F.M. Couto, P. Esteves-Veríssimo, Accurate filtering of privacy-sensitive information in raw genomic data, J. Biomed. Inform. 82 (2018) 1–12.

[60] M. Fernandes, J. Decouchant, M. Völp, F.M. Couto, P. Esteves-Verissimo, DNA-SeAl: sensitivity levels to optimize the performance of privacy-preserving DNA alignment, IEEE J. Biomed. Health Inform. 24 (3) (2020) 907–915.

[61] E. Vayena, U. Gasser, Between openness and privacy in genomics, PLoS Med. 13 (1) (2016) 1–7.

[62] Y. Erlich, A. Narayanan, Routes for breaching and protecting genetic privacy, Nat. Rev. Genet. 15 (2014) 409–421.

[63] J. Baron, K. El Defrawy, K. Minkovich, R. Ostrovsky, E. Tressler, 5pm: secure pattern matching, in: Proceedings of the International Conference on Security and Cryptography for Networks, 2012, pp. 222–240.

[64] M.J. Atallah, F. Kerschbaum, W. Du, Secure and private sequence comparisons, in: Proceedings of the ACM Workshop on Privacy in the Electronic Society, 2003, pp. 39–44.

[65] M. Kantarcioglu, W. Jiang, Y. Liu, B. Malin, A cryptographic approach to securely share and query genomic sequences, IEEE Trans. Inf. Technol. Biomed. 12 (5) (2008) 606–617.

[66] M. Namazi, J.R. Troncoso-Pastoriza, F. Perez-Gonzalez, Dynamic privacy-preserving genomic susceptibility testing, in: Proceedings of the ACM Workshop on Information Hiding and Multimedia Security, 2016, pp. 45–50.

[67] G.S. Çetin, H. Chen, K. Laine, K. Lauter, P. Rindal, Y. Xia, Private queries on encrypted genomic data, BMC Med. Genomics 10 (2) (2017) 45.

[68] D. He, N.A. Furlotte, F. Hormozdiari, J.W.J. Joo, A. Wadia, R. Ostrovsky, A. Sahai, E. Eskin, Identifying genetic relatives without compromising privacy, Genome Res. 24 (4) (2014) 664–672.

[69] S. Namasudra, D. Devi, S. Choudhary, R. Patan, S. Kallam, Security, privacy, trust, and anonymity, in: S. Namasudra, G.C. Deka (Eds.), Advances of DNA Computing in Cryptography, Chapman & Hall/CRC, 2018, pp. 138–150.

[70] S. Namasudra, G.C. Deka, R. Bali, Applications and future trends of DNA computing, in: S. Namasudra, G.C. Deka (Eds.), Advances of DNA Computing in Cryptography, Chapman & Hall/CRC, 2018, pp. 181–192.

[71] Y. Huang, Secure multi-party computation, in: Responsible Genomic Data Sharing, X. Jiang and H. Tang, Eds., Academic Press, 2020, pp. 123–134.

[72] M.M. Al Aziz, M.Z. Hasan, N. Mohammed, D. Alhadidi, Secure and efficient multi-party computation on genomic data, in: Proceedings of the International Database Engineering & Applications Symposium, 2016, pp. 278–283.

[73] H. Cho, D.J. Wu, B. Berger, Secure genome-wide association analysis using multiparty computation, Nat. Biotechnol. 36 (6) (2018) 547–551.

[74] D. Deuber, C. Egger, K. Fech, G. Malavolta, D. Schröder, S.A.K. Thyagarajan, F. Battke, C. Durand, My genome belongs to me: controlling third party computation on genomic data, Proc. Priv. Enhanc. Technol. 2019 (1) (2019) 108–132.

[75] A. Mittos, B. Malin, E.D. Cristofaro, Systematizing genome privacy research: a privacy-enhancing technologies perspective, Priv. Enhanc. Technol. 2019 (1) (2019) 87–107.

[76] K. Learned, A. Durbin, R. Currie, E.T. Kephart, H.C. Beale, L.M. Sanders, J. Pfeil, T.C. Goldstein, S.R. Salama, D. Haussier, O.M. Vaske, I.M. Bjork, Barriers to accessing public cancer genomic data, Sci. Data 6 (98) (2019) 907–915.

[77] Y. Erlich, J.B. Williams, D. Glazer, K. Yocum, N. Farahany, M. Olson, A. Narayanan, L.D. Stein, J.A. Witkowski, R.C. Kain, Redefining genomic privacy: trust and empowerment, PLoS Biol. 12 (11) (2014) 1–5.

[78] C.C. Agbo, Q.H. Mahmoud, J.M. Eklund, Blockchain technology in healthcare: a systematic review, Healthcare 7 (2) (2019) 56.

[79] M. Hölbl, M. Kompara, A. Kamisalic, L. Nemec Zlatolas, A systematic review of the use of blockchain in healthcare, Symmetry 10 (470) (2018).

[80] S. Namasudra, Fast and secure data accessing by using DNA computing for the cloud environment, IEEE Trans. Serv. Comput. 15 (4) (2022) 2289–2300.

[81] F. Rocha, M. Correia, Lucy in the sky without diamonds: stealing confidential data in the cloud, in: Proceedings of the IEEE/IFIP International Conference on Dependable Systems and Networks Workshops, 2011, pp. 129–134.

[82] E.S. Dove, Y. Joly, A.M. Tasse, Public Population Project in Genomics and Society (P3G) International Steering Committee and International Cancer Genome Consortium (ICGC) Ethics and Policy Committee, B.M. Knoppers, Genomic cloud computing: legal and ethical points to consider, Eur. J. Human Genet. 23 (2015) 1271–1278.

[83] X. Zhou, B. Peng, Y.F. Li, Y. Chen, H. Tang, X. Wang, To release or not to release: evaluating information leaks in aggregate human-genome data, in: Proceedings of the European Symposium on Research in Computer Security, 2011, pp. 607–627.

[84] M. Blanton, M.J. Atallah, K.B. Frikken, Q. Malluhi, Secure and efficient outsourcing of sequence comparisons, in: European Symposium on Research in Computer Security, 2012, pp. 505–522.

[85] Y. Chen, B. Peng, X. Wang, H. Tang, Large-scale privacy-preserving mapping of human genomic sequences on hybrid clouds, in: Proceedings of the Network & Distributed System Security Symposium, 2012.

[86] V. Popic, S. Batzoglou, A hybrid cloud read aligner based on MinHash and kmer voting that preserves privacy, Nat. Commun. 8 (15311) (2017) 1–7.

[87] A. Bessani, M. Correia, B. Quaresma, F. André, P. Sousa, DepSky: dependable and secure storage in a cloud-of-clouds, ACM Trans. Storage 9 (4) (2013) 1–33.

[88] R. Mendes, T. Oliveira, V.V. Cogo, N.F. Neves, A.N. Bessani, CHARON: a secure cloud-of-clouds system for storing and sharing big data, IEEE Trans. Cloud Comput. 9 (4) (2021) 1349–1361.

[89] J.L. Raisaro, C. Troncoso, M. Humbert, Z. Kutalik, A. Telenti, J.P. Hubaux, GenoShare: supporting privacy-informed decisions for sharing exact genomic data, EPFL Infoscience (2017) 1–19.

[90] V. Cogo, A. Bessani, Enabling the efficient, dependable cloud-based storage of human genomes, in: Proceedings of the International Symposium on Reliable Distributed Systems Workshops, 2019.

[91] M. Schwarz, S. Weiser, D. Gruss, C. Maurice, S. Mangard, Malware guard extension: using SGX to conceal cache attacks, in: In Proceedings of the International Conference on Detection of Intrusions and Malware, and Vulnerability Assessment, 2017, pp. 3–24.

[92] J. Götzfried, M. Eckert, S. Schinzel, T. Müller, Cache attacks on Intel SGX, in: Proceedings of the European Workshop on Systems Security, 2017, pp. 1–6.
[93] F. Chen, C. Wang, W. Dai, X. Jiang, N. Mohammed, M.M. Al Aziz, M.N. Sadat, C. Sahinalp, K. Lauter, S. Wang, PRESAGE: PRivacy-preserving gEnetic testing via SoftwAre Guard Extension, BMC Med. Genomics 10 (48) (2017) 77–85.
[94] F. Chen, S. Wang, X. Jiang, S. Ding, Y. Lu, J. Kim, S.C. Sahinalp, C. Shimizu, J.C. Burns, V.J. Wright, E. Png, M.L. Hibberd, D.D. Lloyd, H. Yang, A. Telenti, C.S. Bloss, D. Fox, K. Lauter, L. Ohno-Machado, PRINCESS: Privacy-protecting rare disease International Network Collaboration via Encryption through Software guard extensionS, Bioinformatics 33 (6) (2017) 871–878.
[95] C. Lambert, M. Fernandes, J. Decouchant, P. Esteves-Veríssimo, MaskAl: Privacy Preserving Masked Reads Alignment using Intel SGX, in: Symposium on Reliable Distributed Systems (SRDS), 2018.
[96] M. Völp, J. Decouchant, C. Lambert, M. Fernandes, P. Esteves-Verissimo, Enclave-based privacy-preserving alignment of raw genomic information: information leakage and countermeasures, in: Proceedings of the 2nd Workshop on System Software for Trusted Execution, 2017, pp. 1–6.
[97] B. Zubairu, Security risks of biomedical data processing in cloud computing environment, in: Cloud Security: Concepts, Methodologies, Tools, and Applications, Information Resources Management Association, 2019, pp. 1748–1768.
[98] T. Pascoal, J. Decouchant, A. Boutet, P. Esteves-Verissimo, DyPS: Dynamic, Private and Secure GWAS, in: Proceedings on Privacy Enhancing Technologies, Sciendo, 2021.
[99] K. Ayoz, E. Ayday, A.E. Cicek, Genome reconstruction attacks against genomic data-sharing beacons, arXiv preprint:2001.08852 (2020).

About the authors

Maria Fernandes is currently a postdoctoral researcher at University of Oxford. She received a master in Bioinformatics and Computational Biology from University of Lisbon, and a PhD degree in Computer Science from the University of Luxembourg. Her research interests focus on biomedical data storage and analysis, data privacy and human genetics. During her PhD, she worked on designing privacy-preserving approaches for genomic data processing.

Jérémie Decouchant is an assistant professor at Delft University of Technology (NL). Previously, he has been a research scientist at SnT, University of Luxembourg. He received an engineering degree (MSc) from Ensimag, Grenoble, and a PhD degree in computer science from the University of Grenoble-Alpes, France. His research interests revolve around resilient distributed systems and algorithms. In particular, he has been designing privacy-preserving processing workflows for genomic data.

Francisco M. Couto is currently an associate professor with habilitation at Universidade de Lisboa (Faculty of Sciences) and a researcher at LASIGE. He graduated (2000) and has a master (2001) in Informatics and Computer Engineering from the IST. He concluded his doctorate (2006) in Informatics, specialization Bioinformatics, from the Universidade de Lisboa. He was an invited researcher at EBI, AFMB-CNRS, BioAlma during his doctoral studies. His main research contributions cover several key aspects of bioinformatics and knowledge management, namely in proposing and developing: various text mining solutions that explore the semantics encoded in ontologies; semantic similarity measures and tools using biomedical ontologies; and ontology and linked data matching systems. Until August 2022, he published 2 books; was co-author of 10 chapters, 62 journal papers (47 Q1 Scimago), and 32 conference papers (10 core A and A*); and was the supervisor of 10 PhD theses and of 51 master theses. He received the Young Engineer Innovation Prize 2004 from the Portuguese Engineers Guild, and an honorable mention in 2017 and the prize in 2018 of the ULisboa/Caixa Geral de Depósitos (CGD) Scientific Prizes.

CHAPTER THREE

DNA computing in cryptography

Jiechao Gao[a] and Tiange Xie[b]

[a]Department of Computer Science, University of Virginia, Charlottesville, VA, United States
[b]Institute of Information Engineering, Chinese Academy of Sciences, Beijing, China

Contents

Abstract

Cryptography is used to prevent any unintentional party from reading or manipulating any confidential information by converting it from the sender to the receiver and vice versa. It brings ideal security to the e-commerce industry and the Internet. Deoxyribonucleic acid (DNA) computing is a bioscience-inspired technology that uses DNA as an information carrier for secure communications over networks. Adleman proposed the idea of DNA computing in 1994. The molecular structure of DNA and its complex processes such as hybridization or polymerase chain reaction (PCR) provide tremendous parallelism, energy efficiency, and superior storage capacity; it is one of the most advanced forms of information representation. Therefore, there is a trend in cryptography to develop many new algorithms to secure data. In DNA computational cryptography, information is encrypted in a DNA sequence using molecular computation, whereas existing conventional cryptography uses complex mathematical procedures for encryption. This chapter discusses the DNA

Advances in Computers, Volume 129
ISSN 0065-2458
https://doi.org/10.1016/bs.adcom.2022.08.002

computational techniques proposed by different researchers for different encryption goals such as encryption and decryption, authentication, digital signatures, and the challenges of implementing DNA computation in cryptography.

1. Introduction

DNA computing is one of the major research areas in natural computing. In DNA computing, all data or information is encoded as strands of DNA. DNA computing originated from the concept of using molecular biological processes to perform arithmetic and logical operations on data stored in DNA [1]. Leonard Adleman performed the first DNA computing experiments in 1994; his research solved the Hamiltonian path problem. The Hamiltonian path problem, which takes a long time to compute, was chosen to show the computational power of DNA. The physical properties of DNA affect DNA computing in the same way that the physical properties of silicon affect electronic computing. DNA is the storage medium for the genetic data present in every living cellular organism. It consists of nucleotides, chemical groups, and bases. The four nucleotide bases are adenine (A), cytosine (C), guanine (G), and thymine (T) [2]. the DNA strand consists of numerous combinations using these four bases. It is sufficient to encode and store information compared to the 0s and 1s used in electronic computing. The application of DNA computational techniques and concepts to cryptography is a promising direction of research. Cryptography uses DNA computational logic for encryption and decryption (confidentiality), authentication, and digital signatures [3]. Each security service has different encryption algorithms - confidentiality, authentication, and digital signatures as in Fig. 1. encryption

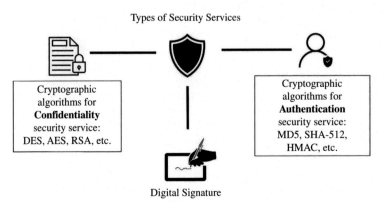

Types of Security Services

Cryptographic algorithms for **Confidentiality** security service: DES, AES, RSA, etc.

Cryptographic algorithms for **Authentication** security service: MD5, SHA-512, HMAC, etc.

Digital Signature

Fig. 1 Taxonomy of cryptographic algorithms.

algorithms like Data Encryption Standard (DES), Advanced Encryption Standard (AES), and Rivest Shamir Adleman (RSA) use a public key exchange to maintain the confidentiality of data by securing the communication network. Hash functions such as Message-Digest Algorithm 5 (MD5), Secure Hash Algorithm-512 (SHA-512) and Hash Message Authentication Code (HMAC) are useful for ensuring data integrity and identifying any unauthorized data changes.

Because of these unique features of DNA, huge storage and massive parallelism. DNA computing contributes to delivering high-security levels. More importantly, DNA computing, traditional cryptographic methods, and algorithms form hybrid cryptographic systems. These systems yield the benefits for both fields. DNA computing can also help to bridge the gap between existing cryptographic methods and novel technologies. DNA computing is efficacious for cryptography for authentic, uninterrupted communication given the extensive parallelism as well as extraordinary storage volume of a DNA molecule. The recurrent methodology followed is to encrypt the data into DNA digital form (DNA sequence). DNA computing in cryptographic algorithms is assumed to make them more powerful, unbreakable, and crucial for computer security [4]. In cryptography, the DNA stores a massive volume of data with the permutation of the four DNA-bases that hold responsible for the arrangement of DNA strands to create hydrogen bonds altogether [5]. In the medical sector, the non-public digital medical images of the patients can be encrypted by employing DNA computing-based techniques [6]. DNA is surveyed as an advanced carrier for data security because it can attain topmost protection and robust security in addition to high capacity, and also it has a low modification rate. Using DNA-based AES cryptography and DNA steganography, a novel data security methodology can be evolved [7]. DNA computing is critical in strengthening the existing cryptographic methodologies for the reliable transmission of confidential data. DNA computing offers new ways for cryptography due to its extensive applications. The four nucleotide bases provide excellent means of executing computations because they can create self-assembly structures. As a result, DNA cryptography has emerged as a combination of computer science along with biological science and information security [8].

Different DNA computing theories are proposed and studied in various other approaches in diverse fields. Bi-directional DNA Encryption Algorithm (BDEA) is a data security technique used for securing the stored data in cloud computing [9]. BDEA is advantageous in accomplishing 2-layer security for American Standard Code for Information Interchange

(ASCII) sets [10]. In a BDEA encryption process, the sender enters the information into a 16bit Unicode plaintext. Then the Unicode plaintext gets converted to ASCII, followed by conversion to a hexadecimal value. This value gets converted to binary via a decimal converter. After obtaining the binary code, the next task is to find its corresponding DNA digital coding. The DNA digital coding for the binary pairs with the four DNA bases is $00 \rightarrow A$, $01 \rightarrow T$, $10 \rightarrow G$ and $11 \rightarrow C$. Lastly, with the help of primer pairs, PCR amplification is performed. The amplified message is then ready to be sent to the receiver. Key-sharing proceeds via the Diffie-Hellman algorithm. The least significant base (LSBase) substitution is a DNA-steganography method for hiding the encryption key in a DNA sequence [11–14]. This method utilizes the characteristics of codons that produce the same amino acids. The last base of the codon is changed, preserving its type. The base to be substituted varies according to the value of the hidden bit. Most schemes use One Time Pad (OTP) with DNA sequencing [15–20]. The OTP cipher is generally unbreakable due to the long length of encryption keys. DNA sequence offers a huge storage space that is useful in storing randomly generated encryption keys of extensive lengths. The longer the encryption key, the more tedious it is to break the cipher. In some of the schemes, the data first gets encrypted using a random DNA sequence. This encrypted sequence is then hidden inside another DNA sequence along with a cover-DNA sequence. Such a technique offers two levels of security-data secrecy and data authenticity. DNA-transcript and DNA-translation are complex processes that are used in developing unbreakable cryptographic protocols that may be immune to common cyberattack methodologies. The data gets stored in a DNA sequence which gets converted to messenger Ribonucleic Acid (RNA)-form of data. This is further transformed into protein-form data depending on a genetic code table and transferred to the recipient with an encryption key. The 3D DNA-level permutation built upon the position-sequence-group is useful for confusing the attacker about the positions of the elements of a 3D DNA-matrix. The random reallocation of the matrix elements to any position helps improve the confusion effect and escalating security levels. This technique also makes an algorithm resistant to known-plaintext attacks.

This chapter discusses DNA computing-based cryptographic and steganographic schemes proposed by other researchers. The chapter also discusses their advantages and disadvantages. The huge storage and massive parallelism properties of the DNA render it useful to be used in developing cryptographic and steganographic schemes. Some of the DNA computing-based techniques are together used with existing cryptographic algorithms like AES to strengthen their resistance to cyber-attacks. The discussed schemes get usually employed for purposes like encryptions, authentications, and digital signatures. The chapter also discusses the existing challenges in the implementation of such schemes. The major contributions of the chapter are:

- To present a detailed overview of various techniques of DNA computing which when incorporated with encryption, authentication, and digital signatures can secure the data against commonly known cyber-attacks.
- To compare the key findings, advantages, and disadvantages of various cryptographic and steganographic schemes proposed by researchers.
- To present how using DNA computing techniques with existing cryptographic algorithms can increase their efficiency, rendering the modern-day cyber-attacks futile.

This chapter comprises eight different sections. Section 2 provides a glimpse into the background of cryptography. Section 3 discusses some of the existing DNA computing techniques for the encryption, authentication, and digital signature processes. Section 4 discusses novel DNA computing techniques using PCR. Existing DNA computing-based cryptography schemes and steganography schemes are discussed separately in section 5 and section 6, respectively. Section 7 of the chapter addresses the various challenges in the implementation of DNA computing in cryptography. Section 8 summarizes all the conclusions of the chapter.

2. Background of cryptography

The development of electromechanical machines in the 19th and 20th century steered a new era for cryptography. Cryptography had become a tool of war. Advanced cryptographic techniques were in use during World WarII, with electromechanical rotor machines like the Enigma or the Lorenz SZ40. Later, Alan Turing accompanying other British cryptanalysts made use of Bletchley Park Bombe to decipher Enigma codes. The invention of computers and the rise of the internet took cryptography to a whole new level. Earlier cryptography would be limited to military, information agencies, and governments. Advances in technology led to its use in various urban sectors, organizations, online commerce transactions, and telecommunication services.

2.1 Types

Over the years, cryptographic methods have evolved from codebooks, and ciphers to cryptographic algorithms. This section discusses these different types of cryptographic techniques:

2.1.1 Codebook

According to the National Institute of Standards and Technology (NIST), "a codebook is a document comprising plaintext and the code equals in a logical arrangement or a method for machine encryption by applying words substitution technique" [21]. Letters, words, and frequently used expressions get listed in one column and their code equivalents (three, four, or five-letter codes) in another column. Codebooks were frequently used during World War II to encrypt messages sent over long distances. They were often said to provide Security by Obscurity [22] because once it was captured or reconstructed, the message's code got easily decoded. An example of a codebook would be Reserve Hand Verfahren (RHV) was used in World War II by the German Navy. RHV procedure had two stages, a transposition stage and a bigram substitution using the bigram tables. In the transposition stage, plaintext would be written into a "cage" or "box" (a shape on a piece of paper) by the cipher clerk. The bigram tables then were used to substitute the pairs of letters [23]. The structure of a codebook has a code, its brief description, its full description, some rules for usage of that code, and examples [24].

2.1.2 Cipher

A cipher is an algorithm (a series of steps) comprising a cipher key for encrypting the plaintext message into the ciphertext concerning defense and discretion. For example, the Caesar cipher is the oldest algorithm used for message encryption. Characters in the plaintext got exchanged with precisely one of the characters from the ciphertext in Caesar cipher [25]. Following steps were taken to establish ciphertext with Caesar cipher [25] illustrated in Fig. 2:

- First, decide the extent of the shift character for forming the ciphertext from the plaintext message.
- Second, redeem characters in the plaintext into ciphertext subjected to the predetermined shift.

Ciphers get classified into block ciphers and stream ciphers based on secret-key cryptography.

(1) Block Cipher: A block cipher illustrated in Fig. 3 is also called a Pseudo-Random Permutation (PRP) [26]. When we run all plausible

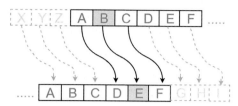

Fig. 2 Caesar cipher scheme.

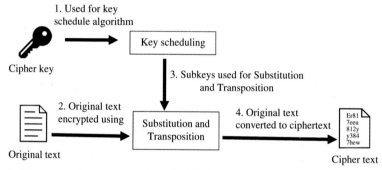

Fig. 3 Diagrammatic representation of a block cipher

inputs via the cipher, we get the result in form of a random permutation of all the data entered (an outcome of the cipher being a bijection) [27]. Block cipher makes use of a single private key shared betwixt both the sender and receiver to both encrypt and decrypt a message. Block ciphers in quotidian use are DES, Blowfish, International Data Encryption Algorithm (IDEA), Rivest Cipher (RC)5, and AES.

(2) Stream Cipher: A stream cipher illustrated in Fig. 4 is a type of symmetric cipher. It runs with a time-varying conversion on plaintext figures separately [28]. Stream cipher can reach higher speeds than block cipher, but it is more vulnerable to attacks. RC4 is a widely used stream cipher. Stream ciphers are suitable for encryption of plaintext messages having a continuous or unknown length, like network streams, and military applications where the cipher stream is availed under a secured arrangement and fed to devices that demand to function in insecure and hostile environments [29].

2.1.3 Algorithms

There are mainly three cryptographic algorithms:

1. **Secret-Key Cryptography:** The encryption/decryption processes require one single, as illustrated in Fig. 5. From the sender's end,

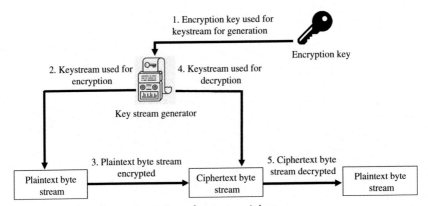

Fig. 4 Diagrammatic representation of a stream cipher.

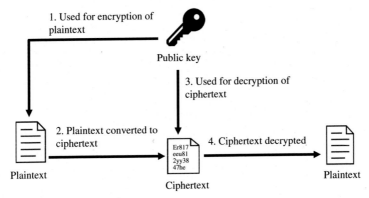

Fig. 5 Secret key cryptography

the plaintext message gets encrypted into ciphertext using a key, after the ciphertext gets to the receiver, who then decrypts the ciphertext back to the plaintext message using the same key [30]. The use of single-key for encryption and decryption is also known as symmetric encryption. This approach got substantially used for privacy and confidentiality. Some quotidian examples of secret-key cryptography include AES, DES, Blowfish, and RC4.

2. **Public Key Cryptography:** It avails two separate keys (public and private keys). The encryption process requires a public key; the decryption process needs a private key as in Fig. 6. In 1975, public-key cryptography got invented by Whitfield Diffie and Martin Hellman. The two keys are related to one another mathematically. The awareness of one of the existing keys is insufficient for someone trying to discover other keys.

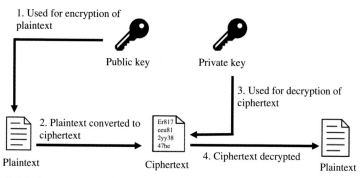

Fig. 6 Public key cryptography.

Table 1 Comparison of asymmetric key algorithms.

Methods	Advantages	Disadvantages
RSA	It is difficult to produce a private key from the public key	It involves a complex key generation process The algorithm is sluggish
Diffie-Hellman	It is comparatively quick because of key length = 256bit	It is vulnerable to man-in-the-middle attack
ECC	It has a low power consumption	It is difficult to implement
DSA	Its implementation is Swift It is resilient to man-in-the-middle-attack	Its computation is time-consuming It involves complex remainder operators in the verification process

The encryption and decryption processes require a set of keys, so the process is called asymmetric cryptography. Some common examples of public-key cryptography in Table 1 include RSA, DSA, Diffie-Hellman algorithm, and Elliptic Curve Cryptosystem (ECC) [31].

3. **Hash functions:** Hash functions generate a fixed-length hash value for the plaintext message for the encryption process. Hash functions deliver a mechanism for verification of any file as it is not practicable to alter its content without changing the hash value assigned to it. The operating systems make use of hash functions to encrypt passwords. For any document, the hash function brings forth a digital fingerprint for the sake of preventing any alterations to it by an intruder or a virus. An ideal cryptographic hash function consists of the following properties [30]:

- The hash values are easy to compute for any message
- Infeasible to generate messages having a hash value

- Infeasible modification to information without modifying the hash value
- Infeasible to encounter any two distinct messages having a similar hash value.

2.2 Key management

Key management is a set of methodologies and stratagem used to facilitate cryptographic techniques by maintaining keying relationships between the sender and the receiver. Key management incorporates procedures and approaches reinforcing [32]:

(1) System users' initialization inside an arena

(2) Production, allocation, installation of the keying material

(3) Regulating usage of keying materials

(4) Updating, withdrawal, demolition of keying materials support

(5) Storing, support/retrieval, documentation of keying materials

Key management proves to be pivotal in securing cryptographic techniques that provide confidentiality, data authentication, data integrity, entity authentication, and digital signatures. The purpose of key management includes the maintenance of keying relationships and keying material such that it clocks pertinent threats, like compromising the secrecy and validity of secret keys or their unlicensed availing [32]. The transfer of keys from the sender to the receiver via a secure channel is as crucial as data encryption.

Key management is an essential constituent of network security and crucial to Delay Tolerant Networks (DTN) security [33]. Key management's foremost purpose is to assist the functioning convenience of keying material for cryptographic purposes. Key management is pivotal for the security of encrypted data shared among authentic users, and for establishing a valid unique encryption key for communication across a secure channel. The key gets approved for data-sharing between two individuals or among a group of people in an organization. Key revocation takes place for circumstances where the key gets compromised. It is needed to notify all the other entities of an organization that was earlier using the key. It is the organization's responsibility to determine the security of data earlier protected by the revoked key.

2.3 Differences between cryptography and steganography

Steganography and cryptography are techniques responsible for maintaining the confidentiality and integrity of the data [34]. The term steganography itself means "covered writing." It deals with concealing the existence of

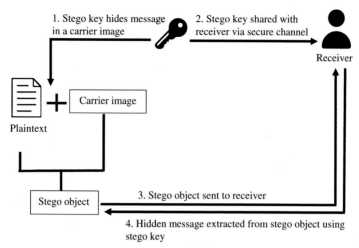

Fig. 7 Basic model of steganography.

Table 2 Comparison analysis of cryptography and steganography.

Property	Cryptography	Steganography
Aim	To protect the data exchanged between a sender and receiver	To hide data transmission taking place between a sender and receiver
No of inputs	One	Minimum two
Output received	Ciphertext	Stego object
Key	Required	Optional
Visibility	Always	Never
Attack	Cryptanalysis	Steganalysis

any secret messages such that it is not possible to keep a record, as also illustrated in Fig. 7. The hidden data is termed a stego object or stego message. A key is also required for identifying the hidden message. During World War I, to hide any code, the magazine letters would get pricked or alphabets dotted with invisible ink, and on reheating, they would reveal the plaintext message.

On the other hand, the term cryptography means "secret writing." Cryptography focuses on converting the plaintext message to ciphertext (unreadable format) rather than hiding the existence of any message [35]. Table 2 brings forth a comparison of both the methodologies concerning security solutions, applications, problems, and additional criteria [34].

Cryptography and Steganography techniques are integrated into various security modules [36], encryption methods [37], and reversible data hiding [38] to enhance the level of data protection.

3. Purposes to use DNA computing in cryptography

DNA computing has emerged as a new computational paradigm because it brought up prospective issues and opportunities to traditional cryptography by employing the potential of considerable parallelism, bulky information of biomolecules, and least-power consumption properties [39] as shown in Fig. 8. DNA computing uses the DNA molecules as an information carrier for encryption, digital signature, and authentication processes [40].

3.1 Encryption and decryption

While synthesizing the DNA molecule [31], the protein part gets extracted to convert to its corresponding nitrogen base. It is applied for encryption/decryption and gets articulated as A, C, T, and G characters. DNA computing-based cryptography includes only A, C, T, and G characters and merging and locating the messages. These A, C, T, and G characters are responsible for creating DNA Sequence S to combine with a message M to put together a new sequence S' and transmit it to the receiver, after which the Sequence S' gets converted back into S.

Despite the countless data encryption methodologies evolved up to the present time, DNA Computing methodologies got an extensive acceptance because of the biological complexity and the computational assets of DNA [41]. Repetition of words or some characters in the English language helped the cryptanalysts to predict the corresponding ciphertext, but DNA sequences invalidate all linguistic properties. Therefore, "first rendering any language to a DNA sequence and then employing cryptography techniques upon it can avert the attacks reliant on frequency analysis" [42].

The following characteristics of a DNA molecule render it a viable contribution to the field of DNA computing [43]:

(1) **Massive Parallelism:** speeds up lengthy and decipherable polynomial-time functions or problems by using only some operations. Like a mix of one thousand eighteen strands of DNA are efficient for working at 10,000 times the rate of an advanced supercomputer.

The binding property of DNA between its nucleotide bases facilitates self-assembly structure creation that delivers a logical method of executing parallel molecular computations. With vast parallelism, DNA computing can tackle various parts of computing problems simultaneously and can break the DES [40].

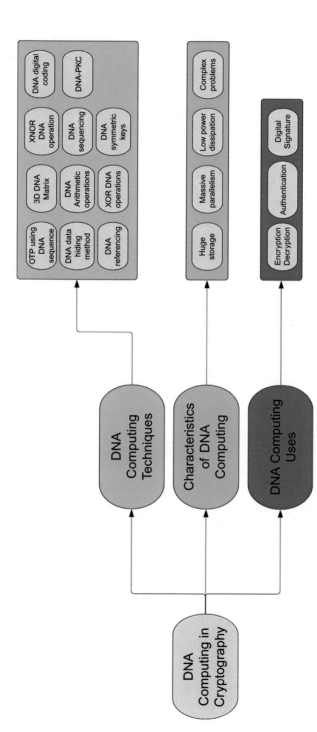

Fig. 8 Increasing cyber security with the application of DNA computing in cryptography.

(2) Huge Storage: A single cubic centimeter of DNA can withhold more data in comparison with a trillion compact disks. The density of data of DNA molecules can reach 18 Mbits/in., but present computer hard drives only store less than 1/100,000 of this data in the same space. In 1 g of DNA molecule, there are 1021 bases equivalent to 108 terabytes. At present, all the data generated can get stored in a few grams of DNA molecules. Due to more than sufficient storage space, lengthy encryption algorithm keys can get generated. Thus, making the algorithm resistant to brute force attacks. For steganography, the huge storage space comes in handy for hiding lengthy OTP keys in the fake DNA sequence.

DNA encryption algorithms must overcome some of the disadvantages of existing encryption algorithms also, listed in Table 1. The list of some requirements considered by any DNA encryption algorithm is as follows:

(1) DNA encoding for comprehensive character sets
(2) Random table generation for encodings
(3) A unique sequence to encode all characters of the original message into the DNA-sequence
(4) Robust encoding
(5) Simulation of biological process
(6) Dynamic encryption procedure

DNA-based computing techniques support the viability as well as the applicability of encryption algorithms. The security offered by the DNA computing-based cryptographic algorithms is adequate for several security purposes available in present network architectures; they are very well capable of resisting any exhaustive attacks or statistical attacks or differential attacks [44]. Following is the description of some of the DNA computing techniques used together with existing cryptographic techniques:

(1) OTP with DNA sequencing: Some algorithms like [15,45] are based on the OTP encryption technique with DNA computing where the randomly generated private key gets applied. The key size can be more than plaintext size. The size limitation of the key was a problem encountered in traditional methods. Here DNA sequencing is useful as a DNA molecule offers huge storage space. For this algorithm, the key gets grouped into separate blocks, each block contains 8characters and is only used once for encryption. OTP cipher is unbreakable because the key is randomly generated and in addition to the DNA sequencing, it can be longer than the given message. The message to be sent gets converted to ASCII value which is converted to binary.

These binary values are encoded into a DNA sequence by applying DNA digital coding. The four binary pairs represent the four bases: $00 \rightarrow A$, $01 \rightarrow C$, $10 \rightarrow G$, $11 \rightarrow T$. A DNA sequence is randomly selected as the key and assembled in separate blocks of 8characters each. The positions of each character in the selected sequence get tabulated. Depending on this table and the selected DNA sequence, the message is converted into an encrypted form. The encrypted message along with the key is then sent to the receiver's end.

(2) **3D DNA Matrix:** In this methodology, the 3D DNA matrix is used for image encryption [46]. The matrix is manipulated based on permutation as well as diffusion of the plain-image. The matrix is obtained using bit plane-splitting, bit plane-recombination and the DNA-encoding of plain-image. Depending on the position sequence group permutation is introduced and the chaotic sequences (produced from chaotic-system) are used for permutating the positions of all the elements present in 3D DNA matrix. For the calculation of initial values of chaotic systems, SHA 256 of the original image gets applied. Then with the help of 3D DNA-level diffusion, the now confused 3D DNA-matrix gets divided into subblocks. Followed by the XOR operation the block gets directed to a sub-DNA matrix and a key DNA matrix is generated via a chaotic system. Lastly, the cipher image is obtained via decryption of the diffused-DNA matrix. Permutation, as well as diffusion, is employed because it serves three advantages. The 3D DNA-level permutation built upon the position-sequence-group is used to confuse the positions of the elements of the matrix. The generation of chaotic sequences is random. The random reallocation of the matrix elements to any position helps improve the confusion effect and escalating security levels. For the diffusion process, the hamming distance of the plain-image is used for the modification of initial values of the chaotic system. The key matrix and DNA encoding get defined by the original image itself. Therefore, this makes the algorithm resistant to known-plaintext attacks effectively.

(3) **DNA module transcript and translation:** This technique is based on one of the crucial doctrines of molecular biology [47]. In DNA transcription, a random DNA segment containing the gene is read from the promoter. Non-coding parts are called "intron" and coding parts are "extron". The introns are removed depending on certain tags and the extrons are re-joined as well as caped. The sequence gets transcribed to single-stranded sequence of messenger RNA. In DNA translation,

the selected messenger RNA sequence gets translated to a sequence of amino acids. The ribosome read the segment of three bases or one codon from messenger RNA and translates them to single amino-acids. DNA computing uses a similar idea for cryptography. As illustrated in Fig. 9, consider that the sender stores the message in binary which then gets converted to DNA form using DNA digital coding. The sender knows about starting code (indicating the beginning of an intron) as well as pattern codes (defining the segments of the frame to be kept and removed). Therefore, the sender knows the location of the intron in DNA-form of data. The sender then cuts out introns depending on the specified pattern and so the DNA-form of data gets converted to the messenger RNA-form of data. This is further transformed into a protein-form of data depending on a genetic code table and conveyed to the recipient. The recipient gets the encryption key from the sender via a secure network. The recipient reverse translates all the conversions using an encryption key to recover the data. Due to the complex conversion process, this method increases the security level and is difficult to crack without the genetic code table and the public key.

(4) DNA arithmetic operations: Some algorithms include DNA operations along with DNA encoding [16,48–50]. DNA operations used in DNA computing can be broadly classified into biological and arithmetic operations. Transcription, translation and PCR come under biological operations whereas binary operations like ADD, AND, XOR, NAND etc. come under arithmetic operations. These binary operations require two input values; therefore, two-bit strings are denoted by two-different DNA single-strands. The first string is the input strand and the other string is the operand strand. For example, the input strand can be

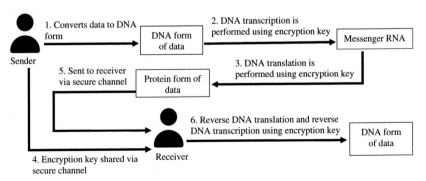

Fig. 9 Flowchart of using DNA transcription and DNA translation

AT→ 0, operand strand can be CG→0. The output of NAND operation on both of these strands will give, 00→1, as the output. One of the
ways can be that the truth tables are encoded in DNA with the help of a
three-level scheme. Level 1 is concerned with the input strand, which
can be made up of any two DNA bases. Level 2 is concerned with
the operand strand, which can be made up of other two remaining bases
along with a base similar to the one used in the input strand. The input is
distinct from the operand, but it still retains a similar base-pairing structure. Different combinations can be made together from input and oper-
and bits. In level 3, the output is calculated depending on the given truth
table of the operator employed. This method is illustrated in Table 3.
Operations like addition and subtraction are performed corresponding to
binary addition/subtraction. Using four DNA bases to encode to four binary
pairs, there exist 24 coding schemes. Though only eight coding schemes fulfill
the Watsonrick-complement rule. There are eight DNA addition as well as
subtraction rules equivalent to eight DNA-encoding schemes.

3.2 Authentication

One of the critical facets of information security is to ensure the authenticity
of any user's identity. The conventional authentication methods usage
included text-based passwords or two-factor schema based on smart cards
and tokens or advanced biometric techniques. The foremost concern while
strengthening any authentication process is the trade-off between usability
and security. Following are the descriptions of some of the DNA computing
techniques used for authentication purposes:

(1) **Authentication using DNA sequencing and DNA operations:**
The schemes comprise two stages. In stage one, the image gets encrypted
by applying DNA sequencing. In stage two, the encrypted image gets
embedded in a watermarked audio [51]. A Pseudo-Random Number
Generator (PRNG) is applied for generating some random bit-sequence
in the permutation-phase. The obtained bit-sequence gets used for

Table 3 Truth table using DNA coding.

Input strand	Operand strand	NAND
AT=0	GC=0	1
	AC=1	1
TA=1	CA=0	1
	CG=1	0

permuting pixels of the original watermarked image. The permuted image gets converted into a DNA-sequence. PRNG has been applied again for generating a bit-sequence in the substitution phase. The newly obtained bit-sequence gets converted to another DNA sequence. Both the DNA sequences are together added (using Galva Field) to get a new DNA-sequence. DNA-XOR operation gets done on this new DNA-sequence. For embedding the sequence, the audio gets decomposed. The obtained signal gets divided into equal-sized frames. Using a key, the sequence gets embedded in the signal to get a watermarked audio.

(2) **Mobile Cloud-based Authentication using a DNA Sequence.** This method provides an authentication model for Mobile-Cloud-based healthcare systems [52]. The method aims at extracting authentication data and reconstructing the DNA-sequence without information losses. The mobile device would process the DNA-sequence and hid secret-authentication data in it to produce a stego-DNA sequence. The stego-DNA sequence gets obtained using a stego-key. This key is transferred to data-consumer via a secure channel. The obtained stego-DNA sequence gets transferred to the cloud for storage. The data-consumer would request for this stego-DNA sequence. The cloud would provide the consumer with the stego DNA-sequence. The authenticity of the DNA-sequence can get verified and it can get reconstructed to obtain the authentication data using the stego-key.

(3) **Wireless Body Area Network (WBAN) Authentication using DNA Sequences:** This authentication procedure consists of three entities- WBAN client, network manager, application provider and three phases- registration, verification and password changing [53]. In the registration phase, the WBAN client submits their ID to the network manager who is responsible for generating random DNA sequences and its complement for the client. The network manager then sends these sequences along with the timestamp back to the WBAN client. The client then needs to check the timestamp within a valid time gap. The client verifies the network manager and then sends the password corresponding to the DNA sequence. The network manager verifies the password and the client. The procedure, therefore, requires verifying both the WBAN client and the network manager. The network manager's details are stored in the client's smart card during the registration phase. This helps in evading server spoof attacks.

(4) DNA Coding with AES Algorithm: DNA encoding techniques can also be used together with an encryption algorithm to determine those authentic users have access to the organization's database or information centers [54]. Encrypted IDs for users can be generated of length greater than 64 bits using characters and a key as inputs. The first step involves reading the characters and converting them to ACII equivalents. The ASCII value then gets converted to binary which is then mapped to the corresponding DNA codes. These DNA codes are then converted to decimal values depending on some specific dictionary. AES encryption is applied, the generation of IDs requires the key as well as the decimal values. An identification database can store the original characters, key and ID.

(5) DNA Genome Encoding: This technique involves encrypting the identity of a client using *E. coli* bacteria genome [55]. Firstly, the ASCII value of the client's identity is determined. Secondly, a specific gene is determined from the genome of the bacteria where the DNA's length is thrice ASCII value's length of the client's identity. The client's identity is encoded in the DNA using multiform-property of codons in a given genetic code table. The encoded bacteriophage should be placed inside a test tube. A test tube consisting of *E. coli* is kept with the server. To authenticate the client's identity, the client needs to add the contents of their test tube to the one kept by the server. The genome mutation will take place resulting comparison between the client's encoded bacteriophage with that of the server. After which the string consisting of binary bits will be generated and converted to its ASCII equivalent. This authentication procedure can ensure security because of the random selection of the genome and the location of the gene where the data gets encoded. Without the location of the gene, the attacker will require to check the whole DNA sequence which is infeasible.

3.3 Digital signature

In the present-day e-commerce domain, digital signatures have significant prospects for facilitating reliable electronic transactions, as in numerous important business processes they occur before or independently of the final transactions [56]. Digital signatures use mathematical techniques or hash functions illustrated in Fig. 10 to validate web pages or documents or messages or software. The hash function generally maps all the information to a

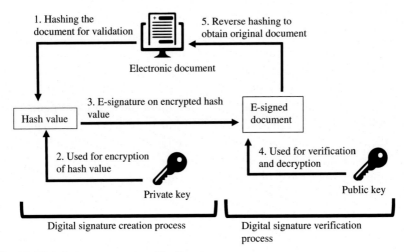

Fig. 10 Digital signature and verification process.

smaller string of finite length (hash value). The proceeding hash value gets encrypted with the help of the private key of the sender. The receiver who gets the information decrypts the signature using the public–key of the sender [55]. Digital signatures allow a recipient to confirm the authenticity of the source of any document; it also ensures veracity as the signature gets refuted if the content signed gets altered at all [57].

The commonly used approach is to sign the genomic structures digitally by DNA sequences in addition to encoding the outcomes of the algorithm embedded in electronic documents. Although this approach increases the security of any transmitted information, it is not plausible because [57]:

(1) Validation requires some cognizance of the author's identity.

(2) It denies a distinct classification of the genomic structures.

(3) The algorithm for signature authorization, invalidates any sequence that differs from what was originally signed.

(4) This rigorous validation may be unnecessary keeping in mind, the occurrence of any unsolicited alterations.

Following is the description of some of the DNA computing techniques used for digital signature purposes:

(1) DNA Symmetric Keys: This technique is implemented for low computation systems or regulated bandwidth, and it involves using DNA coding for the digital signature on the data to be sent [58]. The technique consists of two modules. For the first module, the sender uses a DNA coding sequence for digitally signing the data. For the second module, the sender generates a 128-bit length DNA symmetric key which gets encrypted using DNA public key. Then the symmetric key is shared with the receiver and messages are exchanged. The encryption and decryption of the DNA-coding sequences are done using a non-linear XOR function. The computational time for performing this function corresponds to that of regulated bandwidth systems. With increased complexity in computation, the technique is resistant to brute-force attacks.

(2) **BioGamal Algorithm:** This algorithm combines DNA and Elgamal encryption-decryption algorithms [59]. The DNA encryption-decryption algorithm takes the Hash value of the data as its input. This Hash value gets encoded by first converting to ASCII form, then in its binary equivalent. Using DNA digital coding, the four binary pairs are mapped to the corresponding four DNA bases. 15 key combinations are formed by pairing two DNA bases. For example, key combination: AA, pattern: 0000, value: 0. The values are assigned similarly for different pairs like AC, TA, CC etc. Lastly, using these key combinations and their assigned values, DNA sequencing is applied. Elgamal algorithm is a block cipher where the message from the sender's end gets divided into sub-blocks for encryption. Two keys- public and private key along with a prime number p and a primitive group gets generated. In the encryption process, the public key along with a randomly generated confidential integer k which should be less than or equal to $(p-2)$ encrypts the message. The value of k is used for ASCII conversion of the message. In BioGamal, the digital signature is encoded by applying DNA algorithm, the obtained ciphertext is encoded using the Elgamal algorithm, for decryption these steps are performed vice versa. The BioGamal algorithm is useful in providing two-level security. Level one of security is DNA encryption; level two constitutes Elgamal encryption. BioGamal algorithm also serves as an example of blending DNA computing cryptographic methods with traditional cryptographic algorithms.

(3) **Realtime-Signature Authorization using DNA Encryption:** In this technique, DNA based encryption technique is used for encrypting and later authenticating the digital signature of the sender [60]. The technique also uses IoT for simultaneously reflecting the changes in a document on some other similar file in another location. The sender signs the document using a digital pen. The signature area consists of some specific dot-pattern for tracking the pen's movement while

signing the document. After signing the document, the fingerprints of the sender are also recorded for verification afterward. The dot pattern and the sender's fingerprints are converted to binary format and encoded using DNA encoding. The obtained cipher data then gets conveyed to the receiver.

(4) **DNA–Protein Kinase Catalyst (PKc):** This technique requires two different pairs of cryptographic keys, one is for the encryption process and the other is for the signature [61]. The encryption process uses a public key and the owner has a personal signature key. The signature generated by the key can be verified using a public key but the signature itself cannot be forged. DNA-PKc is not a copy of data instead they are DNA probes. The keys cannot be transferred electronically, a physical exchange of keys is carried out. The keys and ciphertext exist in the form of biological molecules for example, in a mixture or on a chip. In DNA-PKc, there are two steps involved in the generation of the public key. In the first step, encryption and decryption keys are separated as DNA probe sets and probe mixtures respectively. The DNA probes are stored in test tubes separately. The decryption key can be extracted from the probe mixture with the help of probes complementary and then hybridized to the encryption key. In the second step, a set of encryption-decryption keys is selected. The DNA probes are arranged to fabricate the DNA chips. The DNA chips are hybridized along with decryption probes under specific hybridization conditions. This process of key generation is more like a biological experiment and further research can improve its efficiency.

For digital signature verification, the sender can select a pair of verification and signature keys, where the signature key is kept secret and the public key is a verification key. The verification keys can be copied and shared with different verifiers. The signature key is used for DNA chip fabrication for any given message. DNA probes represent binary 0 and 1 and are arranged in this form on the chip. The receiver can use the verification key to verify the sender's signature by hybridizing it with the DNA chip. The other verifiers can also do the same, but no one can forge the sender's signature. This method provides two-level security. Biological carriers of keys and ciphertext are the first level of security. The second level of security consists of computational difficulties. Given the massive parallelism property of a DNA molecule, any attack on DNA-PKc results in computational complexities greater than offered by AES-128. The attacker requires to first break the biological level of security which is dependent on random DNA sequencing. The molecular information of a random DNA probe present on DNA chip is inaccessible by attackers.

4. DNA computing based cryptographic method using PCR

PCR is the rapid DNA magnification technology derived from Watson–Crick complementary model. PCR amplifies the number of DNA strands, the reason being it's laborious to deal with a small number of DNA strands. A primer is defined as a sequence consisting of a few units of nucleotides compatible with a fixed sequence of DNA to be amplified. For performing PCR, it is important to identify the sequence of DNA for augmentation and therefore design an appropriate primer for the same [15]. Biological PCR operations involve a repeated series of temperature changes (thermal cycles), where every cycle has a distinct temperature step [62]. Most of the PCR methods involve the following steps:

(1) Initialization: involves heating of DNA polymerase in a reaction chamber at high temperatures ranging from 94 to 98 °C for 1 to 10 min.

(2) Denaturation: is the first step of a thermal cycle. In this step, the DNA polymerase is heated in a reaction chamber at a temperature of 94–98 °C for 20–30 s. This is done to break the hydrogen bonds existing between the DNA complementary bases and yield two single-strand DNA molecules.

(3) Annealing: is the second step of a thermal cycle. In this step, the temperature gets lowered to 50–65 °C for 20–40 s for annealing the primers to the single strands of DNA obtained from the previous step. The primers used are shorter than the length of the target region. Determining a suitable temperature is crucial to ensure the successful hybridization of primer with the single DNA strand.

(4) Extension/elongation: in this step, the temperature is dependent on the polymerase used. This step involves adding free Deoxyribonucleotide Triphosphate (dNTP)s to the single DNA strand for the synthesis of a new DNA strand. In ideal conditions, the number of DNA target sequences gets doubled after each extension cycle.

(5) Final elongation: is an optional step, performed at 70–74 °C temperature range for 15 for ensuring that any residual single-strand DNA gets entirely elongated.

(6) Final hold: this involves cooling the reaction chamber temperature (4–15 °C) for some indefinite time period for instant storing of any PCR products.

PCR technique is widely in use by clinicians and researchers for diagnosing diseases, cloning and sequencing genes, as well as carrying out advanced quantitative and genomic studies rapidly and delicately [63]. PCR amplification can be affected due to a temperature change. In [64], the PCR technique encodes an image in between primer keys (they are the primer pairs used as keys for encryption and decryption processes), the DNA codons that is encoded by four bases yield 256 combinations of keys responsible for heavy security, and it minimizes ciphertext's size. Primer keys are difficult to replicate since they are biological molecules. They get formed by converting amino acid sequences to DNA nucleotide sequences. The primer keys are useful for changing the DNA sequence. Table 4 tabulates

Table 4 Comparison of cryptographic schemes using PCR methods.

Schemes	Key findings	Advantages	Disadvantages
[18]	Encryption key generation takes place using ultrasonic fragmentation and PCR sequencing	Highly random generation of secure keys Due to the huge storage capacity of DNA, there exists no correlation between any keys generated	It has an excessive modification rate
[19]	Data gets encrypted using ssDNA OTP	Secure long length key generation DNA-based OTP generation	Its time consuming
[20]	PCR process employed in key generation	It is resistant to brute-force attacks	Key sharing needs to be more secured
[65]	Data gets encrypted using DES before the DNA encoding process	Trivial correlation between pointers file makes the scheme robust	As the sequence is shared only between the sender and receiver, there should be a method for users' authentication.
[66]	Use of DNA-XOR operation for data encryption	Secured key sharing It is based on the DNA sequencing-puzzle	It has an excessive modification rate

a comparison between some of the cryptographic schemes that also use PCR method discussed below:

- DNA sequencing along with OTP encryption is also used as cryptographic techniques. A random key is generated using ultrasonic fragmentation and PCR sequencing [18]. Ultrasonic fragmentation is performed on a selected DNA strand for fragmenting it to the needed length. The DNA pieces (of varying lengths) obtained after fragmentation undergoes Gel electrophoresis. Gel electrophoresis process ensures that all DNA pieces are of uniform lengths. A random encryption key gets generated using PCR process. The data to be transferred gets converted to ciphertext using DNA-OTP encryption process.

- PCR can also be used in steganographic techniques [19]. The data gets converted to ASCII to binary. Using ssDNA OTP, the data gets encoded into a DNA sequence. This encrypted data gets among any two primers and together covered inside some microdot. PCR amplifies the hidden data for DNA hybridization. Gel electrophoresis ensures that the length of the primers used are uniform. Hybridization takes place between the primers and the original OTP generated.

- For flanking the data by the primers, PCR method is used [65]. The DNA sequence in which the data is encrypted is placed between two primers. PCR allows the generation of several copies of an exclusively selected DNA sequence. The reference sequence will however get communicated only between the sender's end and receiver's end. For the encryption process, the selected DNA sequence gets searched for identifying starting and terminating primers. The data gets stored between these two primers in a DNA sequence. The data first gets converted to binary form and then to ASCII values. Before DNA encoding, the data gets encrypted using encryption algorithms like DES. The starting and terminating primers help in detecting the length and location of a message in a DNA sequence.

- This scheme uses a dynamic-sequence table for assigning random ASCII values to DNA-sequence table [20]. After some finite number of iterations, OTP is implemented on the binary values (modified). OTP ciphertext is mapped to a DNA-sequence with the help of an amino acid table. Moreover, with the help of PCR amplification, the primers and coding mode are used as keys. Firstly, the plaintext gets converted to its ASCII value and then to binary. The obtained binary values get arranged in a matrix form. The key is generated using PCR process and transmitted physically to the receiver and it is difficult to replicate the key.

- The encryption process consists of two rounds in this scheme [66]. In the first round, the key and the data are converted to ASCII then binary. The obtained binary values are then mapped to a DNA sequence. PCR is used for primer pairs generation which is used as keys that get transmitted through a secure channel. In the second round, the data gets encrypted to DNA segments using the XOR operation.

5. DNA computing based cryptographic schemes

An ideal security model should achieve three main goals- confidentiality, integrity, and availability [67]. Novel DNA cryptographic schemes based on DNA computing techniques should also try to incorporate these factors to improve their implementation. This section discusses the following existing DNA based encryption schemes and Table 5 shows a comparison between these schemes:

For secure image communication, a simple image can be first, permuted with the help of a sequence of Pseudo-Random Number (PRN) and subsequently encrypted with the help of DNA computation [68]. It is a reliable, lightweight cryptographic scheme where a PRNG generates two PRN sequences based on a chaotic logistic map with the help of two key sets. First, it generates a PRN sequence that permutes the image. The second, PRN sequence generates any random DNA sequence. The number of rounds of permutation as well as encryption could increase to upgrade the security of image communication. For an image of size M x N the algorithm requires M x N random numbers, so the complexity becomes $O(n)$. For generating a random DNA sequence of M x N bits use the same chaotic map, therefore the complexity for M x N x 8 number of random bits is $O(n)$. The overall complexity of the algorithm becomes $O(n)$. The algorithm is resistant to statistical attacks because it uses DNA sequencing and a logistic map for the substitution phase of the encryption process. If the value of the key changes, it results in uncorrelated images in comparison to plain image.

Colored images can be encoded using an encryption algorithm apropos of DNA sequencing operations as well as One-Way Coupled-Map Lattices (OCML) [69]. DNA encoding rules get used for the image to be broken down in three gray-level units and randomly converted to three distinct DNA matrices. Second, the XOR operation is operated twice on DNA matrices. Third, according to DNA decoding rules, DNA matrices get

Table 5 Comparison of DNA computing based cryptographic schemes.

Schemes	Key findings	Advantages	Disadvantages
[17]	Use of MLNCM and OTP using DNA computing	Random lattice numbers Encoding of lattice number to DNA Lengthy OTP key	It has an excessive modification rate
[68]	PRNG is employed to generate a PRN-sequence for permuting images and another sequence for random DNA sequencing.	It is resistant to statistical attacks It is lightweight If the value of the key changes, it results in uncorrelated images in comparison to the plain image.	Key sharing not secured Sequence repetition is possible in PRNG
[69]	DNA matrix and OCML	OCML applied during key generation Diffusion is used for image encryption	It has an excessive modification rate
[70]	Using DNA level diffusion and permutation along with hash function	Involves random reallocation of the elements	It has an excessive modification rate No secure key
[71]	Using Mealy machine for DNA sequencing	Secure against known-plaintext-attack, brute force, differential cryptanalysis, cipher-text-only, man-in-the-middle attacks as well as phishing Key is shared after receiver authentication	It has an excessive modification rate Its time consuming
[72]	Use of ECC along with DNA arithmetic operations	The algorithm could avoid brute force, statistical, differential-attacks	Key sharing needs to be secured
[73]	Dynamic key generation using the scrambling technique of DNA computing	Simple, low execution time Low modification rate	Key sharing needs to be secured

Continued

Table 5 Comparison of DNA computing based cryptographic schemes.—cont'd

Schemes	Key findings	Advantages	Disadvantages
[74]	Key generation processes two-level mathematical operations The identity of a user is authenticated	DNA sequence gets scrambled for enhancing the security of encrypted image	It has an excessive modification rate It is time-consuming.
[75]	Key gets created with the help of DNA computing, attributes of the user as well as MAC address of the user, ASCII value by following the DNA bases and complementary rules	Involves complex key generation procedure	It has a low-probability cracking
[76]	Using Morse code and zigzag pattern along with DNA sequence	It is resilient to brute-force attacks, differential attacks and known-plaintext attacks	A slight change in either the DNA-sequence or the zigzag pattern will corrupt the original data
[77]	DNA reference key is added in between DNA bases and then the complementary rule is applied to yield the DNA based key	The attackers cannot replicate the generated key because of the exclusivity of DNA molecule.	It has an excessive modification rate

remodeled to three gray images. Lastly, a diffusion process is additionally applied to modify the image pixel's values by some keystream, to obtain a cipher-image. The keystream generated by OCML is associated with this plain-image.

A compound prototype of DNA computing along with chaotic systems, as well as hash functions can be useful for image encryption [70]. This scheme comprises of DNA level permutation as well as diffusion. DNA level permutation: it is useful in randomly changing the spots of elements within DNA image. On this, a mapping function depending on a logistic map can be functional. DNA level diffusion: DNA operators work upon diffusing the arranged DNA image by a key DNA image. Hence, this led to defining two novel DNA algebraic operators that are DNA-left-circular shift operator and DNA-right-circular shift operator. Using DNA-level permutation and diffusion is advantageous, for confusing the attacker of the element's positions. The random reallocation of the elements further aids in improving security. For every change in the image, there is a change in the DNA-level permutation and diffusion. Moreover, using DNA algebraic operators further complicates the process of locating the elements for the attackers.

Cryptosystems coalescing the aspects of DNA with the randomly formed mealy machine have proven to be efficient and secure against known-plaintext-attack, brute force, differential cryptanalysis, cipher-text-only, man-in-the-middle attacks as well as phishing [71]. The cryptosystem is made up of three different entities- sender, receiver, and key-pair generator. The key-pair generator forms a 256-bit DNA based key depending on the properties of the receiver. To enhance the security of the ciphertext, a Mealy machine (randomly generated) is used for DNA sequencing. The four DNA bases are used as the inputs of the mealy machine. The sender first, registers and requests the key from the key-pair generator. The generated key-pair is provided to the sender. Similarly, the receiver also requests for the key from key-pair generator. The receiver is also provided with the same key-pair. The sender authenticates the receiver via key-pair generator by registering with the receiver's credentials. Once the receiver is authenticated, a data request can be passed to the sender. The data transmitted is in encrypted form using a mealy machine and DNA sequencing.

In Ref. [72], researchers introduced an image encryption schema apropos of ECC enhanced with the help of DNA computing. Within this scheme firstly, an image gets mapped and later encrypted using ECC. Secondly, each point is encoded to obtain a new DNA sequence. Deriving a cipher image is possible with the help of using DNA operations (XOR, +, −). According to the researchers, this model provides two levels of security. ECC focuses on the first levels of security, and the second level gets delivered with the help of DNA Computing. Results gained, after some simulation experiments, performance evaluation as well as assessment with other anticipated encryption algorithms, showed that the algorithm could avoid brute force, or a statistical, as well as a differential-attacks.

For RGB image encryption, the combination of DNA computing and a genetic algorithm is brought forth [73]. This scheme derives from the scrambling technique of DNA computing operations with the help of crossover and mutation processes for establishing dynamic keys depending on a genomic algorithm. Firstly, the decoding of the image GA selected the DNA sequence encoding process and the random key for the three R, G, and B channels, subsequently the process of DNA addition. The decoded DNA gets added to the matrix of output. Finally, to obtain the encrypted image, the XOR-mod procedure is performed on the decoded matrix as well as on a random number of the genomic algorithm.

In Ref. [74], a DNA computing methodology is put forward for encrypting biometric color (face), and gray fingerprint images as these images play a pivotal role in authenticating the identity of an individual. In this method, firstly the values of the images are considered for designation to two distinct matrices. Two-level mathematical operations get applied to the fingerprint image for the generation of the encryption key. DNA computing is performed on the above two matrices to generate the DNA sequence to augment the security and reliability of biometric images. The DNA sequences get scrambled to increase the complexity of the encrypted biometric image.

Some researchers also performed image encryption based upon spatiotemporal chaos of Mixed Linear–Nonlinear Coupled Map Lattices (MLNCML) by employing DNA computing as well as OTP encryption dogma [17]. The plaintext image is used for generating lattice numbers for the encryption process. The generation of random lattice numbers is sensitive to any changes to plain-image. DNA encryption with OTP encryption could further enhance security. The DNA molecule offers massive parallelism which can be used for generating OTP keys of length greater than the message itself. It becomes almost impossible to crack the message by having a lengthy OTP key. Encoding lattice numbers to corresponding DNA bases also enhance the intricacy of the scheme.

A huge amount of stored and transmitted images in cloud-based Closed-Circuit Television (CCTV) systems can be protected, using the image encoding algorithm [78] made upon the hyperchaotic map and DNA coding. It uses the chaotic property of the hyperchaotic maps above DNA computing as shown in Fig. 11. Cloud Computing provides innumerable assistances for both a service provider, and a customer and many enterprise applications and data have migrated to public or hybrid cloud. However, a myriad of security issues accompanies the benefits of cloud computing. Following DNA based encryption schemes are proposed for enhancing the security of a cloud computing environment:

(1) Researchers proposed a DNA data encryption schema for cloud computing environments in [75]. A 1024-bit size key gets created with the help of DNA computing, attributes of the user as well as MAC address of the user, ASCII value by following the DNA bases and complementary rules.

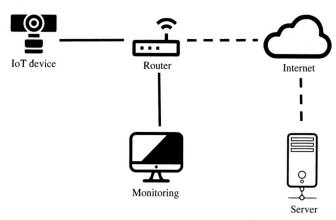

Fig. 11 Cloud apropos CCTV system.

(2) To enhance security measures as well as confidentiality for data storage in a cloud environment, Morse code and zigzag pattern use DNA sequences, aimed at developing an encoding schema [76]. Zigzag pattern used in this scheme is actually a square matrix of integers. It is possible for the matrix to increase sequentially by anti-diagonals. Before uploading data to the cloud, it is encrypted using random DNA sequencing which makes it arduous for the attacker to hack the data stored in the cloud. The original data gets converted to binary form after which it is mapped to corresponding DNA bases using DNA sequencing. The DNA sequence then gets converted to the Morse pattern which is further converted to the zigzag pattern of the square matrix. The zigzag pattern yields the ciphertext which is then uploaded to the cloud. The place for data storage is unknown to the user as well. This scheme is resilient to brute-force attacks, differential attacks, known-plaintext attacks. In case of brute force attack, it is a tedious task to locate the key or original data in its bio-molecular form. If the attacker uses an incorrect key-code, the data will be corrupted. Similarly, for the differential attack, a slight change in the DNA sequence and/or the zigzag pattern will corrupt the data. Given the Morse code conversion, it becomes more difficult to access original data using any chosen set of characters from the data.

(3) DNA computing-based encryption schemes can also be used for securing multimedia in the cloud [77]. For key-generation the multimedia content is divided into n number of 1024-bit size blocks. Each of these blocks get separated into four 256-bit size blocks. Each 8bit binary number gets converted to corresponding ASCII values which in turn are converted to their real binary numbers. Each 2-bit binary pair gets converted to its corresponding DNA bases depending on a given DNA encoding table. DNA reference key is added in between DNA bases and then the complementary rule is applied to yield the DNA-based key. The attackers cannot replicate the generated key because of the exclusivity of the DNA molecule.

6. DNA computing based steganography schemes

Steganography is one such technique in which it is not possible to detect the presence of a secret message, and it proves as a tool for security purposes to transmit confidential information securely. This section discusses some of the DNA based steganography schemes, and Table 6 presents a comparison between these schemes:

DNA sequencing plus a cross-coupled chaotic map is in use for the steganographic scheme [11]. The confidential message gets encoded with the help of DNA computing. After that, it gets embedded onto a cover image with the help of the LSBase substitution technique. In this scheme, the confidential message gets converted into a DNA sequence, and some other arbitrary DNA-sequence gets produced with the help of a chaotic system (cross-coupled). Both DNA sequences get added together by applying DNA addition operation. The LSBs of pixels of the cover image get replaced by a confidential DNA-sequence.

LSBase is a steganographic algorithm that uses codon degeneracy for covering the original data into DNA sequences and causing no changes to the protein structure they get coded for Ref. [12]. The LSBase algorithm alters the third DNA base of codon to either purine or pyrimidine bases depending on whether the secret-bit is binary1 or binary0, respectively. The embedding process gets initiated with the conversion of key-value to binary values. For carrying the secret-key information, a random reference DNA-sequence gets selected. The DNA sequence is then converted to RNA by replacement of the T character with the U character. For the embedding process, the reference sequence gets split in different codon triplets. For all codons, the least-significant base gets checked for pyrimidine when true, it gets altered to U for encoding 0 (secret bit) or C for encoding 1. If the base is purine, it gets modified to A for secret-bit 0 and secret-bit 1, it is altered to G. This step applies to the complete sequence. The modified sequence of RNA gets converted again to a DNA strand. The resultant DNA sequence and the original one both are coded for the same protein and are indistinguishable. As the key-value gets embedded in a DNA

Table 6 Comparison of DNA computing based steganographic schemes.

Schemes	Key findings	Advantages	Disadvantages
[11]	Using a cross-coupled chaotic system for DNA sequence generation	The functionality of the biological DNA gets preserved The rate of DNA-modification is low	It has low capacity because of LSB used in the hiding process
[12]	Using LSBase algorithm	No changes to the protein structure of DNA	It is not easy to implement
[13]	Two secret images are generated using DNA sequencing, conversion to grayscale, and LSB substitution	Two secret images are hidden within an image exclusive of distortion It provides double-hiding layers A secret key has been used	Data not encrypted before hiding
[14]	A three-leveled scheme where DNA-RSA encryption of data is done in the first level. The second and third levels are for data hiding. LBase substitution is used in the second level and LSB substitution in the third level.	It's efficient and secure Data is encrypted with the help of the RSA algorithm before being embedded It provides two hiding layers Public-key is used	The functionality of biological DNA not preserved
[79]	Microdots are used for data encryption	The data is first encrypted and then hidden inside a DNA sequence Low DNA modification rate	Increase the message size applying microdots is time-consuming
[80]	Using DNA-XOR operator and DNA-XOR operator for confidential transmission	It makes use of an encryption key Data to be hidden is divided into 3-secret shares	The length of stego-DNA could be expanded

Continued

Table 6 Comparison of DNA computing based steganographic schemes.—cont'd

Schemes	Key findings	Advantages	Disadvantages
[81]	Information is encrypted using play fair cipher Using the DNA substitution method for data hiding	High capacity of data hiding inside the DNA It has a low-cracking probability It has high scalability	The length of the DNA reference can be expanded after embedding
[82]	Data hiding using DNA-insertion algorithm	Implementation of XORing is difficult to guess It has a low-cracking probability	It has an excessive modification rate
[83]	Information is converted to binary and then embedded in such a way that a fake DNA-sequence is generated	The visibility of the sequences is very low It is resilient to brute force attack Data confidentiality among clients	Data to be sent is not encrypted before hiding There is no procedure for client authentication
[84]	A distinctive palindromic DNA-structure is employed in place of header information to indicate the ending of hidden data.	Data is encrypted with the help of Playfair cipher before being embedded It is robust A secret encryption key has been used	The functionality of amino acid not preserved It has high redundancy It has an excessive modification rate

sequence, it can unsuspectedly get transmitted across any public network. The flowchart of the same is shown in Fig. 12.

For hiding data in colored images, the plain-image can be first converted to grayscale and then encoded into a random DNA-sequence [13]. Firstly, when reading the string from left to right, find the characters closest to A, C, T, and G characters in the difference-chart. Then calculating the corresponding decimal values and storing them in a matrix form is done. Now the secret information is stored in matrix form and converted to a grayscale image. Applying the LSBase substitution technique, a stego image and a cover image get generated. DNA arithmetic operators help embed the data. The key gets generated using the decimal value of the nucleotide bases.

Hiding data into an audio signal which is imperceptible is called audio steganography. Data hiding in audio signals prevents unauthorized access, means of secured transmission and its reception [14]. The scheme has three working levels- one level for encryption and the other two for steganography. In the first level, the data gets encrypted with the help of DNA-based

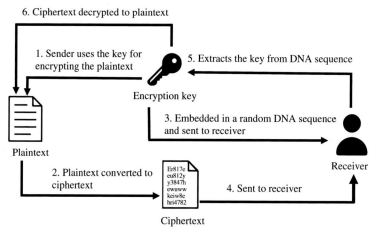

6. Ciphertext decrypted to plaintext

1. Sender uses the key for encrypting the plaintext

5. Extracts the key from DNA sequence

Encryption key

3. Embedded in a random DNA sequence and sent to receiver

Plaintext

Receiver

2. Plaintext converted to ciphertext

Er817e eu812y y3847h ewuww keiw8e hri4782

4. Sent to receiver

Ciphertext

Fig. 12 Flowchart of LSBase steganographic scheme.

RSA encryption. By converting the data into a random DNA-sequence by applying DNA coding. The DNA-form data gets converted to a codon-triplets-sequence which further gets converted into amino acids. Ambiguity numbers respective of the codons are essential for the decryption process. RSA encryption then encrypts the sequence of amino acid. PRNG gets used for generating public as well as private keys for the RSA algorithm. During the second stage, the obtained encoded data gets covered inside a reference DNA-sequence and carried out using the LSBase substitution-technique. Finally, during the third stage, the acquired DNA-sequence from the previous step gets covered within an audio file for concealing the presence of data. WAV format audio files hide the data. Using LSB substitution, an audio file gets read in binary. To avoid any distortions, the length of the hidden DNA sequence gets encoded in the lower halves of the first 32 audio-samples, and the secret sequence itself gets encoded in the lower halves of the remaining audio-samples.

Researchers in Ref. [79] founded a message hiding method derived from DNA steganography. In this method, they designed a few methodologies to firstly encrypt some messages and afterward decompose the corresponding ciphertext into two parts. For sending one of the parts, the sender makes use of DNA steganography. If the microdot gets contaminated, then again, encryption and decomposition of the message would take place until the microdot is no longer assumed.

Researchers in Ref. [80] used the DNA-XOR operator for hiding color images. In this scheme, a truth table of DNA-XOR evaluates all input values and deriving from such evaluations, the highest value of Peak

Signal-to-Noise Ratio (PSNR) gets chosen for classified transmission. These chosen values get embedded in the cover image. The cover image then becomes an encryption key in this secret sharing process. The data is divided into 3-secret shares and then embedded in the red channel, green channel, and blue channel of the cover image for hiding.

DNA strand properties help develop a steganography algorithm to achieve secure and reliable data communication [81]. This method gets employed at two levels. In level one: applying a DNA-based Playfair cipher for encryption of the confidential information/message. In level two: the substitution method hides the encrypted confidential message into some reference DNA. DNA conservative-mutation gets applied to allowing any DNA base substitution to another base carrying two bits. This scheme provides three benefits. First, the use of each codon for hiding two-bits and the preservation of properties of protein sequence allows a high capacity of data hiding inside the DNA. Secondly, the low cracking probability can be attributed to the conservative mutation process along with the biological permutation process. Lastly, the scheme renders high scalability for three megabytes data size.

This scheme uses DNA-insertion algorithm as it is beneficial in enhancing DNA computing-based steganography [82]. A random DNA sequence is selected for this scheme. It then gets converted into binary-form by applying the DNA dictionary-method. The original message to be transmitted gets converted first to ASCII form. The ASCII equivalent gets converted into a binary sequence. This binary sequence gets divided into segments of size 8-bits. In this scheme, two key values are used – K1 and K2. Key-value K1 is for performing the XOR operation on the message to narrow it between 0 and 255. K1 gets converted into bits binary sequence of eight bits. Then it gets XORed with the first 8-bits of the binary form of the original message. The result is another XORed value that again gets XORed with the next 8-bits of the message (binary-form) and henceforth continues. The resultant message bits yielded by XORs get concatenated for forming one binary sequence. Then the initial converted sequence of DNA to binary form gets split into segments with the help of randomly generated K2. Key-value K2 is low to minimize the DNA sequence's length for hiding the message. The split DNA segment gets inserted separately at the starting of every segment of the initial randomly chosen DNA sequence by applying the insertion algorithm. Then the encoded binary sequence gets converted into some fake DNA-sequence by applying the dictionary rule. The implementation of XORing in this scheme is unpredictable. The scheme also achieves less cracking probability.

DNA computing based steganographic methods can also be used while transmitting data to the cloud via unsecured networks, thus ensuring data confidentiality [83]. This scheme involves the embedding phase, where the data gets converted into binary-form and mapped to a random DNA-sequence. The complementary rule gets applied to the generated DNA sequence. This step is essential for increasing the intricacy of the scheme. The clients get a DNA reference sequence. This reference sequence helps in the extraction of the index of each DNA base numerically. Now the embedded data is ready to be sent to the cloud. This scheme increases data confidentiality among clients using a cloud environment. The DNA reference sequence provided to the client is unique, and guessing the correct sequence is impracticable. The visibility of the sequence is very low, making it difficult for the attacker to determine whether it is fake or not.

Using blind extraction technique for data hiding does not require sharing anything between the sender and receiver in advance other than a secret key [84]. In this scheme, data hiding takes place in two phases. The first phase occurs before the embedding process. In the first stage, the data gets encrypted with the application of DNA codons to Playfair cipher. The second phase includes the substitution and insertion phases. For the substitution phase, the stego-DNA gets generated when the encrypted data gets embedded into a random-sequence of DNA. During the insertion phase, the resulting stego-DNA is inserted into a cover DNA to perform the blind extraction process. For signaling the ending of hidden data, a distinctive palindromic DNA-structure is employed. The flowchart for the same is shown in Fig. 13.

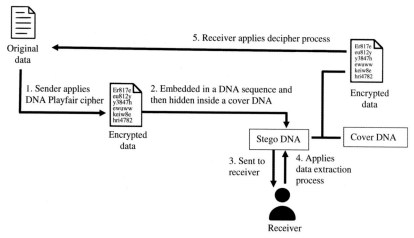

Fig. 13 Blind-extraction technique flowchart.

7. Challenges to implement DNA computing in cryptography

There are many challenges to implementing DNA computing in cryptography:

(1) **Accuracy of the Processes:** DNA computing depends highly on the accuracy of the enzymes used, but the synthesis of a DNA strand is bound to errors like mismatched pairs. In a fully operational computer, the increased number of calculations also exponentially increases the chances of error. Furthermore, the size of the molecules also increases as required by any complicated procedures. Consequently, this increases the probability of shearing the molecules and contributes to error. DNA computing works with a massive amount of data hence exponentially increasing the probability of error. The parts of a computer are probabilistic, and a sub-circuit may provide accurate answers only 90% of the time, and DNA molecules can also denature and get dissolved into nucleotides. By extending computations time, your computer will gradually be "melting" [85].

(2) **Time Consuming:** The massively parallel DNA makes the calculations appear fast. Each process is passive because innumerable solutions get calculated right away, and it acquires the facade of speed. For example, for any problem having multiple possible routes and possibilities, solving each one parallelly together is possible. But for relatively simple computational problems, generating solution sets requires impractically large amounts of memory (i.e. many DNA strands). In DNA computing, the speed of each process is slow in comparison to conventional computing. One more problem is that DNA computations make it challenging to see a solution. At present, there does not exist any credible method to input data from DNA calculation to some electrical output [85]. In the context of well-blended systems, the operations' sequences are a challenge in themselves to achieve [86]. Although DNA performs 1022 computations/s, it is slow and convoluted in setting up the input and output of the data from a biological computer and sorting out conclusive results [87].

(3) **Implementation:** To produce a stable and reliable DNA computing-based cryptography mechanism, ample materials and large-scale biological and lab experiments are required. It becomes crucial to bridge the existing technologies with recent technologies so that it is possible to develop a hybrid cryptographic system that provides unrivaled

confidentiality, and a more secure authentication schema. Biological structures and computer science collectively need to lay a foundation for providing a definite standard for efficient, reliable, and stable DNA computing-based cryptographic algorithms.

DNA cryptography lacks an advanced mathematical background like other cryptographic systems [88]. In some cases, the synthesis of genetic sequences for making fully functional genetic robots would be expensive using current existing methods [89]. At present, some major perplexities of DNA cryptography are the lack of theoretical basis, sophisticated lab requirements, and computation inadequacies [47].

In DNA computation, human interaction is imperative when combining the mixture of the input, the software, and the hardware molecules, interpreting the output molecules. All of this increases the overall time for obtaining some solution to the problem.

Parallel processing requires the amount of DNA that is proportional to the complexity of a problem. All of this could cause a tremendous and inappropriate amount of DNA employed for solving equations comprising numerous variables [90]. The DNA computer may be programmable, but it only performs elementary functions, and significant advances will help implement many complex functionalities [90].

(4) Biological Axioms: Biological axioms are the limitation encountered while encoding data into DNA using techniques, for example, PCR amplification, DNA chips, and probes, DNA splicing, and genome encoding. At present, there exist four biological axioms defined for cryptography [91]:

- Without identifying the correct primer, it becomes tedious to isolate DNA sequences from an unspecified compound of DNA for performing the sequencing process. For PCR amplification that needs two primers, knowing one correct primer is insufficient for conducting a PCR amplification reaction. Only when both primers are accurate, amplification of practical information is possible.

- Without identifying the correct probe, it is tedious to precisely decipher and perform sequencing upon information or data about mixed unspecified DNA probes. This varies simply in nucleotide sequences placed on the DNA chips (or microarrays). For different hybridization conditions, several DNA probes are used and generated. When the hybridization conditions are unknown, it is challenging to decipher and perform sequencing on the data of mixed DNA-probes arranged on a DNA chip.

- Reconstructing and performing extensive DNA-fragment splicing is tedious for a DNA sequence that also contains recurrent sequences.
- Prediction of selenoprotein from the genome and expressing seleno-protein via genetic engineering methodologies is a tedious task.

8. Conclusions

DNA computing has a variety of applications in cryptography. Cryptography has evolved from the use of codebooks to the use of DNA computing techniques. The reliability and performance of DNA encryption algorithms apply to the layered security routines of today's network architectures. Some DNA algorithms can overcome exhaustive or statistical, differential attacks. The DNA computational techniques most commonly used in encryption and steganography schemes include OTP, DNA arithmetic operations (AND, OR, XOR, XNOR), PCR, DNA sequencing, and 3D DNA matrices. Most encryption schemes follow a similar strategy of converting data to ASCII values and further converting them to binary equivalents. These binary pairs are mapped to each of the four DNA bases and stored in a randomly generated DNA sequence. Steganography schemes propose to embed the data into a dummy DNA sequence that is sent to the recipient through any network or medium. DNA computation-based steganography schemes need to achieve higher data hiding ability and minimal cracking probability. When comparing different steganographic schemes, encrypting the data before hiding provides an additional level of security to the whole process. The massive parallelism and storage capabilities of DNA are useful for authentication, encryption processes, and digital signatures. Cryptography techniques based on DNA computing still need extensive research. There are many issues to be addressed such as accuracy of the process, the time consumption of individual procedures, feasible implementation of all mentioned techniques, and biological axioms. The application of DNA computing in cryptography is not limited to encryption of text/messages but extends to providing reliable security mechanisms for image encryption, cloud computing environments, big data, and deep learning. The use of DNA computing in cryptography has significant potential for the development of secure, reliable, and feasible cryptographic algorithms and systems. The concept of DNA computing in cryptography can be extended to secure cloud computing environments, and wireless sensor networks (WSNs). For cloud computing, encryption schemes based on DNA computing can be developed to ensure data confidentiality, security, and client authentication. For WSNs, efficient, attack-resistant schemes that use DNA bases to hide data can be improved. Data can be first converted to ASCII values and then to their binary equivalents. With DNA encoding, binary pairs can be mapped to DNA bases and then encrypted with a key. DNA computing techniques can also be merged with existing encryption algorithms such as AES to enhance the complexity of the resulting cryptosystem.

References

[1] D. Fan, J. Wang, E. Wang, S. Dong, Propelling DNA Computing with Materials' power: recent advancements in innovative DNA logic computing systems and smart bio-applications, Adv. Sci. 7 (24) (2020) 2001766.
[2] L.E. Orgel, The origin of life on the earth, Sci. Am. 271 (4) (1994) 76–83.
[3] S. Chandra, S. Paira, S.S. Alam, G. Sanyal, A comparative survey of symmetric and asymmetric key cryptography, in: In 2014 International Conference on Electronics, Communication and Computational Engineering (ICECCE), IEEE, 2014, pp. 83–93.
[4] A.K. Kaundal, A.K. Verma, DNA based cryptography: a review, Int. J. Inf. Commun. Technol. 4 (7) (2014) 693–698.
[5] T. Anwar, A. Kumar, S. Paul, DNA cryptography based on symmetric key exchange, Int. J. Eng. Technol. 7 (3) (2015) 938–950.
[6] P.T. Akkasaligar, S. Biradar, Selective medical image encryption using DNA cryptography, Inform. Security J. Glob. Perspect. 29 (2) (2020) 91–101.
[7] K.S. Sajisha, S. Mathew, An encryption based on DNA cryptography and steganography, in: International Conference of Electronics, Communication and Aerospace Technology (ICECA), IEEE, Coimbatore, India, 2017, pp. 162–167.
[8] A.E. El-Moursy, M. Elmogy, A. Atwan, DNA-based cryptography: motivation, progress, challenges, and future, J. Softw. Eng. Intell. Syst. 3 (1) (2018) 67–82.
[9] P. Barkha, Implementation of DNA cryptography in cloud computing and using socket programming, in: IEEE International Conference on Computer Communication and Informatics (ICCCI), IEEE, Coimbatore, India, 2016, pp. 1–6.
[10] A. Prajapati, A. Rathod, Enhancing security in cloud computing using bi-directional DNA encryption algorithm, in: I. Computing (Ed.), Communication and Devices, India, Springer, New Delhi, 2015, pp. 349–356.
[11] B. Mondal, A secure steganographic scheme based on chaotic map and DNA computing, Micro-Electron. Telecommun. Eng. 106 (1) (2020) 545–554.
[12] A. Khalifa, LSBase: a key encapsulation scheme to improve hybrid crypto-systems using DNA steganography, in: 8th International Conference on Computer Engineering & Systems (ICCES), IEEE, Cairo, Egypt, 2013, pp. 105–110.
[13] M.S. Chakraborty, S.K. Bandyopadhyay, Data hiding by image steganography applying DNA sequence arithmetic, Int. J. Adv. Inform. Sci. Technol. (IJAIST) 44 (44) (2015) 1–3.
[14] R.M. Tank, H.D. Vasava, V. Agrawal, DNA-based audio steganography, Oriental J. Comput. Sci. Technol. 8 (1) (2015) 43–48.
[15] K. Gupta, S. Singh, DNA based cryptographic techniques: a review, Int. J. Adv. Res. Comput. Sci. Softw. Eng. 3 (3) (2013) 607–610.
[16] T. Li, J. Shi, X. Li, J. Wu, F. Pan, Image encryption based on pixel-level diffusion with dynamic filtering and DNA-level permutation with 3D Latin cubes, Entropy 21 (3) (2019) 319.
[17] Y.Q. Zhang, X.Y. Wang, J. Liu, Z.L. Chi, An image encryption scheme based on the MLNCML system using DNA sequences, Opt. Lasers Eng. 82 (1) (2016) 95–103.

[18] Y. Zhang, X. Liu, M. Sun, DNA based random key generation and management for OTP encryption, Biosystems 159 (1) (2017) 51–63.

[19] M. Borda, O. Tornea, DNA secret writing techniques, in: 8th IEEE International Conference on Communications, IEEE, Bucharest, Romania, 2010, pp. 451–456.

[20] E.M.S. Hossain, K.M.R. Alam, M.R. Biswas, Y. Morimoto, A DNA cryptographic technique based on dynamic DNA sequence table, in: 19th IEEE International Conference on Computer and Information Technology (ICCIT), Dhaka, Bangladesh, IEEE, 2016, pp. 270–275.

[21] NIST Codebook. Available: https://csrc.nist.gov/glossary/term/codebook [Accessed on 14 November 2020]. 2020.

[22] Codebook. Available: https://www.cryptomuseum.com/crypto/codebook/index.htm [Accessed on 14 November 2020]. 2020.

[23] Enigma. Available: https://enigma.hoerenberg.com/index.php?cat=Reservehandver fahren&page=RHV%20Part%201 [Accessed on 14 November 2020]. 2020.

[24] K.M. MacQueen, E. McLellan, K. Kay, B. Milstein, Codebook development for team-based qualitative analysis, Cam J. 10 (2) (1998) 31–36.

[25] B. Oktaviana, A.P.U. Siahaan, Three-pass protocol implementation in Caesar cipher classic cryptography, IOSR J. Comput. Eng. 18 (4) (2016) 26–29.

[26] J. Paul, A. Saju, L. Nair, Data based transposition to enhance data avalanche and differential data propagation in advanced encryption standard, Int. J. Comput. Appl. 67 (1) (2013) 6–9.

[27] T. St Denis, Advanced encryption standard, in: T. St Denis (Ed.), Cryptography for Developers, first ed., Elsevier, 2006, pp. 139–200.

[28] A. Canteaut, Stream cipher, in: H.C.A. van Tilborg, S. Jajodia (Eds.), Encyclopedia of Cryptography and Security, Springer, Boston, MA, 2011, pp. 1263–1265.

[29] C. Manifavas, G. Hatzivasilis, K. Fysarakis, Y. Papaefstathiou, A survey of lightweight stream ciphers for embedded systems, Security Commun. Netw. 9 (10) (2016) 1226–1246.

[30] Cryptography. Available: https://www.itu.int/en/ITU-D/Cybersecurity/Documents/ 01-Introduction%20to%20Cryptography.pdf [Accessed on 15 November 2020]. 2020.

[31] S. Hariram, R. Dhamodharan, A survey on DNA based cryptography using differential encryption and decryption algorithm, J. IOSR 10 (5) (2015) 14–18.

[32] Key Management Techniques. Available: http://cacr.uwaterloo.ca/hac/about/chap13. pdf [Accessed on 15 November 2020] 2020.

[33] S.A. Menesidou, V. Katos, G. Kambourakis, Cryptographic key management in delay tolerant networks: a survey, Future Internet 9 (3) (2017) 26–47.

[34] M.S. Taha, M.S.M. Rahim, S.A. Lafta, M.M. Hashim, H.M. Alzuabidi, Combination of steganography and cryptography: a short survey, IOP Conf. Ser. Mater. Sci. Eng. 518 (5) (2019) 052003.

[35] M.K.I. Rahmani, K. Arora, N. Pal, A crypto-steganography: a survey, Int. J. Adv. Comput. Sci. Appl. 5 (1) (2014) 149–154.

[36] D.K. Sarmah, N. Bajpai, Proposed system for data hiding using cryptography and steganography, Int. J. Comput. Appl. 8 (9) (2010) 7–10.

[37] S. Usha, G.A.S. Kumar, K. Boopathybagan, A secure triple level encryption method using cryptography and steganography, in: Proceedings of 2011 International Conference on Computer Science and Network Technology, IEEE, Harbin, China, 2011, pp. 1017–1020.

[38] N. Rashmi, K. Jyothi, An improved method for reversible data hiding steganography combined with cryptography, in: 2nd International Conference on Inventive Systems and Control (ICISC), IEEE, Coimbatore, India, 2018, pp. 81–84.

[39] G. Cui, C. Li, H. Li, X. Li, DNA computing and its application to information security field, in: Fifth International Conference on Natural Computation, IEEE, Tianjin, China, 2009, pp. 148–152.

[40] G. Cui, L. Qin, Y. Wang, X. Zhang, Information security technology based on DNA computing, in: International Workshop on Anti-Counterfeiting, Security and Identification (ASID), IEEE, Xiamen, China, 2007, pp. 288–291.

[41] S.C. Sukumaran, M. Mohammed, PCR and bio-signature for data confidentiality and integrity in mobile cloud computing, J. King Saud Univ. Comput. Inform. Sci. 32 (10) (2018) 12–17.

[42] A. Singh, R. Singh, Information hiding techniques based on DNA inconsistency: an overview, in: 2nd International Conference on Computing for Sustainable Global Development (INDIACom), IEEE, New Delhi, India, 2015, pp. 2068–2072.

[43] DNA Computing. Available: https://cs.stanford.edu/people/eroberts/courses/soco/projects/2003-04/dna-computing/evaluation.htm [Accessed on 15 November 2020]. 2020.

[44] G. Jacob, DNA based cryptography: an overview and analysis, Int. J. Emerg. Sci. 3 (1) (2013) 36–42.

[45] N.S. Kolte, K.V. Kulhalli, S.C. Shinde, DNA cryptography using index-based symmetric DNA encryption algorithm, Int. J. Eng. 10 (1) (2017) 810–813.

[46] X. Chai, Z. Gan, Y. Lu, Y. Chen, D. Han, A novel image encryption algorithm based on the chaotic system and DNA computing, Int. J. Modern Phys. C 28 (05) (2017) 1750069.

[47] S. Sadeg, M. Gougache, N. Mansouri, H. Drias, An encryption algorithm inspired from DNA, in: International Conference on Machine and Web Intelligence, IEEE, Algiers, Algeria, 2010, pp. 344–349.

[48] D.A. Zebari, H. Haron, S.R.M. Zeebaree, D.Q. Zeebaree, Multi-level of DNA encryption technique based on DNA arithmetic and biological operations, in: International Conference on Advanced Science and Engineering (ICOASE), IEEE, Duhok, Iraq, 2018, pp. 312–317.

[49] X. Li, C. Zhou, N. Xu, A secure and efficient image encryption algorithm based on DNA coding and spatiotemporal chaos, IJ Netw. Security 20 (1) (2018) 110–120.

[50] M.G.A. Malik, Z. Bashir, N. Iqbal, M.A. Imtiaz, Color image encryption algorithm based on hyper-chaos and DNA computing, IEEE Access 8 (1) (2020) 88093–88107.

[51] B. Mondal, T. Mandal, T. Choudhury, D.A. Khan, Use of 'a light weight secure image encryption scheme based on chaos and DNA computing' for encrypted audio watermarking, Int. J. Adv. Intell. Paradigms 13 (1–2) (2019) 67–79.

[52] M.S. Rahman, I. Khalil, X. Yi, A lossless DNA data hiding approach for data authenticity in mobile cloud-based healthcare systems, Int. J. Inf. Manag. 45 (2019) 276–288.

[53] E.E. Özkoç, DNA-based user authentication schemes for wireless body area network, e-Society (2018) 217–223.

[54] Z.H. Thenon, A.S. Elameer, S. Hassan, Proposed secret authentication method for e-government purposes, in: The 1st International Conference on Information Technology (ICoIT'17), Iraq, 2017, pp. 418–428.

[55] R. Dastanian, A. Karimi, H.S. Shahhoseini, A novel multi-client authentication method using infection of bacteria, Int. J. Inform. Electron. Eng. 2 (5) (2012) 790.

[56] A. Gupta, Y.A. Tung, J.R. Marsden, Digital signature: use and modification to achieve success in next generational e-business processes, Inf. Manag. 41 (5) (2004) 561–575.

[57] J.E. Gallegos, D.M. Kar, I. Ray, J. Peccoud, Securing the exchange of synthetic genetic constructs using digital signatures, ACS Synth. Biol. 9 (10) (2020) 2656–2664.

[58] G. Madhulika, C.S. Rao, Generating digital signature using DNA coding, in: Proceedings of the 3rd International Conference on Frontiers of Intelligent Computing: Theory and Applications (FICTA), vol. 328, no. 1, 2015, pp. 21–28.

[59] R. Kasodhan, N. Gupta, A new approach of digital signature verification based on BioGamal algorithm, in: 3rd International Conference on Computing Methodologies and Communication (ICCMC), IEEE, Erode, India, 2019, pp. 10–15.

[60] S.A. Zachariah, D. Rajasekar, L. Agilandeeswari, M. Prabukumar, IoT-based real time signature authentication and transfer from document to document with DNA encryption, in: 2nd International Conference on Next Generation Computing Technologies (NGCT), IEEE, Dehradun, India, 2016, pp. 01–08.

[61] X. Lai, M. Lu, L. Qin, J. Han, X. Fang, Asymmetric encryption and signature method with DNA technology, SCIENCE CHINA Inf. Sci. 53 (3) (2010) 506–514.

[62] B.B. Raj, V.C. Sharmila, An survey on DNA based cryptography, in: International Conference on Emerging Trends and Innovations in Engineering and Technological Research (ICETIETR), IEEE, Ernakulam, India, 2018, pp. 1–3.

[63] L. Garibyan, N. Avashia, Research techniques made simple: polymerase chain reaction (PCR), J. Invest. Dermatol. 133 (3) (2013) 1–4.

[64] P. Vinotha, D. Jose, VLSI implementation of image encryption using DNA cryptography, Intell. Commun. Technol. Virtual Mobile Netw. 33 (1) (2019) 190–198.

[65] A. Aggarwal, P. Kanth, Secure data transmission using DNA encryption, Int. J. Adv. Res. Comput. Sci. 5 (6) (2014) 57–61.

[66] Y. Zhang, D. Zhang, P. Sun, F. Guo, DNA sequencing puzzle based DNA cryptography algorithm, in: 7th IASTED International Conference on Modelling, Simulation and Identification, IASTED, Calgary, Canada, 2017, pp. 1–9.

[67] S. Namasudra, D. Devi, S. Choudhary, R. Patan, S. Kallam, Security, privacy, trust, and anonymity, in: S. Namasudra, G.C. Deka (Eds.), Advances of DNA Computing in Cryptography, Taylor & Francis, 2018, pp. 138–150.

[68] B. Mondal, T. Mandal, A light weight secure image encryption scheme based on chaos & DNA computing, J. King Saud Univ. Comput. Inform. Sci. 29 (4) (2017) 499–504.

[69] X. Wu, J. Kurths, H. Kan, A robust and lossless DNA encryption scheme for color images, Multimed. Tools Appl. 77 (10) (2018) 12349–12376.

[70] E.Z. Zefreh, An image encryption scheme based on a hybrid model of DNA computing, chaotic systems and hash functions, Multimed. Tools Appl. 79 (33) (2020) 24993–25022.

[71] P. Pavithran, S. Mathew, S. Namasudra, P. Lorenz, A novel cryptosystem based on DNA cryptography and randomly generated mealy machine, Comput. Secur. 104 (1) (2021) 102160.

[72] S. Bendaoud, F. Amounas, E.H. El Kinani, A new image encryption scheme based on enhanced elliptic curve cryptosystem using DNA computing, in: Proceedings of the 2nd International Conference on Networking, Information Systems & Security, Association for Computing Machinery, New York, USA, 2019, pp. 1–5.

[73] H.H. Hussien, DNA computing for RGB image encryption with genetic algorithm, in: 14th International Conference on Computer Engineering and Systems (ICCES), IEEE, Cairo, Egypt, 2019, pp. 169–173.

[74] S. Arunpandian, S.S. Dhenakaran, DNA based computing encryption scheme blending color and gray images, in: International Conference on Communication and Signal Processing (ICCSP), IEEE, Chennai, India, 2020, pp. 0966–0970.

[75] S. Namasudra, D. Devi, S. Kadry, R. Sundarasekar, A. Shanthini, Towards DNA based data security in the cloud computing environment, Comput. Commun. 151 (1) (2020) 539–547.

[76] A. Murugan, R. Thilagavathy, Cloud storage security scheme using DNA computing with morse code and zigzag pattern, in: IEEE International Conference on Power, Control, Signals and Instrumentation Engineering (ICPCSI), IEEE, Chennai, India, 2017, pp. 2263–2268.

[77] S. Namasudra, R. Chakraborty, A. Majumder, N.R. Moparthi, Securing multimedia by using DNA based encryption in the cloud computing environment, ACM Trans. Multimed. Comput. Commun. Appl. 16 (3) (2020) 1–19.

[78] T. Wu, X. Fan, K. Wang, C. Lai, N. Xiong, J.M. Wu, A DNA computation-based image encryption scheme for cloud CCTV systems, IEEE Access 7 (1) (2019) 181434–181443.

[79] Z. Wang, X. Zhao, H. Wang, G. Cui, Information hiding based on DNA steganography, in: IEEE 4th International Conference on Software Engineering and Service Science, Beijing, China, IEEE, 2013, pp. 946–949.

[80] T. Tuncer, E. Avci, A reversible data hiding algorithm based on probabilistic DNA-XOR secret sharing scheme for color images, Displays 41 (1) (2016) 1–8.

[81] A. Khalifa, A. Atito, High-capacity DNA-based steganography, in: 8th International Conference on Informatics and Systems (INFOS), IEEE, Cairo, Egypt, 2012, pp. 76–80.

[82] P. Malathi, M. Manoaj, R. Manoj, V. Raghavan, R.E. Vinodhini, Highly improved DNA based steganography, Procedia Comput. Sci. 115 (1) (2017) 651–659.

[83] M.R. Abbasy, B. Shanmugam, Enabling data hiding for resource sharing in cloud computing environments based on DNA sequences, in: IEEE World Congress on Services, IEEE, Washington, USA, 2011, pp. 385–390.

[84] A. Khalifa, A. Elhadad, S. Hamad, Secure blind data hiding into pseudo DNA sequences using playfair ciphering and generic complementary substitution, Appl. Math. 10 (4) (2016) 1483–1492.

[85] DNA Computing. Accessible: https://www.cs.uaf.edu/2010/fall/cs441/proj1/dna/DNAComputingHTMLNotes.html [Accessed on 15 November 2020]. 2020.

[86] N. Aubert, Y. Rondelez, T. Fujii, M. Hagiya, Enforcing logical delays in DNA computing systems, Nat. Comput. 13 (4) (2014) 559–572.

[87] S. Baskiyar, Simulating DNA computing, in: Proceedings of the 9th International Conference on High Performance Computing (HiPC '02), Springer, Berlin, Germany, 2002, pp. 411–419.

[88] M. Babaei, A novel text and image encryption method based on chaos theory and DNA computing, Nat. Comput. 12 (1) (2013) 101–107.

[89] K. Hameed, DNA computation-based approach for enhanced computing power, Int. J. Emerg. Sci. 1 (1) (2011) 23–30.

[90] DNA Computers. Accessible: https://thetartan.org/2010/3/22/scitech/dna_computers#:~:text=DNA%20computers%20require%20some%20amount,also%20interpret%20the%20output%20molecules [Accessed on 15 November 2020] 2020.

[91] Y. Zhang, Z. Wang, Z. Wang, X. Liu, X. Yuan, A DNA-based encryption method based on two biological axioms of DNA chip and polymerase chain reaction (PCR) amplification techniques, Chem. Eur. J. 23 (54) (2017) 13387–13403.

About the authors

Jiechao Gao received the BSc degree from Jilin University, China in 2016, and the MSc degree from Columbia University, USA, in 2018. He is currently a Ph.D. student in the Department of Computer Science of the University of Virginia. His research interests include distributed networks, Internet of Things, Cloud computing, federated learning, machine learning algorithms and applications. He has published several papers in well-known venues so far, such as IPDPS, ICPP, IPSN, IEEE trans on service computing and etc. He also served in academic works such as reviewing papers from several well-known venues such as multimedia and tools, soft computing, Peer J, Electronics, Sensors, IEEE sensors journal, ACM ToSN, IEEE TITS, IEEE IoT journal, etc, he also serves as guest editor in International Journal of Distributed Sensor Networks, TPC of INFOCOM, and NeurIPS workshops.

Tiange Xie received the BSc degree from Peking University, China in 2016, and the MSc degree from Columbia University, USA, in 2018. He is currently a Ph.D. student in the Institute of Information Engineering, Chinese Academy of Sciences, Beijing, China. His research interests include cybersecurity and IoT security.

CHAPTER FOUR

A novel image encryption and decryption scheme by using DNA computing

Chiranjeev Bhaya[a], Arup Kumar Pal[a], and SK Hafizul Islam[b]

[a]Department of Computer Science and Engineering, Indian Institute of Technology (Indian School of Mines), Dhanbad, Jharkhand, India
[b]Department of Computer Science and Engineering, Indian Institute of Information Technology, Kalyani, West Bengal, India

Contents

Abstract

This chapter presents the application of DNA computing in the field of image crypto-systems. Since images are multimedia data with high redundancy, conventional algorithms for encrypting text data prove to be impractical. Cryptosystems for images have to be designed in such a way as to make sure that they remove both visual and statistical content from the original data, for which DNA computing stands as an efficient tool. Several such cryptosystems designed for natural, color, and medical images have been reviewed in this chapter. Researchers have developed various DNA-based operations to be used in image cryptosystems, which have been elaborately discussed. The basic algorithms used in most of the cryptosystems have been discussed, and a cryptosystem has been proposed to illustrate the working of the various steps involved. The essential security parameters required for any image cryptosystem have been presented, and performance evaluation of the proposed method on various standard images has been shown. In the end, the chapter throws some light on the possible improvements, motivating the readers to add some significant contributions to this exciting subject.

Abbreviations

ACM	Arnold's Cat MAP
ASCII	American Standard Code for Information Interchange
CDF	Cumulative Density Function
CML	Coupled Map Lattice
CPU	Central Processing Unit
CT	Computed Tomography
DCT	Discrete Cosine Transformation
DICOM	Digital Imaging and Communications in Medicine
DNA	Deoxyribonucleic Acid
FRFT	Fractional Fourier Transform
HSV	Hue Saturation Value
JPEG	Joint Photographic Expert Group
KS	Key Sensitivity
LSS	Logistic Sine System
MD-5	Message-Digest algorithm 5
MRI	Magnetic Resonance Imaging
NPCR	Number of Pixels Change Rate
PSNR	Peak Signal-To-Noise Ratio
PWLCM	Piecewise Linear Chaotic Map
RAM	Random Access Memory
RGB	Red Green Blue
RNA	Ribonucleic Acid
SHA	Secure Hash Algorithm
UACI	Unified Average Changing Intensity
USC-SIPI	University of Southern California—Signal and Image Processing Institute
XOR	Exclusive OR

1. Introduction

In the era of digitalization, we are surrounded by advanced technical equipment like high-performance computers, high-speed Internet connections, high-resolution cameras, sophisticated smartphones, easy and fast means of communication and entertainment, etc., which have become an integral part of our lives. The technology industry has significantly impacted almost all fields, starting from personal message and multimedia sharing in social media to highly secured information sharing in the army and government. A large part of the digital technology involves the sharing of multimedia data, especially images. Images in the field of medicine are widely used for the diagnosis of diseases. These include Computed Tomography (CT) Scan, X-Ray Scans, mammograms, Magnetic Resonance Imaging (MRI), etc., which are stored digitally [1]. Just like any other data, these images may contain sensitive information and must be prevented against malicious use. Hence, the need for image encryption comes into the picture to ensure their confidentiality in a world that is full of adversaries.

Cryptography is the art of sending secured messages by overcoming adversaries present in the network. The main objective of information security is to ensure confidentiality, integrity, and availability of data [2]. Confidentiality is done by encrypting the original data into an unintelligible format so that unauthorized users cannot see the meaningful content sent by the sender. This unintelligible data is called the *cipher* and is obtained with the help of *cryptographic keys* in an encryption algorithm. Based on the type of key used, the encryption algorithm may be *symmetric* or *asymmetric* [3]. In *symmetric-key* cryptosystems (Fig. 1A), the same key is used for encryption as well as decryption algorithm. However, the exchange of the key is a challenge in these types of algorithms, which has to be kept secret. This problem is solved by using different keys for encryption and decryption, and such cryptosystems are called *asymmetric-key cryptosystems*. In these cryptosystems (Fig. 1B) every user has a public and a private key. The public key is known to everyone, and the private key is kept secret. Encryption is done with the public key of the receiver, and decryption can be done only by its private key. However, these algorithms are based on highly complex mathematical computations and are inefficient for large multimedia data like images. In most of the image cryptosystems, the symmetric key is used.

Images are multimedia data with massive size and high redundancy. Owing to these facts, traditional cryptosystems used for encrypting text data prove futile for images because the cipher image formed from these

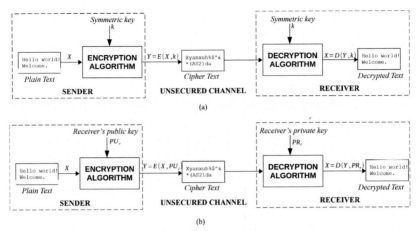

Fig. 1 Types of cryptosystems based on key: (A) symmetric-key cryptosystem and (B) asymmetric-key cryptosystem.

cryptosystems fails to hide the visual information [4]. Cryptosystems for images are, therefore, more sophisticated and complex compared to those for text data. For multimedia data like images, it is essential to remove the correlation between the plain and the cipher image and the visual and statistical information from its local and global regions. Conventional image encryption algorithms mainly operate on pixel level and bit level. However, they have some limitations. In pixel-level encryption, every pixel of the image is considered as a single unit; whereas in bit level, every bit of every pixel is considered as a single unit. Askar et al. [5] proposed a pixel-level confusion–diffusion-based image cryptosystem where row-shifting and column-shifting operations were carried out for confusion and XOR operation was carried out for diffusion of pixels using a 1D-discrete chaotic map. This method had a low tolerance for a differential attack as it failed to give proper UACI values. Ye et al. [6] proposed a permutation rewriting method combining confusion–diffusion process into a single unit. However, to attain the required security parameters, more than one round of iteration was required. Fu et al. [7] proposed a method of permutation using pixel swapping strategy. This method, too, required multiple iterations. Moreover, the permutation algorithm required preiteration involving complex calculations, thus making the system more complicated. Zhu et al. [8] proposed a bit-level confusion method where the image was decomposed into its bit–planes, and *Arnold's Cat Map* (ACM) was applied on every bit-plane independently. Pixel-level diffusion was then carried out after confusion. Xu et al. [9] proposed chaos cyclic shift and XOR operations for bit-level diffusion on each of the bit-planes. The diffused planes were then divided into two groups of odd and even planes, and confusion was carried independently on both the groups. Both Refs. [8] and [9] had to undergo multiple rounds of iterations for secured encryption. Cao et al. [10] proposed a two-dimensional logistic ICMIC cascade map and used it for encrypting images by performing bit-level circular shift permutation and bit-level XOR diffusion. Hasheminejad et al. [11] used SHA-1 on the plain image for key generation. However, this may not be a good implementation due to the weak unsecured properties of the SHA-1 algorithm. The encryption technique and limitations of the conventional image cryptosystems have been summarized in Table 1.

To deal with such a vast amount of data, researchers have used the high parallelism and vast data storage capacity of DNA molecules to design efficient cryptosystems [12]. The concepts of molecular biology, hardware, and bioinformatics are used to design DNA-based algorithms to solve various computational problems [13], image encryption is one of them. These encryption algorithms consist of DNA sequence operations [14] like DNA

Table 1 Conventional image cryptosystems and their limitations.

Authors	Year	Encryption technique	Limitations
Askar et al. [5]	2017	Row-shifting and column-shifting for confusion and XOR for diffusion using 1D-discrete chaotic map.	Low tolerance for differential attack due to low UACI value
Ye et al.[6]	2018	Permutation rewriting method combining permutation-diffusion process into a single unit	Multiple rounds of iteration required
Fu et al. [7]	2018	Permutation using pixel swapping technology	Permutation algorithm required complex preiteration calculation
Zhu et al. [8]	2011	Bit-level confusion using ACM followed by pixel-level diffusion	Multiple rounds of iteration required
Xu et al. [9]	2016	Chaos cyclic shift and XOR for bit-level confusion, separate diffusion in odd and even bit planes	Multiple rounds of iteration required
Hasheminejad et al. [11]	2019	SHA-1 on plane image for key-generation	Weak unsecured properties of SHA-1

encoding, decoding, addition, subtraction, XOR, etc. These algorithms often use various chaotic maps to generate pseudo-random sequences that are highly sensitive to initial parameters [15]. Some authors have also used the concept of nucleic acids obtained from DNA molecules in their encryption process [16]. The use of DNA strengthens the cryptosystems' security and ensures that they are resistant against various statistical and cryptographic attacks. These studies have been briefly described in this chapter.

The main contributions of this chapter are as follows.

- This chapter reviews some of the recent image cryptosystems using DNA computing that have been designed by various authors across the globe.
- A vivid description on digital image, chaotic maps, and DNA operations have been discussed.
- The basic procedure of designing DNA-based image cryptosystem has been discussed. A method illustrating the same has also been proposed.
- The basic security parameters which any image cryptosystem must fulfill have been graphically and mathematically presented.
- The performance analysis of the proposed cryptosystem has been presented, showing that it fulfills the criteria of a secured cryptosystem.

Hereafter, we move to the mainstream of the chapter. Section 2 discusses some recent cryptosystems using DNA-computing on various types of images developed by researchers from all over the world. Section 3 discusses some preliminaries related to images, chaotic maps, and DNA operations. Section 4 explains the basic design of DNA-based image cryptosystems along with a proposed method, illustrating the importance of the various steps of the process. Section 5 discusses the general security parameters that any image cryptosystem must follow. Section 6 shows the performance evaluation of the proposed cryptosystem on some natural images. Eventually, the conclusion has been presented in Section 7, giving a summary of the chapter and throwing some light on future prospects.

2. Literature review

Over the recent years, researchers all over the globe have developed numerous cryptosystems to encrypt images. The use of DNA in these algorithms has enabled them to enhance security and performances. Images may be of different types, so different cryptosystems have been designed for each one of them. This section discusses some of the recently designed state-of-the-art DNA-based image cryptosystems.

2.1 Cryptosystems for natural images

A typical natural grayscale image is a matrix of pixels each having a depth of 8-bits. Cryptosystems designed for these images tend to remove visual and statistical information to create random noise-like chaos. Often, these cryptosystems use high-dimensional chaotic maps to generate pseudo-random sequences for various operations like selecting DNA rules for encoding and decoding or to perform operations like XOR or addition–subtraction. Some of these cryptosystems also use confusion–diffusion architectures. Patel et al. [17], for example, used the 3D logistic chaotic map (Eq. 5) to generate multilevel encryption. To initialize the three dimensions of the map, Chebyshev, prime, and ASCII keys were deduced from the symmetric key parameters. A preprocessing was applied to convert color images to gray, and the first two dimensions of the key were used for row and column permutations, respectively. DNA encoding was applied for pixel-based diffusion by using the eight rules, and DNA XOR was applied as a logical operation to change the pixel values. After DNA decoding, the cipher image was obtained.

A similar concept was used by Dongming et al. [18] in which they used *Piecewise Linear Chaotic Map* (PWLCM) to generate two pseudo-random matrices. The images and the matrices were both converted into DNA sequences from the encoding rules and were XORed to form noise like intermediate images. The PWLCMs were further converted into two *Random Phase Masks* to perform *Double Random Phase Encoding* to generate the final cipher image. A significant contribution of the paper was the design of DNA XOR gates to speed up the encryption process. They used both optical parameters as well as the DNA cryptography technique to generate an efficient encryption algorithm.

Wang et al. [19] proposed an improved version of logistic map by introducing *Coupled Map Lattice* (CML) to generate the required pseudo-random sequences. They used cryptographic hash (SHA-256) to generate the initial parameters for CML and performed both pixel and DNA level scrambling and diffusion by using operations like DNA addition, subtraction, XOR, and DNA complementary rules to generate the cipher image. They introduced the concept of Hamming distance-based cyclic shifts to increase the security of the algorithm.

Hu et al. [20] used CML in *Logistic Sine System* (LSS) to develop improved chaotic sequences. This LSS-based CML was used to generate DNA-based key streams and key image which was required for encryption. The key image was XORed with the binary image and the result was converted to DNA sequences by encoding algorithm. DNA insertion and deletion operations were applied using the values from the key streams and the resultant cipher image was obtained after DNA decoding.

Yadollahi et al. [16] devised a method to increase the efficiency of DNA-based image encryption by introducing *Ribonucleic Acid* (RNA) phase. RNAs are single-strand molecules composed of the same nucleotides as those of DNAs except for Thiamine (T) which is converted into *Uracil* (U). In their algorithm, the DNA phase consisted of encrypting the plain image by DNA encoding, DNA XOR, and DNA decoding by using a logistic map. The intermediate result obtained after this phase was used to convert to RNA and was encoded into groups of three nucleotides to form *codons*. The combination of 64 codons was used for shuffling and the resultant RNA phase cipher-image was highly secured.

The algorithms for DNA-based cryptosystems for natural images have been summarized in Table 2.

2.2 Cryptosystems for medical images

Medical images are different from natural images as they have lower entropy and useful information is restricted to only a small region, an example of which can be seen in Fig. 2. Moreover, they contain very minute details and require more efficient cryptographic algorithms because any loss of sensitive information cannot be encouraged while decrypting. Akkasalingar and

Table 2 DNA-based cryptosystems for natural images.

Authors	Year	Encryption technique
Hu et al. [20]	2017	Coupled Map Lattice in Logistic Sine System to generate key image. Binary XOR followed by DNA encoding, DNA insertion and deletion, and DNA decoding
Patel et al. [17]	2020	Multilevel encryption using 3D logistic chaotic map. Pixel-based diffusion using DNA XOR
Dongming et al. [21]	2020	Double random phase encoding using piecewise linear chaotic map using DNA XOR gates
Wang et al. [19]	2020	Pixel- and DNA-level scrambling using DNA addition, subtraction, XOR, and complementary rules. Coupled Map Lattice for generating pseudo-random sequence
Yadollahi et al. [16]	2020	Using both DNA and RNA phase for enhanced security. DNA XOR using logistic map in DNA phase and codon-based shuffling in RNA phase

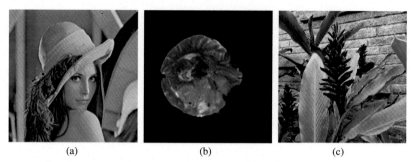

(a)   (b)   (c)

Fig. 2 Different types of images: (A) a natural image where information is spread throughout the region, (B) a medical image where information is present only in a confined region, and (C) a color image.

Biradar [22] proposed a cryptosystem that first selected the significant pixels from the image and applied permutation, diffusion, and DNA XOR after encoding them into DNA sequences. The remaining of the pixels were encoded into DNA sequences and only DNA XOR was applied. These were then combined and decoded into binary sequences to generate a chaotic image. A *dual hyperchaotic map* was used for the processes and selection of the rules for DNA encoding and decoding. Dividing the image into significant and nonsignificant pixels ensured fast encryption with enhanced security.

Belazi et al. [1] proposed a medical image encryption algorithm by using *Logistic-* and *Sine-Chebyshev maps* whose initial values were obtained from SHA-256 hash functions. They used multiplication over the prime fields to perform block-level permutation and S-boxes to perform substitution. The resultant matrix was encoded into DNA sequences by using one of the eight rules from the maps and DNA complement operation was used to further shuffle the contents followed by DNA decoding. Bitwise XOR was again performed on the obtained matrix and the entire process was repeated once again for the second round. The proposed algorithm was computationally efficient, having simple operations that had linear time complexity.

Another similar cryptosystem was proposed by Dagadu et al. [23] in which they used *Bernoulli Shift* and *Zigzag* maps to produce two key matrices which were used for row-by-row diffusion and DNA XOR operation. A separate *Logistic map* was used to determine the rules for DNA encoding and decoding. The initial values of these maps were obtained from MD-5 hash function and the algorithm could be applied on color medical images as well. The algorithm required only one round in the entire encryption process and the results were robust against various cryptographic and statistical attacks.

Stalin et al. [4] devised a fast medical encryption algorithm using a *four-dimensional nonlinear logistic map* with the help of which they generated multiple keys and performed pixel substitution. The image was divided into 16-blocks and each block was encoded into DNA sequences using the encoding and complement rules. Further, DNA subtraction and bitwise XOR were also performed to obtain the final encrypted cipher image.

A novel technique of encrypting medical images in the frequency domain was proposed by Banu and Amtitharajan [24] in which they used *Integer Wavelet Transform* to divide the image into four subbands. One of the subbands contained the main visual information of the image and was further divided into four subparts and were interchanged to remove the visual information from the image. To further increase the security of the obtained image, operations like DNA XNOR and DNA complement were performed in the DNA domain, the values and rules for which were obtained from *Logistic* and *Lorentz chaotic maps*. The algorithm could encrypt DICOM images very efficiently ensuring high security in them.

The algorithms for DNA-based cryptosystems for medical images have been summarized in Table 3.

Table 3 DNA-based cryptosystems for medical images.

Authors	Year	Encryption technique
Belazi et al. [1]	2019	Logistic and Sine-Chebychev maps for generating pseudo-random sequences. Multiplication over prime field for permutation and S-box-based diffusion followed by DNA complement and bitwise XOR
Dagadu et al. [23]	2019	Logistic map to determine rules for DNA encoding and decoding, and Bernoulli shift and Zigzag maps for row-by-row diffusion and DNA XOR operation
Stalin et al. [4]	2019	Pixel substitution using multiple keys generated by four-dimensional nonlinear logistic map. Use of DNA complement and subtraction operations in DNA phase and bitwise XOR in non-DNA phase
Akkasalingar and Biradar [22]	2020	Separate encryption of significant and nonsignificant image regions using permutation, diffusion, and DNA XOR. Dual hyperchaotic map for selecting DNA encoding and decoding rules
Banu and Amititharajan [24]	2020	Encryption in integer wavelet transformed domain using DNA XNOR and DNA complement operations. Selection of rules from logistic and Lorentz chaotic maps

2.3 Cryptosystems for color images

Unlike grayscale images which are composed of a single channel, typical color images are generally composed of three channels, each of which forms a channel matrix consisting of pixels which are 8-bit deep. Therefore, they are more voluminous in size and have greater information content. Algorithms for color image cryptosystems must ensure encryption of all the three channels of the image and remove visual and statistical information from them. Niyat and Moattar [25] used *3D-Chen chaotic system* to generate the key stream and random image for image permutation by using the data of each dimension for each of the three channels of the cryptosystem. Each channel went through permutation diffusion operation in DNA domain, the encoding and decoding rules of which were decided by *3D-logistic map*. The image was divided into a number of blocks and DNA XOR was applied at block level to obtain the cipher image.

Chai et al. [26] used a *4D-Hyperchaotic system* to generate the chaotic sequences required for encryption of color images. The initial values for the map were obtained from SHA 384 hash function, and three chaotic matrices one for each channel were obtained from the map, with the help of which simultaneous permutations were performed. They devised a dynamic DNA-based diffusion mechanism that was based on random numbers obtained from plain text and use the eight possible DNA-encoding and -decoding rules to perform diffusion which increased their security.

Another similar color image encryption algorithm based on *4D-Hyperchaotic system* was proposed by Liu et al. [27] in which they decomposed the color image into three DNA matrices, one for each color channel using the possible combinations of DNA encoding rules. The initial key was a combination of hash obtained from the image and an external key stream using SHA-512 algorithm. They proposed three different DNA-based diffusion rules to be acted upon the DNA sequences of the image which would be chosen dynamically from the map values. Confusion and diffusion was independently applied on the three DNA matrices and eventually were combined together and decoded to form the cipher image.

To enhance the efficiency of DNA-based color image cryptosystem, Farah et al. [15] used *Fractional Fourier Transform* (FRFT) along with DNA sequences to generate the cipher images. In their encryption algorithm, the substitution phase consisted of DNA encoding, DNA XOR, and DNA decoding. To make the substituted image more chaotic, they applied FRFT and XORed the result with chaotic sequences obtained from *Lorenz* maps, repeating over three times. The eventually obtained chaotic image was highly random and could resist cryptographic attacks.

Table 4 DNA-based cryptosystems for color images.

Authors	Year	Encryption Technique
Chai et al. [26]	2019	Dynamic DNA-based diffusion based on random numbers of plain text. 4D-hyperchaotic system for generating pseudo-random sequences
Liu et al. [27]	2019	Decomposition of color image into three independent matrices and use different DNA-based diffusion rules to be chosen dynamically from 4D-hyperchaotic system map values
Niyat and Moattar [25]	2020	3D-Chen chaotic system to generate key stream and random image for permutation. 3D-logistic map for DNA encoding, DNA XOR, and DNA decoding at block level
Farah et al. [15]	2020	Use of Fractional Fourier Transform and DNA computing. DNA XOR with the chaotic sequences obtained from Lorentz maps in the transformed domain
Dong et al. [21]	2020	Compression-based encryption technique. Compression using block-based discrete cosine transform, and encryption using DNA-based operations using Chen hyperchaotic system and logistic map

Dong et al. [21] proposed a compression-based encryption algorithm in which the color images are first reduced in size by using block-based *Discrete Cosine Transformation* (DCT). Quantification and data processing are then done to make the resultant matrix compatible for DNA encoding. *Chen hyperchaotic system* are used to obtain the pseudo-random sequences for DNA-based operations in the encryption process and *logistic map* for pixel-level scrambling. This algorithm ensured that the amount of data required for encryption was reduced because of the compression stage, which ensured the overall reduction in time and space of the cryptosystem.

The algorithms for DNA-based cryptosystems for color images have been summarized in Table 4.

3. Preliminaries

To understand the concept of the use of DNA in image cryptosystems, one must have some preliminary ideas on digital images, chaotic maps, and various DNA operations that are used in almost every cryptosystem. These concepts are briefly discussed in this section.

3.1 Anatomy of a digital image

A *digital image* is a projection of a three-dimensional view of the external environment into a two-dimensional digital plane, captured by a camera containing light sensors and projected on a screen or printed on a paper.

A digital image is a two-dimensional matrix of a specific size. Every individual unit of an image is called a *pixel*. A pixel is composed of a number of bits called the *depth of the pixel*. Images can be composed of several *channels*. For example, color images are composed of three channels, whereas grayscale images are composed of a single channel. There are various formats for representation of color images [28]. The most common format is the RGB format, where the image is composed of three channels, each representing the intensities of red, green, and blue colors, respectively. Other formats for color representation like YCbCr, L*a*b*, HSV, etc. [29] are also used

Images are also stored in different formats. Since the images' information is highly redundant, they are stored in both lossless and lossy manner depending on their purpose. The naive approach of storing images in *bitmap format* where for grayscale images and color, every pixel is stored with an intensity value of 8-bit depth and 24-bit depth, respectively. Bitmap is a lossless format for image storage, which although gives high clarity but also requires huge size for storage. Another popular format for image storage is JPEG *(Joint Photographic Expert Group)*, which uses discrete cosine transform to images in a compressed form [30]. Although this is a lossy format of image storage, there is very little loss in visual perception and is widely used in digital photography as it effectively compresses the image to a substantial size. Medical images require a very high amount of information and cannot tolerate any loss. They are mostly stored in DICOM *(Digital Imaging and Communications in Medicine)* format, which enables the medical experts to view and have a close analysis of the features [31]. In DICOM images, pixels have a higher bit depth of 16-bits, and all the medical information and history of the patient is stored in the image itself.

3.2 Chaotic maps

To encrypt voluminous data like images, there is a need for generating a pseudo-random sequence which is highly sensitive to initial conditions (which can be obtained from the encryption key) and highly ergodic. Chaotic maps prove to be an efficient solution to this problem. A *chaotic map* for a variable x is a periodic function governed by the difference relation

$x_{n+1} = F(x_n, c)$ for some initial control parameter c and integer n [32]. When extended to a larger number of variables, these become *multidimensional chaotic maps*. These functions are designed in such a way that they show chaotic behavior for some values of n. One of such chaotic maps is the *Logistic Map* [33] and is defined as

$$x_{n+1} = \mu x_n (1 - x_n) \tag{1}$$

where $x \in [0, 1]$ and μ is the initial control parameter. Fig. 3 shows the bifurcation diagram of the map, from which it can be seen that chaos begins from $\mu \approx 3.56995$ onward. Similarly some other one-dimensional chaotic maps are
- *Tent Map* [34]

$$x_{n+1} = \begin{cases} \mu x_n, & x_n < 1/2 \\ \mu(1-x_n), & 1/2 \leq x_n \end{cases} \tag{2}$$

   which is chaotic for $\mu \in [1, 2]$.
- *Sine Map* [35]

$$x_{n+1} = \mu \sin(\pi x_n) \tag{3}$$

where $\mu \in [0.87, 1]$ shows chaotic behavior.

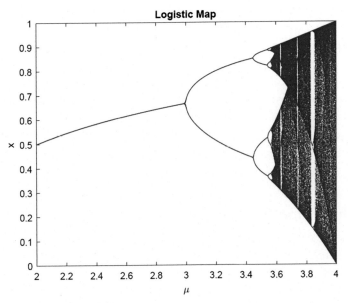

Fig. 3 Bifurcation diagram of logistic map

Some higher dimensional chaotic maps are defined as follows.

- *2D-Henon Map* [36]

$$x_{n+1} = 1 - \alpha x_n^2 + y_n$$
$$y_{n+1} = \beta x_n \tag{4}$$

where $\alpha = [1.06, 1.22] \cup [1.27, 1.29] \cup [1.31, 1.42]$ and $\beta = 0.3$.

- *3D-Chaotic Logistic Map* [17]

$$x_{n+1} = \alpha x_n(1 - x_n) + \beta y_n^2 x_n + \gamma z_n^3$$
$$y_{n+1} = \alpha y_n(1 - y_n) + \beta z_n^2 y_n + \gamma x_n^3 \tag{5}$$
$$z_{n+1} = \alpha z_n(1 - z_n) + \beta x_n^2 z_n + \gamma y_n^3$$

where $\alpha \in (3.53, 3.81)$, $\beta \in (0, 0.022)$, and $\gamma \in (0, 0.015)$ shows chaotic behavior.

- *4D-Hyperchaotic Map* [26]

$$x_{n+1} = \alpha x_n + \beta y_n z_n$$
$$y_{n+1} = \gamma y_n + \delta x_n z_n$$
$$z_{n+1} = \eta x_n y_n + \kappa z_n + \mu x_n w_n \tag{6}$$
$$w_{n+1} = \nu y_n$$

where $\alpha = 8$, $\beta = -1$, $\gamma = -40$, $\delta = 1$, $\eta = 2$, $\mu = 1$, $\nu = -2$, and $\kappa \in (-26.2, -7.1)$ shows hyperchaotic behavior.

3.3 DNA complementary rules

A DNA strand is made of four nitrogenous bases, namely Adenine (*A*), Cytosine (*C*), Guanine (*G*), and Thymine (*T*), the sequence of which determines the information stored in them. The *DNA Complementary Rules* [18], also known as *Watson–Crick Complement*, states that the nucleotide pairs (*A*, *T*) and (*C*, *G*) must be complement to each other. These rules are important to determine the encoding and decoding rules for DNA conversion.

3.4 DNA encoding and decoding rules

DNA encoding is basically a mapping f from binary bits to DNA nucleotides, i.e., $f: \{0, 1\}^* \rightarrow \{A, C, T, G\}^*$. To map two bits of binary data to a single nucleotide, a total of 24 different combinations are possible. However, following the DNA Complementary Rules gives only a set of eight valid combinations [4]. These rules are summarized in Table 5.

Table 5 DNA encoding rules.

Binary value	Rule 1	Rule 2	Rule 3	Rule 4	Rule 5	Rule 6	Rule 7	Rule 8
00	A	A	C	G	C	G	T	T
01	C	G	A	A	T	T	C	G
10	G	C	T	T	A	A	G	C
11	T	T	G	C	G	C	A	A

Table 6 DNA decoding rules.

Nucleotide	Rule 1	Rule 2	Rule 3	Rule 4	Rule 5	Rule 6	Rule 7	Rule 8
A	00	00	01	01	10	10	11	11
C	01	10	00	11	00	11	01	10
G	10	01	11	00	11	00	10	01
T	11	11	10	10	01	01	00	00

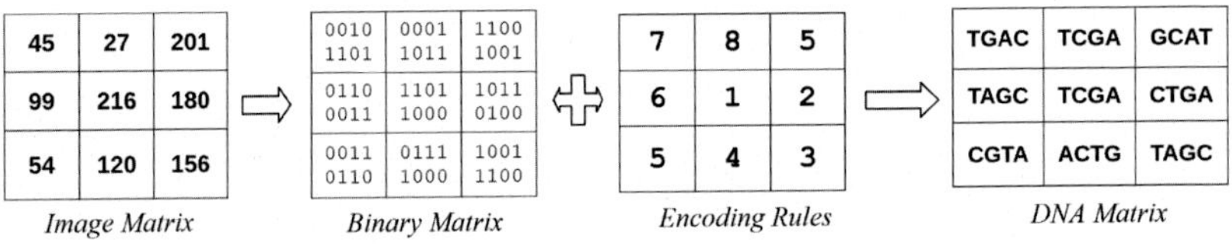

Fig. 4 Example of DNA Encoding.

Corresponding to the eight encoding rules, there are eight decoding rules to convert the nucleotide into a binary sequence. These decoding rules are summarized in Table 6.

Fig. 4 shows an example of DNA encoding using the different encoding rules. An *image matrix* of dimension 3×3 is converted to *binary matrix* by converting the values from each cell to its 8-bit binary equivalent. Each cell is then encoded into its corresponding DNA value by using the corresponding DNA encoding rule to form the *DNA matrix*. The beauty of this encoding step is that without the knowledge of the encoding rules, it is not possible to determine the values of the image matrix from the DNA matrix itself. For example, the values 99 and 156 map to the same DNA sequence *TAGC* using different encoding rules (*Rule 6* and *Rule 3*, respectively).

3.5 DNA XOR, addition, and subtraction

To use DNA nucleotides as a means of computation, several DNA-based operations like XOR [18], addition, and subtraction [14] have been defined which are similar to algebraic operations. Corresponding to the eight encoding and decoding rules, there will be eight combinations of these operations (where some of them may be similar). These operations are summarized in Tables 7–9, respectively.

Table 7 DNA XOR.

⊕	A	C	G	T
A	A	C	G	T
C	C	A	T	G
G	G	T	A	C
T	T	G	C	A

(a) Rule 1

⊕	A	C	G	T
A	A	C	G	T
C	C	A	T	G
G	G	T	A	C
T	T	G	C	A

(b) Rule 2

⊕	A	C	G	T
A	C	A	T	G
C	A	C	G	T
G	T	G	C	A
T	G	T	A	C

(c) Rule 3

⊕	A	C	G	T
A	G	T	A	C
C	T	G	C	A
G	A	C	G	T
T	C	A	T	G

(d) Rule 4

⊕	A	C	G	T
A	C	A	T	G
C	A	C	G	T
G	T	G	C	A
T	G	T	A	C

(e) Rule 5

⊕	A	C	G	T
A	G	T	A	C
C	T	G	C	A
G	A	C	G	T
T	C	A	T	G

(f) Rule 6

⊕	A	C	G	T
A	T	G	C	A
C	G	T	A	C
G	C	A	T	G
T	A	C	G	T

(g) Rule 7

⊕	A	C	G	T
A	T	G	C	A
C	G	T	A	C
G	C	A	T	G
T	A	C	G	T

(h) Rule 8

Table 8 DNA addition.

+	A	C	G	T
A	A	C	G	T
C	C	G	T	A
G	G	T	A	C
T	T	A	C	G

(a) Rule 1

+	A	C	G	T
A	A	C	G	T
C	C	A	T	G
G	G	T	C	A
T	T	G	A	C

(b) Rule 2

+	A	C	G	T
A	T	A	C	G
C	A	C	G	T
G	C	G	T	A
T	G	T	A	C

(c) Rule 3

+	A	C	G	T
A	T	G	A	C
C	G	T	C	A
G	A	C	G	T
T	C	A	T	G

(d) Rule 4

+	A	C	G	T
A	C	A	T	G
C	A	C	G	T
G	T	G	A	C
T	G	T	C	A

(e) Rule 5

+	A	C	G	T
A	G	T	A	C
C	T	A	C	G
G	A	C	G	T
T	C	G	T	A

(f) Rule 6

+	A	C	G	T
A	G	T	C	A
C	T	G	A	C
G	C	A	T	G
T	A	C	G	T

(g) Rule 7

+	A	C	G	T
A	C	G	T	A
C	G	T	A	C
G	T	A	C	G
T	A	C	G	T

(h) Rule 8

Table 9 DNA subtraction.

−	A	C	G	T
A	A	T	G	C
C	C	A	T	G
G	G	C	A	T
T	T	G	C	A

(a) Rule 1

−	A	C	G	T
A	A	C	T	G
C	C	A	G	T
G	G	T	A	C
T	T	G	C	A

(b) Rule 2

−	A	C	G	T
A	C	A	T	G
C	G	C	A	T
G	T	G	C	A
T	A	T	G	C

(c) Rule 3

−	A	C	G	T
A	G	T	A	C
C	T	G	C	A
G	C	A	G	T
T	A	C	T	G

(d) Rule 4

−	A	C	G	T
A	C	A	G	T
C	A	C	T	G
G	T	G	C	A
T	G	T	A	C

(e) Rule 5

−	A	C	G	T
A	G	C	A	T
C	T	G	C	A
G	A	T	G	C
T	C	A	T	G

(f) Rule 6

−	A	C	G	T
A	T	G	C	A
C	G	T	A	C
G	A	C	T	G
T	C	A	G	T

(g) Rule 7

−	A	C	G	T
A	T	G	C	A
C	A	T	G	C
G	C	A	T	G
T	G	C	A	T

(h) Rule 8

3.6 DNA insertion and deletion

Hu et al. [20] used the concept of pseudo-DNA operations to define *DNA Insertion* and *DNA Deletion* which imitate the processes of *DNA cutting* and *DNA ligation*. In DNA cutting, a long nucleotide strand is divided into two smaller substrands from a point, and in ligation, two smaller strands are merged to form a larger strand. These two processes can be used together for DNA insertion and deletion.

Let X be a long DNA strand from which l nucleotides have to be removed from a position c read from the first nucleotide from the left. This operation is called DNA deletion, written as *Del* which results in a new strand X' and is defined in Eq. (7) as

$$X' = Del(X, l, c) \tag{7}$$

Similarly, let Y be another DNA strand which has to be inserted after c places in X. This is the insertion process *Ins* and will result in a new strand X', defined in Eq. (8) as

$$X' = Ins(X, Y, c) \tag{8}$$

The process for DNA deletion and insertion can be seen in Fig. 5. In DNA deletion, the DNA strand X is cut at two points, one at c and other at $c + l$ to give three strands X_1, X_2, and X_3. The substrands X_1 and X_3 are ligated together to form the deleted DNA strand X'. In DNA insertion,

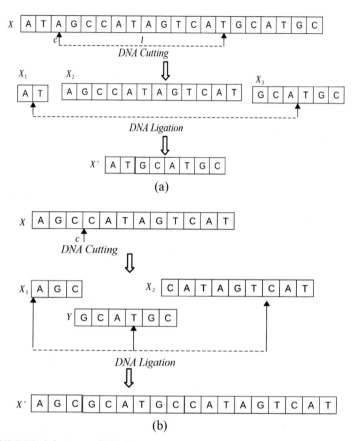

Fig. 5 (A) DNA deletion and (B) DNA insertion.

the DNA strand X is cut at one point at c which results in two substrands X_1 and X_2. They are then merged with Y in the order X_1, Y, X_2 which are ligated to form X'.

4. Designing DNA-based image cryptosystems

Although different researchers use different techniques to construct their cryptosystems, there lies some similarities and underlying principles among them. Image encryption algorithms aim at removing visual and statistical information in the images to generate the cipher images. For this, large pseudo-random sequences are used, especially with the help of chaotic maps, the initial conditions for which are obtained by the symmetric key. Hence, key scheduling plays an important role in the cryptosystem.

This section describes the general underlying principles between most of the DNA-based cryptosystems and proposes a method for designing a DNA-based image cryptosystems using chaotic maps.

4.1 Basic procedure

Most of the image cryptosystems follow a *confusion–diffusion* model where the confusion step is used to scramble the pixels to change the visual information of the image and the diffusion step to alter the statistical information. These steps may be repeated over themselves a number of times till a secure cipher image is obtained. The basic block diagram of a confusion–diffusion model is shown in Fig. 6.

In most of the DNA-based image cryptosystems, the confusion and diffusion processes are embedded in some *DNA-based* and *non-DNA-based operations*, as shown in Fig. 7. The operations which are non-DNA based are the basic conventional cryptographic functions like pixel-level or bit-level permutation, scrambling, etc. DNA-based operations are those which are carried on the DNA-encoded nucleotides of the corresponding images and include operations which have already been discussed in Sections 3.5 and 3.6. To carry out these operations, the chaotic maps are expanded to the required size and converted into the corresponding data types which give the required values. These processes may be iterated over several times

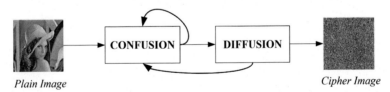

Plain Image   *Cipher Image*

Fig. 6 Block diagram of confusion–diffusion model.

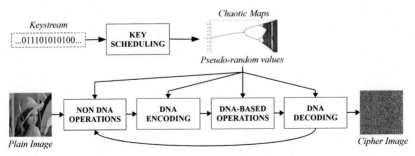

Fig. 7 Block diagram of basic DNA-based encryption algorithms.

to enhance the quality of the cipher image. The resultant DNA chaotic sequences can be synthesized and stored in the form of DNA dust using certain DNA data storage techniques [37] or can be converted into binary equivalent by using DNA decoding rules and transmitted along a communication channel. The sequence of the operations may also differ for different cryptosystems.

The important steps in DNA-based encryption algorithms are discussed as follows.

4.1.1 Key scheduling

The key scheduling algorithm is used to provide the initial values for the chaotic maps and also the initial control parameters. As discussed in Section 3.2, the values generated by the chaotic maps lie in $[0, 1]$ and the control parameters in some $[\alpha, \beta]$ to show chaotic behavior. Consider a chaotic map having k_1 variables whose initial values $x_0^{(1)}, x_0^{(2)}, \ldots, x_0^{(k_1)}$ have to be determined and k_2 number of control parameters $c_1, c_2, \ldots, c_{k_2}$. The keystream $\mathcal{K}$ of length n is divided into $(k_1 + k_2)$ parts and is operated upon some mathematical function f_i to convert it into values in the range acceptable for the variable or the control parameters. Algorithm 1 presents a generalized key-scheduling algorithm.

ALGORITHM 1 Key scheduling.

Input:	Symmetric key $\mathcal{K}$
Output:	Initial values of the *variables* $x_0^{(1)}, x_0^{(2)}, \ldots, x_0^{(k_1)}$ and *control parameters* $c_1, c_2, \ldots, c_{k_2}$ of the chaotic map

1   Divide $\mathcal{K}$ into $k_1 + k_2$ number of groups to form $B_1, B_2, \ldots, B_{(k_1+k_2)}$
2   **foreach** *variable* $x_0^{(i)}$, $i = 1$ to k_1 **do**
3      $x_0^{(i)} = \mathtt{frac}(f_i(B_i))$, for some function f_i // $\mathtt{frac}(x) = x - \lfloor x \rfloor$
4   **end**
5   **foreach** *control parameter* c_i, $i = 1$ to k_2 **do**
6   Let $[\alpha_i, \beta_i]$ be the range in which c_i shows chaotic behavior
7      $c_i = \alpha_i + (f_{i+k_1}(B_{i+k_1})) \times (\beta_i - \alpha_i)$, for some function f_{i+k_1}
8   **end**
9   **return** $x_0^{(1)}, x_0^{(2)}, \ldots, x_0^{(k_1)}, c_1, c_2, \ldots, c_{k_2}$

4.1.2 Pseudo-random values from chaotic maps

The chaotic maps have excellent properties of generating large number of pseudo-random values. These values generated lies in $[0, 1]$ and can be converted to required range and size. Let the chaotic map having k_1 variables and k_2 control parameters be defined by the function $\{x_{n+1}^{(1)}, x_{n+1}^{(2)}, \ldots,$ $x_{n+1}^{(k_1)}\} = F(x_n^{(1)}, x_n^{(2)}, \ldots, x_n^{(k_1)}, \ c_1, c_2, \ldots, c_{k_2})$. To get m pseudo-random sequences, this function can be iterated over m times and then converted of the required data type. For example, to get sequence for choosing one of the eight DNA rules, the values obtained from the chaotic maps can be transformed as $1 + x_n \times 8$ and the integer value can be chosen. The algorithm for obtaining pseudo-random values from chaotic maps is described in Algorithm 2.

ALGORITHM 2 Pseudo-random values from chaotic maps.

Input:	Chaotic map function with *initial values* and *control parameters*, required size of pseudo-random sequences m with values in $[\alpha, \beta]$
Output:	Pseudo-random sequences $\mathcal{R}$

1   Let $\{x_{n+1}^{(1)}, x_{n+1}^{(2)}, \ldots, x_{n+1}^{(k_1)}\} = F(x_n^{(1)}, x_n^{(2)}, \ldots, x_n^{(k_1)}, c_1, c_2, \ldots, c_{k_2})$ be the chaotic function
2   Set the *initial values* and *control parameters*
3   **for** $n = 1$ to m **do**
4   **foreach** *dimension* d of the chaotic map **do**
5   Calculate $x_n^{(d)}$ according to the chaotic map equation
6       $\mathcal{R}[n, d] = \alpha + x_n^{(d)} \times (\beta - \alpha)$
7   **end**
8   **end**
9   **return** $\mathcal{R}$

4.1.3 DNA encoding and decoding

The idea behind DNA encoding and decoding and the corresponding rules have already been discussed in Section 3.4. To convert an image matrix into a DNA matrix, corresponding rules for each pixel are required, which can be obtained from chaotic maps, as discussed in Algorithm 2. Then a mapping is done for every two bits of the image pixel to a DNA nucleotide according to the corresponding rule, as shown in Table 5. Similarly, in the DNA decoding step, the nucleotide sequences are converted back into an image matrix by converting the values of each pixel using a rule obtained by the chaotic map. A single nucleotide is converted into two bits according to the corresponding rule, as shown in Table 6. The algorithms for DNA encoding and decoding are presented in Algorithms 3 and 4, respectively.

ALGORITHM 3 DNA encoding.

Input: Image matrix I of dimension $m \times n$ and pixel depth p, Rule sequence $\mathcal{R}$ obtained from chaotic map using Algorithm 2 of size $m \times n$ having integral values in $[1,8]$

Output: DNA matrix D

1 $\quad dna = \begin{bmatrix} A & A & C & G & C & G & T & T \\ C & G & A & A & T & T & C & G \\ G & C & T & T & A & A & G & C \\ T & T & G & C & G & C & A & A \end{bmatrix}$ // index of 1st element is [1,1]

2   **for** $i = 1$ to m **do**
3   **for** $j = 1$ to n **do**
4        $b[1...p] = \text{DecimalToBinary}(I(i,j), p)$
5   **for** every two bits (b_k, b_{k+1}) in b **do**
6          $val = 2 \times b_k + b_{k+1} + 1$
7          $d = dna[val, \mathcal{R}(i,j)]$ // Convert $I(i,j)$ to p-bit binary
8   Append d to $D(i,j)$
9   **end**
10   **end**
11   **end**
12   **return** D

ALGORITHM 4 DNA decoding.

Input: Image matrix I of dimension $m \times n$ and pixel depth p, Rule sequence $\mathcal{R}$ obtained from chaotic map using Algorithm 2 of size $m \times n$ having integral values in $[1,8]$

Output: DNA matrix D

1   Initialize all pixels of I to 0

2   $dec = \begin{bmatrix} 0 & 0 & 1 & 1 & 2 & 2 & 3 & 3 \\ 1 & 2 & 0 & 3 & 0 & 3 & 1 & 2 \\ 2 & 1 & 3 & 0 & 3 & 0 & 2 & 1 \\ 3 & 3 & 2 & 2 & 1 & 1 & 0 & 0 \end{bmatrix}$ // index of 1st element is [1,1]

3   **for** $i = 1$ to m **do**
4   **for** $j = 1$ to n **do**
5   **for** every nucleotide d in $D(i,j)$ **do**
6   **if** $d == A$ **then**
7           $val = dec[1, \mathcal{R}(i,j)]$
8   **end**
9   **else if** $d == C$ **then**
10          $val = dec[2, \mathcal{R}(i,j)]$
11   **end**
12   **else if** $d == G$ **then**
13          $val = dec[3, \mathcal{R}(i,j)]$
14   **end**
15   **else if** $d == T$ **then**
16          $val = dec[4, \mathcal{R}(i,j)]$
17   **end**
18         $I(i,j) = 4 \cdot I(i,j) + val$
19   **end**
20   **end**
21   **end**
22   **return** I

4.2 Methodology

This section presents a DNA-based image encryption algorithm as shown in Fig. 8. The proposed cryptosystem uses *2D Chaotic Logistic Maps* (2D-CLM) [17] to generate the pseudo-random sequences for various steps. The algorithm uses a *pixel permutation* algorithm which shuffles the pixels of the image according to a *permutation matrix* obtained from the chaotic map. This forms the non–DNA algorithm and is the *confusion* step. The permuted image is encoded into DNA sequences using the *encoding rules* obtained from the chaotic maps. Another chaotic DNA sequence is obtained and is XORed with the encoded DNA sequence of the permuted image using DNA XOR rule as shown in Table 7. The resultant DNA sequence is decoded using the *decoding rules*. These steps are repeated over a number of times to enhance the quality of the final cipher image produced. Experimentally, it has been found that four rounds of encryption results in a high-quality cipher image.

4.2.1 Chaotic map and key scheduling

The proposed cryptosystem uses 2D–CLM [17], which is defined in Eq. (9) as

$$x_{n+1} = \alpha_1 x_n (1 - x_n) + \beta_1 y_n^2$$
$$y_{n+1} = \alpha_2 y_n (1 - y_n) + \beta_2 (x_n^2 + x_n y_n) \tag{9}$$

where x and y are the variables holding values in $[0, 1]$ and $\alpha_1, \alpha_2, \beta_1,$ and β_2 are the parameters showing chaotic behavior for $2.75 < \alpha_1 \leq 3.4$, $2.7 < \alpha_2 \leq 3.45$, $0.15 < \beta_1 \leq 0.21$, and $0.13 < \beta_2 \leq 0.15$.

The key scheduling algorithm shown in Algorithm 1 is used to obtain the initial values of the variables x_0, y_0 and the parameters from the secret key $\mathcal{K}$.

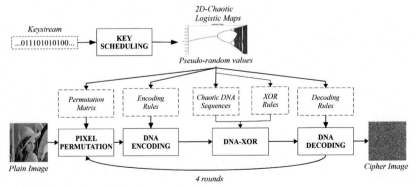

Fig. 8 Flow-diagram of the proposed method.

The secret key is 192-bit long and is divided into 6 blocks $B_1, B_2, \ldots, B_6$, each block of size 32 bits as shown in Fig. 9. Let the bits of each block B_i be represented as $b_1 b_2 \ldots b_{32}$. They are converted into floating-point values using a function f defined in Eq. (10) as

$$f(B_i = b_1 b_2 \ldots b_{32}) = \sum_{i=1}^{32} 2^{-i} b_i \tag{10}$$

The initial values of the map are obtained as shown in Eq. (11).

$$
\begin{aligned}
x_0 &= f(B_1) \\
y_0 &= f(B_2) \\
\alpha_1 &= 2.75 + f(B_3) \times (3.4 - 2.75) \\
\alpha_2 &= 2.7 + f(B_4) \times (3.45 - 2.7) \\
\beta_1 &= 0.15 + f(B_5) \times (0.21 - 0.15) \\
\beta_2 &= 0.13 + f(B_6) \times (0.15 - 0.13)
\end{aligned}
\tag{11}
$$

For every round of the encryption process, four such chaotic maps are required. The maps are expanded to the size of the image using Eq. (9). The last value of the kth map $(x_f^{(k)}, y_f^{(k)})$ becomes the initial value of the $(k+1)^{th}$ map $(x_0^{(k+1)}, y_0^{(k+1)})$.

4.2.2 Generating pseudo-random values
For every round r of encryption, a *permutation matrix, chaotic DNA sequences* and *encoding, decoding,* and *XOR rules* are required for the various processes of encryption. These are obtained from the four chaotic maps $M_1^{(r)}, M_2^{(r)}$, $M_3^{(r)}$, and $M_4^{(r)}$ for the rth round using Algorithm 2 as follows.

- *Permutation Matrix*: The chaotic map $M_1^{(r)}$ is chosen. For an $m \times n$ image, both the dimensions of the maps are given as input to Algorithm 2 with $[\alpha, \beta] = [0, m]$ and $[0, n]$, respectively. The integral values of the sequences obtained as an output, say, $\mathcal{R}_r$ and $\mathcal{R}_c$ are used to determine the index of row and column for the permuted matrix. In case of collision, the next empty cell is chosen as the index.

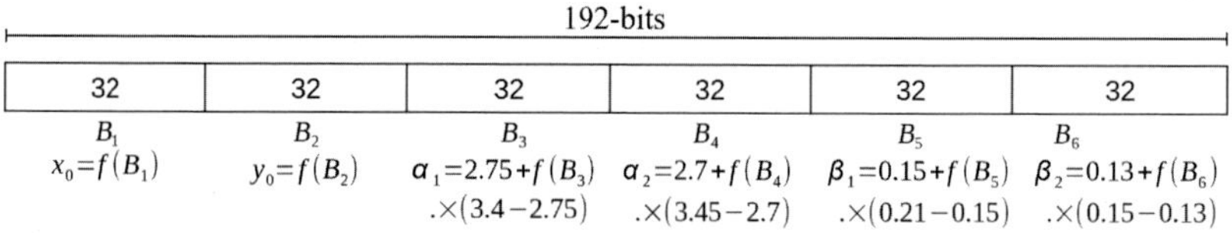

192-bits					
32	32	32	32	32	32
B_1	B_2	B_3	B_4	B_5	B_6
$x_0 = f(B_1)$	$y_0 = f(B_2)$	$\alpha_1 = 2.75 + f(B_3)$	$\alpha_2 = 2.7 + f(B_4)$	$\beta_1 = 0.15 + f(B_5)$	$\beta_2 = 0.13 + f(B_6)$
		$. \times (3.4 - 2.75)$	$. \times (3.45 - 2.7)$	$. \times (0.21 - 0.15)$	$. \times (0.15 - 0.13)$

Fig. 9 Key scheduling for the proposed method.

- *Encoding, Decoding, and XOR rules*: The two dimensions of $M_2^{(r)}$ and one dimension of $M_3^{(r)}$ are used for obtaining the encoding, decoding, and XOR rules, respectively, from Algorithm 2 with the values of $[\alpha, \beta]$ = [1, 8], rounding them off to their closest integral values.
- *Chaotic DNA Sequences*: The pseudo-random values are obtained from the map $M_4^{(r)}$ using Algorithm 2, with $[\alpha, \beta]$ = [0, 255]. The obtained values are rounded off to nearest integers and converted into DNA sequences using Algorithm 3.

4.2.3 Encryption

A single round of encryption has four steps: *pixel permutation, DNA encoding, DNA-XOR,* and *DNA decoding*. These steps are described as follows.

- *Pixel Permutation*: This step shuffles the pixels of the image so that its visual content is lost. Indices for pixel permutation are obtained from the permutation matrix. Let p_r and p_c be the values of the permutation matrix at position (x, y). Then the value of the pixel of the image at that position, say $I(x, y)$, is shifted to $I'(p_r, p_c)$, where I' is the shuffled image. The reverse of this process can be applied to get back the original positions of the pixels while decryption.
- *DNA Encoding*: The pixels of the shuffled image are encoded into DNA sequences using Algorithm 3. The rules are chosen from the encoding rules obtained from the chaotic map sequences. Since eight different rules are used randomly for each pixel, it adds a nonlinear layer to the encryption algorithm, making it difficult to decrypt without the knowledge of the correct key and the chaotic map sequences.
- *DNA XOR*: The DNA-encoded image sequences are XORed with the chaotic DNA sequences by using the XOR rules which are chosen differently for different pixels according to the values of the XOR rules. The XOR operations of DNA nucleotides are presented in Table 7. This step is a substitution step and leads to diffusion of the DNA-encoded permuted image pixels.
- *DNA Decoding*: The DNA sequences obtained from the previous steps are converted back into binary sequences using Algorithm 4 using the decoding rules. The output of this step is a cipher image which is random and noise-like.

Fig. 10 illustrates an example of one round of the cryptosystem for a 3×3 image matrix. It can be well observed that there is no linear relationship between the values in the original image matrix and the obtained

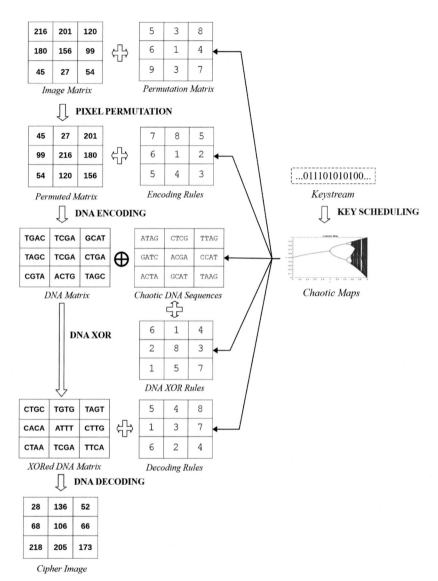

Fig. 10 Illustration of one round of the proposed DNA-based image cryptosystem of Fig. 8.

cipher matrix. This is because the DNA encoding and decoding add a nonlinear layer and the DNA-based XOR is not straightforward as bitwise XOR. The use of DNA, therefore, ensures better confusion and diffusion than conventional encryption algorithms and therefore, increases the overall efficiency of the cryptosystem.

5. Security analysis on cryptosystems

Any secured cryptosystem should be robust against various attacks which may be brute-force, statistical, or differential. Moreover, after image encryption, the cipher images obtained should be random noise-like and must not contain any visual information so that the original image can be guessed by simply seeing the cipher image. This section discusses some of the important parameters which any secured image cryptosystem must fulfill to ensure its robustness against adversaries.

5.1 Types of attacks on cryptosystems

Cryptosystems are attacked by cryptanalyst or attackers to obtain information about the original message or the key that is transmitted. To attack a cryptosystem, the attacker must have some information about the system and the cipher message, and the type of attack mainly depends on the amount of information. These have been described in Ref. [3]. The simplest attack is *ciphertext only* attack where the attacker knows the encryption algorithm along with the ciphertext. In *known plaintext* attack, the attacker has additional knowledge of one or more plaintext–ciphertext pairs formed from the same secret key. In *chosen plaintext* attack, the plaintext message is generated by the attacker from which the ciphertext can be obtained from the same key. In *chosen ciphertext* attack, the ciphertext can be chosen by the attacker and the corresponding plaintext can be known. The most powerful attack is *chosen text* attack where all the information of the aforementioned attacks is available to the attacker.

Generally, the cryptosystems are designed in such a way as to resist these attacks for a certain amount of time, which is large enough, say a thousand years, after which the importance of the information is lost. Such cryptosystems are called *computationally secure*. However, if a cryptosystem is resistant to the attacks even after giving an infinite amount of time with an infinite amount of computation power, the algorithms are *unconditionally secure*. Cryptosystems designed for images focus mainly on computational security because the latter has very little significance in this context.

5.2 Randomness of cipher image

The primary and most important parameter for any cryptosystem is to generate a random noise-like cipher image from which the visual information of

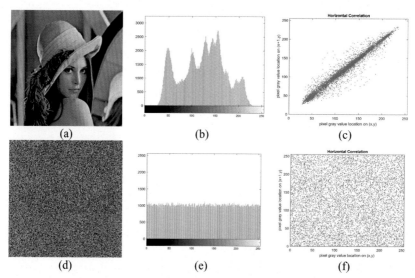

Fig. 11 Analysis of plain and cipher images: (A) plain image, (B) histogram of plain image, (C) correlation of plain image pixels, (D) cipher image, (E) histogram of cipher image, and (F) correlation of cipher image pixels.

the original image must be lost. Fig. 11A and D show a typical plain and cipher image, respectively. It can be very well seen that the cipher image is homogeneously noisy having no texture or pattern, contrary to the plain image which has specific structure and patterns. Any attacker, therefore, cannot guess the content of the plain image by merely obtaining the cipher image signals while transmission.

5.3 Brute-force attacks

In a brute-force attack, the attacker tries to guess the plain image by trying all possible key combinations on the attacked cipher images until an intelligible plain image is obtained. For an encryption algorithm to resist brute-force attack, keyspace and key sensitivity parameters are analyzed [16].

5.3.1 Key space analysis

Any cryptosystem should have a large-sized encryption key so that by merely trying all possible combinations, the original message cannot be reconstructed until the lifetime of the data. To resist such attacks, a key-space larger than 2^{100} is desirable [20]. A keyspace is the number of different combinations of keys possible [16]; however, on average, it requires around half the key-space attempts to guess the encryption key. One may correctly argue that a larger keyspace implies better resistance against a brute-force attack, but the use of symmetric keys in image cryptosystems also requires secured key sharing, which may be highly disadvantageous for very long keys. Therefore, an efficient cryptosystem should have optimal keyspace to ensure both security and implementability.

5.3.2 Key sensitivity analysis

Key sensitivity is the measurement of the change in the number of pixels of the cipher image on changing one bit of the encryption key. Any secured cipher image should be decryptable unless the exact value of the entire key sequence is known. The cryptosystem should prevent the knowledge of the original image or similar images from which the content of the original image can be guessed, even when some part of the encryption key is correctly known [14].

Consider a plain image $\mathcal{P}$ which is encrypted by an algorithm E using a key stream k to form cipher image $\mathcal{C} = E(\mathcal{P}, k)$. One bit of the key is changed to obtain the key k' and is used to encrypt the same plain image to obtain a different cipher image $\mathcal{C}' = E(\mathcal{P}, k')$. Let the dimension of both the plain and the cipher image be $m \times n$ with a pixel depth p. Then the key sensitivity KS is defined in Eq. (12) as [1]

$$KS = \frac{1}{m \times n \times p} \sum_{i=1}^{m} \sum_{j=1}^{n} \sum_{k=1}^{p} \mathcal{C}(i, j, k) \oplus \mathcal{C}'(i, j, k) \tag{12}$$

where $\oplus$ is the bitXOR operation and $\mathcal{C}(i, j, k)$ represents the kth bit of the pixel of image $\mathcal{C}$ in the ith row and j^{th} column.

One of the remarkable features of chaotic maps is their high sensitivity to the initial parameters which are obtained from the keystreams for image encryption. This ensures that if there is a slight change in the initial conditions of the maps, the resultant chaotic sequences widely vary, and therefore, the images encrypted with these sequences will vary to a very large extent, especially in those subroutines where the chaotic sequences are used to perform algebraic or encoding–decoding operations.

5.4 Statistical attacks

It may very well happen that the images after encryption are noise-like and lose all their visual information. However, since the pixels of images are highly redundant and correlated, the statistical information in them might be retained in some form or the other. To prevent such attacks, the pixels

Table 10 Security analysis parameters of various cryptosystems.

Method	Key space	Entropy	Correlation	NPCR	UACI
Cryptosystems for Natural Images					
Hu et al. [20]	2^{249}	7.9975	− 0.0043	99.5991	33.4650
Wang et al. [19]	2^{494}	7.9971	− 0.0003	99.5956	33.3638
Patel et al. [17]	2^{128}	7.9996	− 0.0027	99.6994	31.5592
Yadollahi et al.[16]	2^{128}	7.9994	0.0010	99.6091	33.2718
Cryptosystems for Medical Images					
Akkasalingar et al. [22]	2^{399}	7.8466	0.0181	99.6800	33.5883
Belazi et al. [1]	2^{716}	7.9993	0.0027	99.6173	33.4759
Dagadu et al. [23]	2^{128}	7.9972	− 0.0003	99.6100	33.3800
Stalin et al. [4]	2^{200}	7.9982	− 0.0005	99.6800	33.4880
Banu et al. [24]	2^{790}	7.9980	0.0024	99.6800	33.4700
Cryptosystems for Color Images					
Niyat et al. [25]	$2^{24 \times m \times n}$	7.9974	0.0004	99.6000	33.4316
Chai et al. [26]	2^{614}	7.9971	− 0.0012	99.6100	33.5000
Liu et al. [27]	2^{301}	7.9993	− 0.0018	99.6100	30.4166
Farah et al. [15]	NA	7.9991	0.0354	99.5676	33.4352
Dong et al. [21]	2^{455}	7.9970	− 0.0007	99.6058	33.4845

in the cipher image must lose all the correlation among the pixels and uniform distribution. To assess these parameters, histogram, correlation, and entropy analyses are performed on the cipher images. These analyses are briefly discussed in the following sections and the parametric values for different cryptosystems are summarized in Table 10.

5.4.1 Histogram analysis

The histogram gives a count of the number of pixels having a particular intensity value. A typical natural image consists of pixels having 8-bit depth and therefore each pixel can have an intensity value from 0 to 255 [26]. Fig. 11B and E shows the histograms of a plain image and its corresponding cipher image, respectively. The plain image has a nonuniform histogram distribution where some particular intensity levels have higher frequencies of occurrence than others. However, a good cipher image should have a uniform histogram, to remove the statistical information about the intensity of the pixels in their corresponding plain images.

Whereas the histogram gives only a visual interpretation of the pixel distribution, the statistical value can be measured by variance analysis [26]. Consider an image having a pixel depth of p, then it will have pixels in the range of $[0, 2^p - 1]$. Let f_i denote the frequency of pixels with intensity i. The variance $var(X)$ of a histogram is defined in Eq. (13) as

$$var(X) = \frac{1}{(2^p)^2} \sum_{i=0}^{2^p-1} \sum_{j=0}^{2^p-1} (f_i - f_j)^2 \tag{13}$$

Since the histograms of plain images are nonuniform, they will have a higher variance than the cipher images. For example, in Chai et al. [26], the variances of the original color image of Lena were reported to be 6.24×10^4, 2.64×10^4, and 8.50×10^4 for the red, green, and blue channels, respectively. In the corresponding cipher images, however, the variance was reduced to 247.78, 279.62, and 265.71 for each of the channels, respectively.

5.4.2 Entropy analysis

Claude Shannon gave the measurement of the amount of information contained in data and he termed it as entropy [27]. Consider an image of pixel depth p which will have intensities in $[0, 2^p - 1]$. Let the probability of a pixel having an intensity i be P_i. Then the entropy H of the image is defined in Eq. (14) as

$$H = -\sum_{i=0}^{2^p-1} P_i \log_2(P_i) \tag{14}$$

Plane images have a nonuniform distribution of pixel intensities, so their entropy is quite low. For example, the Lena image shown in Fig. 11A has an entropy of around 7.5788 bits/symbol [19]. However, cipher images have almost equal probability of distribution of pixel intensities; therefore, their entropy should lie close to 8. Table 10 shows the entropy of the cipher image obtained after encrypting the plane images using various algorithms, all of which are close to 7.99, therefore implying that the cipher images are highly random.

5.4.3 Correlation analysis

In a plain image, the adjacent pixels are highly correlated, which means that they have almost the same pixel intensity values most of the time. However, in cipher images, this correlation has to be removed so that the image becomes noisy. This can be visualized very well from Fig. 11C and F which shows a plot of the relationship between the pixels and their adjacent pixels in the horizontal direction for the plain and cipher images, respectively, where the plot shows a linear trend for plain image and is scattered and uncorrelated for the cipher image.

Statistically the correlation can be measured in terms of correlation coefficient that was proposed by Pearson [22]. Let $X = \{x_1, x_2, ..., x_N\}$ and $Y = \{y_1, y_2, ..., y_N\}$ be two data sets having N elements each. The correlation coefficient $r(x, y)$ between X and Y is defined in Eq. (15) as

$$r(x, y) = \frac{cov(x, y)}{\sigma_x \sigma_y}$$
$$= \frac{\sum_{i=1}^{N} x_i y_i - \frac{1}{N} \sum_{i=1}^{N} x_i \sum_{i=1}^{N} y_i}{\sqrt{\sum_{i=1}^{N} x_i^2 - \frac{1}{N} \left(\sum_{i=1}^{N} x_i \right)^2} \sqrt{\sum_{i=1}^{N} y_i^2 - \frac{1}{N} \left(\sum_{i=1}^{N} y_i \right)^2}}$$

(15)

where $cov(x, y)$ is the covariance of x and y and σ_x and σ_y denote the standard deviation of x and y, respectively.

The value of r lies between -1 and 1. A value close to 1 indicates a high positive correlation. For example, the correlation reported for Lena image (Fig. 11A) as reported in Wang et al. [19] is 0.9676, 0.9685, and 0.9714 along horizontal, vertical, and diagonal directions, respectively, showing that the adjacent pixels of a plain image are highly positively correlated. However, the average correlation values of the cipher images, as presented in Table 10, are all close to zero, indicating a very low correlation.

5.5 Differential attack

In a differential attack, the attacker tries to guess the key by making some changes in a given plain text and observing the changes that happen to the ciphertext with the same key. It is a type of chosen plain-text attack [19] which was first described by Birham and Shamir [38]. For an image encryption algorithm to resist differential attack, a small change in one pixel of the plain image should bring out a large change in the cipher image when encrypted with the same key. This practically means that the information of every pixel should be propagated throughout the image in the encryption algorithm. To study the resistance of a cryptosystem against differential attack, *Number of Pixels Change Rate* (NPCR) and *Unified Average Changing Intensity* (UACI) tests are carried out [4].

Let $\mathcal{P}$ be a plain image that is encrypted with a key k using an encryption algorithm E to obtain the cipher image $\mathcal{C} = E(\mathcal{P}, k)$. One of the pixels of $\mathcal{P}$ is changed to obtain the modified plain image $\mathcal{P}'$ which is then encrypted using the same algorithm and key to obtain cipher image $\mathcal{C}' = E(\mathcal{P}', k)$. If the cryptosystem resists a differential attack, there should be a large difference in the corresponding pixels of $\mathcal{C}$ and $\mathcal{C}'$. Let $\delta(x, y)$ be the binary count of whether the corresponding pixels at positions (x, y) in both the cipher images are different or not. Its value is obtained from Eq. (16) as

$$\delta(x, y) = \begin{cases} 0, & \text{if } \mathcal{C}(x,y) = \mathcal{C}'(x,y) \\ 1, & \text{if } \mathcal{C}(x,y) \neq \mathcal{C}'(x,y) \end{cases} \tag{16}$$

For an image of dimension $m \times n$ and pixel depth of p bits, NPCR and UACI are mathematically defined in Eqs. (17) and (18), respectively, as [39]

$$\text{NPCR} : \mathcal{N}(\mathcal{C}, \mathcal{C}') = \frac{1}{m \times n} \sum_{x=1}^{m} \sum_{y=1}^{n} \delta(x, y) \times 100\% \tag{17}$$

$$\text{UACI} : \mathcal{U}(\mathcal{C}, \mathcal{C}') = \frac{1}{(2^p - 1) \cdot (m \times n)} \sum_{x=1}^{m} \sum_{y=1}^{n} |\mathcal{C}(x, y) - \mathcal{C}'(x, y)| \times 100\% \tag{18}$$

The ideal values for NPCR and UACI were studied in Ref. [40]. Let $P = 2^p - 1$, which is the maximum intensity value a pixel can have. For an ideally encrypted image, the value of NPCR must be higher than its critical value $\mathcal{N}_\alpha^*$, which is given in Eq. (19) as

$$\mathcal{N}_\alpha^* = \mu_{\mathcal{N}} - \Phi^{-1}(\alpha)\sigma_{\mathcal{N}} \tag{19}$$

where $\mu_{\mathcal{N}}$ and $\sigma_{\mathcal{N}}$ are the expectation and standard deviation, respectively, and are calculated using Eqs. (20) and (21) as

$$\mu_{\mathcal{N}} = \frac{P}{P + 1} \tag{20}$$

$$\sigma_{\mathcal{N}}^2 = \frac{P}{mn(P + 1)^2} \tag{21}$$

and Φ^{-1} is the inverse of *Cumulative Density Function* (CDF) of standard normal distribution and α is the level of significance. Similarly, the value of UACI must lie between $\mathcal{U}_\alpha^{*-}$ and $\mathcal{U}_\alpha^{*+}$, which are calculated using Eq. (22) as

$$\mathcal{U}_\alpha^{*-} = \mu_\mathcal{U} - \Phi^{-1}\left(\frac{\alpha}{2}\right)\sigma_\mathcal{U}$$

$$\mathcal{U}_\alpha^{*+} = \mu_\mathcal{U} + \Phi^{-1}\left(\frac{\alpha}{2}\right)\sigma_\mathcal{U} \tag{22}$$

where $\mu_\mathcal{U}$ and $\sigma_\mathcal{U}$ are defined in Eqs. (23) and (24), respectively, as

$$\mu_\mathcal{U} = \frac{P+2}{2P+3} \tag{23}$$

and

$$\sigma_\mathcal{U}^2 = \frac{(P+2)(P^2+2P+3)}{18(P+1)^2 mnP} \tag{24}$$

The critical values for NPCR and UACI for various dimensions of image and pixel depth and values of α are summarized in Table 11. Table 10 shows the NPCR and UACI values obtained by different proposed cryptosystems.

5.6 Occlusion and noise attacks

When the cipher image is transmitted to the receiver via wired or wireless channel, it may happen that due to transmission error, some part of the image is lost. This is called *occlusion attack* [27]. However, the cryptosystems may be able to recover some approximations of the image, even after some part of it has been lost. In such a case, the cryptosystem is said to resist the occlusion attack.

Similarly, while transmission, the cipher image may be prone to noise which alters some of the pixel intensities [21]. However, the decoding algorithm may still be able to retrieve a substantial part of the image and resist noise attack.

An example of occlusion attack using the algorithm proposed in Ref. [41] is shown in Fig. 12. Twenty-five percent of the cipher image was lost in the occlusion attack as shown in Fig. 12C; however, the decryption algorithm could recover back an approximation of the original image which still has enough visual information to give an idea what the plain image was.

Similarly, an example of noise attack using the same algorithm [41] is shown in Fig. 13. The cipher image shown in Fig. 13C was attacked with *salt and pepper noise* and was decrypted to obtain the image shown in Fig. 13D.

Table 11 Critical values for NPCR and UACI.

Pixel depth	Dimension	Significance level	Critical NPCR	Critical UACI	
p	$m \times n$	α	$\mathcal{N}_\alpha^*$	$\mathcal{U}_\alpha^{*-}$	$\mathcal{U}_\alpha^{*+}$
8	128×128	0.05	99.529217	33.101211	33.825872
8	128×128	0.01	99.496006	32.987359	33.939725
8	128×128	0.001	99.458780	32.855235	34.071848
8	256×256	0.05	99.569296	33.282376	33.644707
8	256×256	0.01	99.552690	33.225450	33.701633
8	256×256	0.001	99.534077	33.159389	33.767695
8	512×512	0.05	99.589335	33.372959	33.554124
8	512×512	0.01	99.581033	33.344496	33.582587
8	512×512	0.001	99.571726	33.311465	33.615618
8	1028×1028	0.05	99.599394	33.418427	33.508657
8	1028×1028	0.01	99.595259	33.404250	33.522833
8	1028×1028	0.001	99.590624	33.387799	33.539284
16	128×128	0.05	99.993454	32.972924	33.694760
16	128×128	0.01	99.991375	32.859515	33.808169
16	128×128	0.001	99.989044	32.727907	33.939777
16	256×256	0.05	99.995964	33.153383	33.514301
16	256×256	0.01	99.994924	33.096679	33.571005
16	256×256	0.001	99.993759	33.030875	33.636809
16	512×512	0.05	99.997219	33.243612	33.424071
16	512×512	0.01	99.996699	33.215260	33.452424
16	512×512	0.001	99.996116	33.182358	33.485326
16	1028×1028	0.05	99.997849	33.288903	33.378781
16	1028×1028	0.01	99.997590	33.274782	33.392902
16	1028×1028	0.001	99.997300	33.258395	33.409289

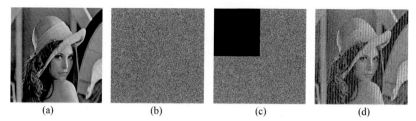

Fig. 12 Occlusion attack: (A) plain image, (B) cipher image, (C) occlusion attack on cipher image, and (D) decrypted image from the occluded cipher.

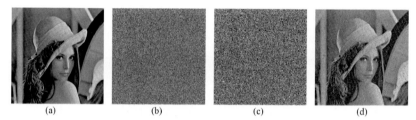

Fig. 13 Noise attack: (A) plain image, (B) cipher image, (C) noise attack on cipher image with *salt and pepper noise*, and (D) decrypted image from noisy cipher.

The decrypted image, although noisy, has a lot of similarity with the plain image, and therefore, the algorithm can resist noise attack.

5.7 Peak signal-to-noise ratio

The performance of a cryptosystem concerning occlusion and noise attack can be evaluated using *Peak Signal-to-Noise Ratio* (PSNR). Let $\mathcal{P}$ be the plain image which is enciphered with an encoding algorithm E using a key k to obtain the cipher image $\mathcal{C} = E(\mathcal{P}, k)$. The cipher image undergoes occlusion or noise attack to obtain the image $\mathcal{C}'$, which is decrypted with the corresponding decoding algorithm D using the same key k to obtain the decrypted image $\mathcal{D} = D(\mathcal{C}', k)$. If the image has dimension $m \times n$ and pixel depth p, the PSNR is defined in Eq. (25) as [24]

$$\text{PSNR} = 10 \times \log_{10}\left[\frac{(2^p - 1)^2}{\frac{1}{m \times n}\sum_{x=1}^{m}\sum_{y=1}^{n}\{\mathcal{D}(x, y) - \mathcal{P}(x, y)\}^2}\right] \quad (25)$$

For an image to be visually perceptible, it should have a high PSNR (preferably above 30 dB). Cipher images, on the other hand, have low PSNR values (usually less than 10 dB) because they are random noise–like and have a high mean squared error in the corresponding pixels.

6. Results and discussions

6.1 Dataset description

The proposed cryptosystem has been tested on a number of images, some of which have been shown in Section 6.3. All these images have been taken from USC-SIPI image database [42]. The images in the database are classified into four categories: *textures, aerials, sequences,* and *miscellaneous.* The performance evaluation of four such images, one for each category, has been shown in Table 12.

6.2 Experimental environment

All the experiments have been run on a personal laptop with Intel(R) Core(TM) i5-8250U CPU 1.60 GHz 1.80 GHz. The CPU has 1 socket, 4 cores, and 8 logical processors with enabled virtualization. It has three caches: L1, L2, and L3 with memory of 256 KB, 1 MB, and 6 MB, respectively. The primary memory consists of a 4 GB DDR4 RAM (3.9 GB usable) with a speed of 2400 MHz. The system has a 64-bit Windows 10 home single language operating system and x64-based processor. The code has been written and executed in MATLAB R2019a.

6.3 Results and analysis

The proposed cryptosystem uses a symmetric key of 192 bits. Therefore, it has a keyspace of 2^{196} which is larger than 2^{100} and is enough to resist any brute-force attack [20].

The performance analysis of the proposed cryptosystem on some images has been presented in Table 12. The corresponding cipher images obtained by the proposed algorithm are random noise-like and have no statistical or visual similarity with the original images.

Entropy measures the amount of randomness in the image and Eq. (14) is used to calculate the entropy of the original and the cipher images by considering all the image pixel intensities. The original images have a low entropy, whereas the cipher images have entropies close to 8, which verify the randomness of the latter.

Histogram analysis gives a visual interpretation of the distribution of the pixel intensities of the images. The histogram is drawn by computing the frequency of the image's pixel intensity distribution. In natural images, the distribution of intensities are nonuniform and are concentrated to certain

Table 12 Performance evaluation of the proposed cryptosystems on various images.

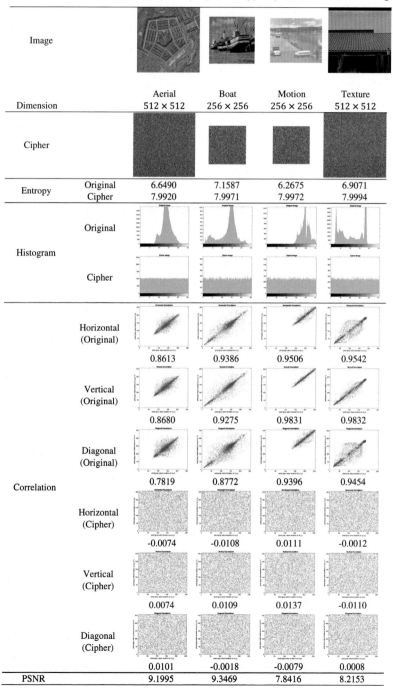

		Aerial	Boat	Motion	Texture
Image					
Dimension		512 × 512	256 × 256	256 × 256	512 × 512
Cipher					
Entropy	Original	6.6490	7.1587	6.2675	6.9071
	Cipher	7.9920	7.9971	7.9972	7.9994
Histogram	Original				
	Cipher				
Correlation	Horizontal (Original)	0.8613	0.9386	0.9506	0.9542
	Vertical (Original)	0.8680	0.9275	0.9831	0.9832
	Diagonal (Original)	0.7819	0.8772	0.9396	0.9454
	Horizontal (Cipher)	-0.0074	-0.0108	0.0111	-0.0012
	Vertical (Cipher)	0.0074	0.0109	0.0137	-0.0110
	Diagonal (Cipher)	0.0101	-0.0018	-0.0079	0.0008
PSNR		9.1995	9.3469	7.8416	8.2153

values, as seen from the histograms in the table. However, the encryption process removes this statistical information and the histogram of the cipher images is uniform.

The correlation coefficients are used to analyze the similarity between the adjacent pixels of the image. For calculating the correlation, 10, 000 random pixels are chosen from the images and a plot is drawn by considering the adjacent horizontal, vertical, and diagonal pixels. The coefficient of correlation is calculated from Eq. (15). The adjacent pixels of a natural images are highly correlated, which can be observed from the graphs and the values of the coefficient, which is close to 1. This correlation is removed in the cipher images, and their corresponding correlation coefficients are close to 0.

PSNR is the measurement of the quality of the transmitted cipher image and a low value indicates it to be highly noisy. To calculate the PSNR, the square of the difference of the corresponding pixel intensities between the original and the cipher image is considered and is calculated using Eq. (25). The PSNR between the original and the cipher images is less than 10, indicating that the cipher images are highly noisy and the original images cannot be guessed by mere visual or statistical processes.

7. Conclusion and future scope

This chapter presents a study on the different types of DNA computing-based image cryptosystems for different types of images that have been recently developed by researchers all over the globe. The use of DNA encoding, decoding, and various DNA-based operations has been vividly discussed in this chapter. A brief discussion has been done on the basic architecture of DNA-based image cryptosystem along with a proposed methodology, illustrating the importance of DNA computing in the field of image cryptography. Moreover, the important parameters that a cryptosystem must fulfill to be called secure have been elaborately discussed by providing mathematical definitions and illustrative examples from some of the state-of-the-art cryptosystems present in the literature. A brief analysis of the proposed cryptosystem has also been presented. The use of DNA computing in image cryptosystem is an interesting field of study and research, where there is always a scope for improvement. Researchers can design cryptosystems which are faster and more secured, probably requiring lesser number of steps and improved efficiency. The parallelism and high storage capacity of DNA can be utilized to design time and storage-efficient cryptosystems. New DNA-based operations along with DNA-based chaotic maps can be designed to perform better than the existing state-of-the-art architectures.

References

[1] A. Belazi, M. Talha, S. Kharbech, W. Xiang, Novel medical image encryption scheme based on chaos and DNA encoding, IEEE Access 7 (2019) 36667–36681, https://doi.org/10.1109/ACCESS.2019.2906292.

[2] B. Guttman, E.A. Roback, An Introduction to Computer Security: The NIST Handbook, Diane Publishing, 1995, https://doi.org/10.6028/NIST.SP.800-12r1.

[3] W. Stallings, Cryptography and Network Security, 4/E, Pearson Education India, 2006.

[4] S. Stalin, P. Maheshwary, P.K. Shukla, M. Maheshwari, B. Gour, A. Khare, Fast and secure medical image encryption based on non linear 4D logistic map and DNA sequences, J. Med. Syst. 43 (8) (2019) 267, https://doi.org/10.1007/s10916-019-1389-z.

[5] S.S. Askar, A.A. Karawia, F.S. Alammar, Cryptographic algorithm based on pixel shuffling and dynamical chaotic economic map, IET Image Process. 12 (1) (2017) 158–167, https://doi.org/10.1049/iet-ipr.2016.0906.

[6] G. Ye, C. Pan, X. Huang, Q. Mei, An efficient pixel-level chaotic image encryption algorithm, Nonlinear Dyn. 94 (1) (2018) 745–756, https://doi.org/10.1007/s11071-018-4391-y.

[7] C. Fu, G.-y. Zhang, M. Zhu, Z. Chen, W.-m. Lei, A new chaos-based color image encryption scheme with an efficient substitution keystream generation strategy, Secur. Commun. Netw. 2018 (2018) 1–13, https://doi.org/10.1155/2018/2708532.

[8] Z.-l. Zhu, W. Zhang, K.-w. Wong, H. Yu, A chaos-based symmetric image encryption scheme using a bit-level permutation, Inform. Sci. 181 (6) (2011) 1171–1186, https://doi.org/10.1016/j.ins.2010.11.009.

[9] L. Xu, Z. Li, J. Li, W. Hua, A novel bit-level image encryption algorithm based on chaotic maps, Opt. Lasers Eng. 78 (2016) 17–25, https://doi.org/10.1016/j.optlaseng.2015.09.007.

[10] C. Cao, K. Sun, W. Liu, A novel bit-level image encryption algorithm based on 2D-LICM hyperchaotic map, Signal Process. 143 (2018) 122–133, https://doi.org/10.1016/j.sigpro.2017.08.020.

[11] A. Hasheminejad, M.J. Rostami, A novel bit level multiphase algorithm for image encryption based on PWLCM chaotic map, Optik 184 (2019) 205–213, https://doi.org/10.1016/j.ijleo.2019.03.065.

[12] C. Bhaya, M.S. Obaidat, A.K. Pal, S.K.H. Islam, Encrypted medical image storage in DNA domain, in: ICC 2021-IEEE International Conference on Communications, IEEE, 2021, pp. 1–7, https://doi.org/10.1109/ICC42927.2021.9500718.

[13] S. Namasudra, G.C. Deka, Introduction of DNA computing in cryptography, in: Advances of DNA Computing in Cryptography, Chapman and Hall/CRC, 2018, pp. 1–18.

[14] Q. Zhang, L. Guo, X. Wei, Image encryption using DNA addition combining with chaotic maps, Math. Comput. Model. 52 (11–12) (2010) 2028–2035, https://doi.org/10.1016/j.mcm.2010.06.005.

[15] M.A.B. Farah, R. Guesmi, A. Kachouri, M. Samet, A novel chaos based optical image encryption using fractional Fourier transform and DNA sequence operation, Opt. Laser Technol. 121 (2020) 105777, https://doi.org/10.1016/j.optlastec.2019.105777.

[16] M. Yadollahi, R. Enayatifar, H. Nematzadeh, M. Lee, J.-Y. Choi, A novel image security technique based on nucleic acid concepts, J. Inform. Secur. Appl. 53 (2020) 102505, https://doi.org/10.1016/j.jisa.2020.102505.

[17] S. Patel, K.P. Bharath, R. Kumar, Symmetric keys image encryption and decryption using 3D chaotic maps with DNA encoding technique, Multimed. Tools Appl. 79 (2020) 1–19, https://doi.org/10.1007/s11042-020-09551-9.

[18] D. Huo, D.-f. Zhou, S. Yuan, S. Yi, L. Zhang, X. Zhou, Image encryption using exclusive-OR with DNA complementary rules and double random phase encoding, Phys. Lett. A 383 (9) (2019) 915–922, https://doi.org/10.1016/j.physleta.2018.12.011.

[19] X. Wang, Y. Wang, X. Zhu, C. Luo, A novel chaotic algorithm for image encryption utilizing one-time pad based on pixel level and DNA level, Opt. Lasers Eng. 125 (2020) 105851, https://doi.org/10.1016/j.optlaseng.2019.105851.

[20] T. Hu, Y. Liu, L.-H. Gong, S.-F. Guo, H.-M. Yuan, Chaotic image cryptosystem using DNA deletion and DNA insertion, Signal Process. 134 (2017) 234–243, https://doi.org/10.1016/j.sigpro.2016.12.008.

[21] H. Dong, E. Bai, X.-Q. Jiang, Y. Wu, Color image compression-encryption using fractional-order hyperchaotic system and DNA coding, IEEE Access 8 (2020) 163524–163540, https://doi.org/10.1109/ACCESS.2020.3022398.

[22] P.T. Akkasaligar, S. Biradar, Selective medical image encryption using DNA cryptography, Inf. Secur. J. A Global Perspect. 29 (2) (2020) 91–101, https://doi.org/10.1080/19393555.2020.1718248.

[23] J.C. Dagadu, J.-P. Li, E.O. Aboagye, Medical image encryption based on hybrid chaotic DNA diffusion, Wirel. Pers. Commun. 108 (1) (2019) 591–612, https://doi.org/10.1007/s11277-019-06420-z.

[24] A. Banu S, R. Amirtharajan, A robust medical image encryption in dual domain: chaos-DNA-IWT combined approach, Med. Biol. Eng. Comput. 58 (2020) 1–14, https://doi.org/10.1007/s11517-020-02178-w.

[25] A.Y. Niyat, M.H. Moattar, Color image encryption based on hybrid chaotic system and DNA sequences, Multimed. Tools Appl. 79 (1–2) (2020) 1497–1518, https://doi.org/10.1007/s11042-019-08247-z.

[26] X. Chai, X. Fu, Z. Gan, Y. Lu, Y. Chen, A color image cryptosystem based on dynamic DNA encryption and chaos, Signal Process. 155 (2019) 44–62, https://doi.org/10.1016/j.sigpro.2018.09.029.

[27] Z. Liu, C. Wu, J. Wang, Y. Hu, A color image encryption using dynamic DNA and 4-D memristive hyper-chaos, IEEE Access 7 (2019) 78367–78378, https://doi.org/10.1109/ACCESS.2019.2922376.

[28] R.C. Gonzalez, R.E. Woods, S.L. Eddins, Digital Image Processing Using MATLAB, Pearson Education India, 2004.

[29] E. Cliffs, A.K. Jain, Fundamentals of Digital Image Processing, Prentice-Hall, Hoboken, New Jersey, 1989.

[30] J. Minguillon, J. Pujol, JPEG standard uniform quantization error modeling with applications to sequential and progressive operation modes, J. Electron. Imaging 10 (2) (2001) 475–486, https://doi.org/10.1117/1.1344592.

[31] P. Mildenberger, M. Eichelberg, E. Martin, Introduction to the DICOM standard, Eur. Radiol. 12 (4) (2002) 920–927, https://doi.org/10.1007/s003300101100.

[32] E.A. Jackson, A. Hübler, Periodic entrainment of chaotic logistic map dynamics, Phys. D 44 (3) (1990) 407–420, https://doi.org/10.1016/0167-2789(90)90155-I.

[33] S.C. Phatak, S.S. Rao, Logistic map: a possible random–number generator, Phys. Rev. E 51 (4) (1995) 3670, https://doi.org/10.1103/PhysRevE.51.3670.

[34] M.A. Khan, V. Jeoti, Modified chaotic tent map with improved robust region, in: 2013 IEEE 11th Malaysia International Conference on Communications (MICC), IEEE, 2013, pp. 496–499, https://doi.org/10.1109/MICC.2013.6805880.

[35] J. Griffin, The sine map, 4 (2013) 2018 (Retrieved May).

[36] M. Cai, Complex dynamics in generalized Henon map, Discret. Dyn. Nat. Soc. 2015 (2015) 1–18, https://doi.org/10.1155/2015/270604.

[37] S.M.H.T. Yazdi, H.M. Kiah, E. Garcia-Ruiz, J. Ma, H. Zhao, O. Milenkovic, DNA-based storage: trends and methods, IEEE Trans. Mol. Biol. Multi-Scale Commun. 1 (3) (2015) 230–248, https://doi.org/10.1109/TMBMC.2016.2537305.

[38] E. Biham, A. Shamir, Differential cryptanalysis of DES-like cryptosystems, J. Cryptol. 4 (1) (1991) 3–72, https://doi.org/10.1007/BF00630563.

[39] P. Mishra, C. Bhaya, A.K. Pal, A.K. Singh, A novel binary operator for designing medical and natural image cryptosystems, Signal Process. Image Commun. 98 (2021) 116377, https://doi.org/10.1016/j.image.2021.116377.

[40] Y. Wu, J.P. Noonan, S. Agaian, NPCR and UACI randomness tests for image encryption, Cyber J. 1 (2) (2011) 31–38.
[41] P. Mishra, C. Bhaya, A.K. Pal, A.K. Singh, A medical image cryptosystem using bit-
level diffusion with DNA coding, J. Ambient. Intell. Humaniz. Comput. 12 (2021)
1–22, https://doi.org/10.1007/s12652-021-03410-7.
[42] SIPI, SIPI Image Database, http://sipi.usc.edu/database/, 2021. (accessed 01.06.21).

About the authors

Chiranjeev Bhaya received his BTech degree in Computer Science and Engineering from Govt. College of Engineering and Textile Technology, Serampore, West Bengal, India in the year 2018 and his MTech degree in Computer Science and Engineering with Specialization in Information Security from Indian Institute of Technology (Indian School of Mines), Dhanbad in the year 2020. His research interests include applications of DNA computing in various fields such as Coding Theory, Data and Image Cryptography, Machine Learning, etc.

Arup Kumar Pal is currently working as an associate professor in the Department of Computer Science and Engineering, Indian Institute of Technology (ISM) Dhanbad, India. Before joining this institute, he was a lecturer in the Department of Computer Science and Engineering, NIT Jamshedpur, from April 2011 to December 2011. He received his BTech in Computer Science and Engineering from Govt. College of Engineering and Textile Technology, Berhampore, India, in 2006. He did his PhD in Computer Science and Engineering from IIT (ISM) Dhanbad in 2011. He has more than 10 years of teaching and research experience. He has contributed over 100 research papers in several journals and conference proceedings of national and international reputes. His main research interest includes Data Compression, Multimedia Security, and CBIR. He has organized several FDPs in the area of Image Processing and Cryptography. He has served as an Advisor Committee member and Technical Programme Committee (TPC) member in various conferences/workshops.

SK Hafizul Islam (SM'18) received MSc degree in applied mathematics from Vidyasagar University, Midnapore, India, in 2006 and MTech degree in Computer Application and PhD degree in Computer Science and Engineering in 2009 and 2013, respectively, from the Indian Institute of Technology [IIT (ISM)] Dhanbad, Jharkhand, India, under the INSPIRE Fellowship PhD Program (funded by the Department of Science and Technology, Government of India). He is currently an assistant professor in the Department of Computer Science and Engineering, Indian Institute of Information Technology Kalyani (IIIT Kalyani), West Bengal, India. Before joining IIIT Kalyani, he was an assistant professor at the Department of Computer Science and Information Systems, Birla Institute of Technology and Science, Pilani (BITS Pilani), Rajasthan, India. He has more than 10 years of teaching and 13 years of research experience. He has authored or co-authored 130 research papers in journals and conference proceedings of international reputes. His research interests include Cryptography, Information Security, Neural Cryptography, Lattice-based Cryptography, IoT & Blockchain Security, and Deep Learning.

He is an associate editor for "IEEE Transactions on Intelligent Transportation Systems," "IEEE Access," "International Journal of Communication Systems (Wiley)," "Telecommunication Systems (Springer)," "IET Wireless Sensor Systems," "Security and Privacy (Wiley)," "Array-Journal (Elsevier)," "Wireless Communications and Mobile Computing (Hindwai)," and "Journal of Cloud Computing (Springer)." He received the University Gold Medal, the S. D. Singha Memorial Endowment Gold Medal, and the Sabitri Parya Memorial Endowment Gold Medal from Vidyasagar University in 2006. He also received the University Gold Medal from IIT(ISM) Dhanbad in 2009 and the OPERA award from BITS Pilani in 2015. He is a senior member of IEEE and a member of ACM.

CHAPTER FIVE

Hiding information in an image using DNA cryptography

P. Bharathi Devi[a], P. Ravindra[b], and R. Kiran Kumar[c]

[a]Department of Computer Science, S.K.B.R. Government Degree College, Macherla, Andhra Pradesh, India
[b]Department of Computer Science, K.B.N. College, Vijayawada, Andhra Pradesh, India
[c]Department of Computer Science, Krishna University, Machilipatnam, Andhra Pradesh, India

Contents

Abstract

In this digital era, all communications are executed over the internet. As there are numerous users and attackers over the internet, confidential data face many security issues. Nowadays, Deoxyribonucleic Acid (DNA) computing is an advanced field for improving data security. In DNA cryptography, data are converted into DNA bases, namely Adenine(A), Guanine(G), Cytosine(C) and Thymine(T), instead of using binary values 0 and 1. To improve data security, in this chapter, a novel two layers scheme

Advances in Computers, Volume 129
ISSN 0065-2458
https://doi.org/10.1016/bs.adcom.2022.08.003

has been proposed by using DNA computing. The first layer converts the secret data into DNA bases, while in the second layer, these are wrapped in an image making it more secure. The second layer is mainly responsible for hiding the data known as steganography. The proposed scheme not only hides the data, but also encrypts the secret data by using a DNA ASCII Table. The proposed scheme is secured against masquerade attack, brute force attack, ciphertext only attack, known plaintext attack and man-in-the-middle attack. To evaluate the performance of the proposed scheme, Peak Signal-to-Noise Ratio (PSNR), Mean Square Error (MSE), and Structural Similarity Index (SSIM) values are considered. Results and discussion prove the efficiency of the proposed scheme.

Abbreviations

A	adenine
AES	advanced encryption standard
ASCII	American Standard Code for Information Interchange
C	cytosine
CIA	Confidentiality, Integrity and Availability
DES	data encryption standard
DNA	deoxyribonucleic acid
DoS	denial of service
ECC	elliptic curve cryptography
FR	full reference
G	guanine
HH	high high
HL	high low
IDEA	international data encryption algorithm
LH	low high
LL	low low
LSB	least significant bit
LWT	lifting wavelet transform
mRNA	messenger RNA
MSE	mean square error
NR	no reference
OTP	one time pad
PCR	polymerase chain reaction
PSNR	peak signal-to-noise ratio
PXORSS	probabilistic XOR secret sharing
RNA	ribonucleic acid
RSA	Rivest-Shamir-Adleman
RR	reduced reference
SSIM	structural similarity index
SSL	secured socket layer
T	thymine

1. Introduction

Improving data security is always a key aspect in the field of computer science. With the advancement of technologies, such as cloud computing, big data and many more, users always communicate their confidential data through the internet. Here, the main concern is that the exchanging of data must not breach data security. Sometimes, attackers hack the confidential data. They hack the secret key, and they also even replace the sensitive data by their fake data. To solve such critical issues, researchers are continuously developing novel cryptographic and steganographic algorithms, to address data security issues. By using efficient cryptographic and steganographic algorithms unauthorized data access can be prevented [1–4]. Cryptography is a technique in which plaintext of data are transformed into a secret form called as ciphertext by using an algorithm. Many people use cryptography to make their information unreadable. However, it must satisfy the CIA (Confidentiality, Integrity and Availability) triad. On the other hand, steganography conceals the existence of the data. Numerous algorithms are available based on symmetric and asymmetric cryptosystem. The symmetric cryptosystem contains only one key that is used for both data encryption and data decryption. The single key is shared between both the sender and receiver through a secure channel before applying the algorithm. Some examples of symmetric algorithms are Data Encryption Standard (DES), Advanced Encryption Standard (AES), Blowfish, International Data Encryption Algorithm(IDEA), etc. In the asymmetric cryptosystem, both parties use two keys, namely, private key and public key. The public key can be shared in the public domain, and the private key is always kept secret. The process of managing both the keys has several steps like generating key, distribution of key and key revocation. When a key is strong and complex, it improves data security. Some popular examples of asymmetric algorithms are RSA (Rivest-Shamir-Adleman) algorithm, Diffie Hellman algorithm, etc. Though asymmetric cryptosystem is much popular, sometimes, it is vulnerable to many attacks like chosen-plaintext attack, man-in-the middle-attack, etc. In addition, it gives a low performance. In Elliptic Curve Cryptography (ECC) algorithm, which is an asymmetric cryptosystem, it is difficult to find the positions of the curve to data encrypt, and as result, it creates confusion for the receiver to decrypt the information. Similarly, the RSA algorithm has the problem to find the value of prime factorization [5]. Currently, many attacks are being developed by hackers to retrieve the confidential information. Therefore, researchers are focusing on quantum cryptography, which is based on quantum mechanics. Here, data is sent by encrypting the photon value, and the identification of photon is possible only at a distance of 90 miles [6–9]. It is based on the laws of physics. This technique is secured; however, it costs high for implementation. In steganography, the data is usually embedded into a text, image, audio, video, etc. This technique hides the secret information to prevent it from unauthorized users. Short encrypted messages can be sent to the receiver by covering the message with an image, which improves data security. The goal of steganography is to provide the sealed information. Henceforth, it is hard for the third person or attackers to know about the presence of the masked message in the cover media like images, audios, videos, etc. [10–13]. The cover media that is used as a data carrier is the most important aspect to conceal the data. The existing models use text, image, audio, video, DNA, etc. as covering media. To apply steganography for text data, modification is done in the paragraph layout. The identification of hidden text becomes easy by doing steganalysis. Even though it occupies less memory, it is not preferable for security aspects. Images are also used as a carrier to keep the short messages either in the place of iris or other facial parts. Audios are used as a carrier by placing the message inaudible frequency positions or converting the frequencies into binary format, and then, apply some XOR functions by using key [14,15].Video is also used as a data carrier, where the video is divided into several frames, and then, an insertion technique is applied for embedding data in any one of the frames [16,17].

To solve the above mentioned problem, currently, researchers are using DNA computing, which was first introduced by L M. Adleman in 1994. As mentioned earlier, each DNA molecule contains four nitrogenous bases, namely Adenine, Guanine, Cytosine and Thymine. A and G are called purines and Cand Tare called pyrimidines [18,19]. The nucleotides in DNA are structured in a long strand, one strand from $5'$ to $3'$ direction and another strand from $3'$ to $5'$ direction; this is how the structure of double helix is formed [20]. DNA has the ability to replicate itself, i.e., it produces its own identical copies. The human body consists of 3 billion bases. In DNA, one base is paired with another, i.e., "A" is paired with "T" and "C" is paired with "G," which is known as Watson-Crick complementary pair rule. In DNA computing, basic concepts of DNA molecules and hardware are used for computation. DNA computing is popular due to its advanced features, such as storage and massively parallel capabilities and many more. The massive parallelism of DNA computing is one of the key factors, which supports executing a complex operation in a small amount of time. Approximately 700 TB of information can be stored in 1 g DNA bases [21]. Thus, few grams DNA can store the data of the world. DNA encoding plays a key role while transmitting the data from its plaintext into the corresponding DNA sequence or strand. Here, each DNA base can have a binary value like A can be used for representing 00, C can be used for representing 01. In the same way, G can be used for representing 10 and T for 11. However, anyone can assign any rule to convert the binary values, i.e., plaintext to the DNA bases. As there are four DNA bases, there can be 4! i.e., 24 processes of assigning DNA bases to its corresponding binary values. So, to achieve enhanced security, different values can be assigned to each DNA base. However, it should be remembered that the complementary pair rule must be maintained [22].

In the last few years, several models are developed based on DNA computing to improve data security. Namasudra et al. [23] have proposed a scheme for securing multimedia by using DNA based encryption in the cloud environment. Here, a 1024-bit secret key and user's attributes are used to encrypt any multimedia file. In 2020, a novel fuzzy DNA computing based image encryption and steganography technique model was proposed [24]. Here, a simple chaotic map is used along with DNA computing, and to get the final encrypted image, the wavelet fusion algorithm is used for fusing the resulting four fuzz-DNA encoded images. Alsaffar et al. [25] have proposed a steganography model using DNA computing and AES algorithm to hide a compressed text in a color image. In this scheme, the size of data is reduced to 75% after the encryption process. In 2021, Setiadi [26] have suggested a measurement tool for the quality assessment of steganography. Based on the results of this study, Structural Similarity Index (SSIM) is a better measure of imperceptibility in all aspects. In [27], Abdelfattah et al. have suggested a secure image encryption scheme based on DNA computing. In this scheme, a new multi chaotic map is introduced for wireless area networks to improve security. This scheme uses the patient medical images and hides the patient information in that image using the Least Significant Bit (LSB) technique. It supports that the information is visually unavailable to unauthorized persons or hackers. Verma et al. [28] have proposed a model for image steganography based on LSB. Security analysis of this scheme is done by using histogram analysis and the quality of an image is tested based on two factors, namely, Peak Signal-to-Noise Ratio (PSNR) and Mean Square Error (MSE). Swain et al. [29] have proposed a novel steganography technique by mapping words with a LSB array. In this scheme, steganographic and start indices are noted down based on the different words of the secret messages and they are mapped on the chosen array. All these schemes are unable to achieve high security and involve many complex operations. In many cases, the performance of a system is degraded.

To solve the problems of the existing schemes, in this chapter, a novel scheme is proposed in which a DNA strand is used as a data carrier to hide the information. In the proposed scheme, the information is first converting into DNA bases by using DNA cryptography [30,31]. Here, DNA sequence is used as a data carrier in addition to the existing carriers like image, audio or video, which improves data security. The proposed work consists of two layers. The first layer deals with the encryption of the secret message by using a DNA-based American Standard Code for Information Interchange (ASCII) table. Here, the message is converted into a DNA sequence and a random key is also generated from the input image. This process strengthens the proposed scheme because when the encrypted secret message is hidden in the cover media, the probability of hacking the information becomes less. The second layer is responsible for hidden the encrypted data in an image in the form of DNA bases. Identifying the hidden information from the image becomes difficult for the intruders. The combination of cryptography and steganography in the proposed approach improves data security significantly.

The major contributions of the proposed work are as follows:

(1) In this chapter, an ovel model is designed for data encryption by using a DNA based ASCII table.

(2) To hide the existence of encrypted data, it is embedded in an image in the form of DNA bases, which improves data security significantly.

(3) Security analysis and experimental results of the proposed methodology are tabulated to prove its efficiency.

Section 2 discusses several existing works with their drawbacks. Section 3 explains the background study of the proposed scheme, i.e., protein synthesisation and the use of a structured codon table in the protein synthesis process. Then, Section 4 describes the entire proposed work in detail. In Section 5, the security analysis of the proposed scheme is presented, and Section 6 discusses the results and discussion of the proposed scheme. At last, Section 7 concludes the overall chapter.

2. Literature reviews

Gehani et al. [32] have developed a model by using a substitution method based on the distinct One-Time Pad (OTP) technique and DNA computing. This scheme uses basic principles of DNA computing suggested by L. M. Adleman [33–36]. Gehani et al. have used DNA computing for data encryption and they have maintained a codebook, which is helpful to convert the block of plaintext to the encrypted data. In 1999, Clelland et al. [37] have proposed a steganography model that uses DNA microdots. In this model, the encryption process is executed by using Polymerase Chain Reaction(PCR) and primers. Then Clelland et al. have applied microdots to hide the information, thus, the data security has been improved. Amin et al. [38] have implemented another encryption scheme based on the genomes available in the GenBank. They have downloaded the genomes of Canisfamiliaris from the GenBank and applied a searching algorithm to find the place of quadruple DNA in it. Borda et al.[39]have proposed another encryption technique by generating an OTP of the length of the number of binary bits multiplied by 10. The key points used in this scheme are indexed based on DNA computing and XOR operation. They have applied both encryption and steganography techniques by assigning a binary value for each oligonucleotide sequence, which is complementary to the randomly generated sequence.

Gao et al. [40] have proposed a DNA-based encryption framework utilizing biological letters, i.e., codon representation letters of the protein synthesisation process. A codon is a DNA sequence that consists of three adjacent DNA bases out of four bases. This scheme uses primers and the distributed algorithm of this scheme holds one secret codebook shared between sender and receiver through a secure channel. In this scheme, the plaintext is converted into the binary form, and then, it is converted into a DNA sequence. However, Gao et al. have not represented practical implementation of their scheme.

Ning [41] has proposed a novel DNA cryptography technique based on the central dogma of molecular biology. In this scheme, the splicing system is utilized to remove the frank DNA sequences and a random exon is added to the ciphertext, which is in the form of DNA bases. This exon is basically a single-stranded Ribonucleic Acid (RNA) known as Messenger RNA(mRNA). This mRNA is then changed into protein. At last, it is sent to the receiver of the data. The recipient executes the reverse processes and gets the plaintext. In this scheme, the key is generated randomly from the plaintext and a different key is created for each data communication process.

Cui et al. [42] have introduced an encryption algorithm by utilizing advanced DNA coding, i.e., PCR enhancement and DNA amalgamation, and then, utilized the conventional cryptography algorithm. Their algorithm mainly focuses to generate key, and PCR primers are considered as the key pairs. Here, both the sender and receiver consensually decide aprimer sequence. In this scheme, the plaintext is changed into a hexadecimal number and the information is scrambled utilizing the recipient's public key. Then, it is converted into a long DNA strand. Here, the sender sends the information to the receiver by combining the ciphertext with certain fake DNA sequences. The receiver utilizes the primer sequence and other credentials to extract the original information from the ciphertext. This scheme is slow because of many complex operations.

Rahman et al. [43] have developed an encryption scheme based on DNA computing. This scheme is mainly based on the encoding algorithm and encryption algorithm in the primary level and subsequent level, respectively. At the primary level, the plaintext is converted into a DNA sequence by using a DNA encoding table. In the subsequent level, the encryption process is executed for many rounds, which is also based on DNA computing. Then, XOR operation is performed between the resultant DNA sequence and key followed by converting into mRNA.

Babu et al. [44] have proposed a pseudo biotic DNA cryptography scheme. This scheme is mainly based on the splicing mechanism and the concept of the central dogma. The splicing mechanism is used to generate a key value for the encryption process and the key is generated from the plaintext by using a key generation algorithm. Here, an XOR operation is performed between the randomly generated DNA sequence and randomly generated mRNA sequence. This scheme is slow and not secured against the password guessing attack.

Tuncer and Avci [45] have implemented a steganography model using DNA computing. In this scheme, a grayscale image is utilized as the cover picture and the ciphertext is separated from secret shares by utilizing probabilistic XOR secret sharing (PXORSS). The cover picture is divided into four groups, namely, Low-Low (LL), Low-High (LH), High-Low (HL) and High-High (HH), by utilizing Lifting Wavelet Transform (LWT). The secret information can be separated in this scheme into four secret shares with a solitary LWT level. Here, various LWT levels can be utilized for more secret shares.

Malathi [46] has proposed a highly improved steganography scheme using DNA computing, and they have used text as the data carrier. For improving data security, the author has used a fake DNA sequence. In this scheme, two keys are used. Here, the primary key $K1$ is used to XOR the plaintext message and it is ranges from 0 to 255. Then, $K1$ is changed into another binary value and XOR operation is performed between the newly generated binary value and the resultant of the XOR operation of the previous step. Then, this sequence is converted into a DNA sequence. The second key $K2$ is actually a DNA sequence, which is mixed up with the fake DNA sequence. Then, again XOR operation is performed between the mixed DNA sequence and DNA sequence, which is already generated. Due to the execution of many operations, this technique consumes time.

Namasudra et al. [47] have proposed a novel scheme for improving data security using DNA cryptography in the cloud environment. In this scheme, a long 1024-bit secret key is used, which is generated based on the user's attributes. Namasudra et al. have presented both security analysis and results and discussion. This scheme is secured against password guessing attack, insider attack, Denial of Service (DoS) attack and many more. It is better than many existing schemes. However, the time to access the data from the cloud server is more in this scheme, so in many cases, the users need to pay more for using the numerous cloud services.

Singh et al. [48] have developed a novel scheme for securing the mobile network. The authors have modified the text based on DNA computing, which is transmitted through the network. Here, two keys are used. The main key (i.e., DNA sequence) is called an intron, and it is chosen arbitrarily from the DNA sequence. Further, it eliminates all the DNA sequences, which are the same as intron, from the ciphertext in terms of DNA sequence. The subsequent key converts the codon into amino acid and stores the flag value.

Patnala and Kumar [49] have proposed a model based on amino acids, and they have prepared a lookup table that contains the 64 amino acids mapped with $\{a...z, A...Z, 0...9\}$. However, this mapping table is restricted to only 64 characters. To solve this drawback, Kumar and Patnala [50] have proposed another model by extending the character set to an ASCII set and proposed a DNA computing based ASCII table, which contains $4*4*4*4$, i.e., 256 possible DNA sequences by arranging the DNA bases in four positions. This DNA computing based ASCII table acts as a key dictionary of this scheme, where mapping of each character of the ASCII table is done with the corresponding DNA sequence of length 4. This scheme improves data security as the table is randomly generated during each data transmission.

Table 1 represents the summary of many existing DNA computing based encryption and decryption schemes.

Table 1 Comparative study on DNA computing-based cryptographic schemes.

Scheme	Key findings	Advantages	Disadvantages
Amin et al. [38]	• This scheme is mainly based on genome sequence	• This scheme uses OTP • It has high throughput	• This scheme is not secured and can be easily improved using a larger DNA strand
Borda et al. [39]	• Here, indexing is based on DNA computing and XOR	• The length of OTP is large	• Processing time is high is this scheme
Gao et al. [40]	• DNA primers are used for secure communication	• Biological alphabet, i.e., codebook is maintained	• Here, practical implementation is not given
Ning [41]	• Central dogma of molecular biology is used • Key is generated from the plaintext	• Using public key cryptosystem for encryption	• The more complex keys, the more complex the decryption process
Cui et al. [42]	• PCR primers are considered as key in this scheme	• It supports high security and PCR amplification is used in this scheme	• This scheme is slow because of the presence of complex operation
Rahman et al. [43]	• The plaintext is converted to DNA sequence using DNA encoding table • This scheme supports two levels security	• This scheme is secured against many attacks	• The proposed scheme is limited to 64 permutations of DNA bases
Babu et al. [44]	• In this scheme, splicing system is used to improve the security	• Here, key is generated randomly	• It is slow and not secured against password guessing attack

Table 1 Comparative study on DNA computing-based cryptographic schemes.—cont'd

Scheme	Key findings	Advantages	Disadvantages
Tuncer and Avci [45]	• A grayscale image is used in this scheme as a cover image	• Various LWT levels can be used for secret shares	• This scheme consumes more time for encrypting any data
Malathi [46]	• Many XOR operations improve the data security • Plaintext is converted into protein	• This scheme supports enhanced security	• Here, execution time is high due to the presence of many operation
Namasudra et al. [47]	• A 1024-bit DNA-based password is used to encrypt the confidential data	• This scheme is highly secured	• In this scheme, data accessing time is high, so customer may need to pay more for using the cloud service
Singh et al. [48]	• This scheme uses a DNA sequence as a key • Here, the key is randomly chosen	• Splicing mechanism is used in this scheme	• Computational ability of the nodes in mobile networks is still an issue
Patnala and Kumar [49]	• This scheme is based on amino acid • Here, codons are used for transmitting the information	• This scheme randomly arranges lookup table for each transmission	• The lookup table is limited to 64 characters only
Kumar and Patnala [50]	• DNA ASCII table is used for encryption process and it is generated randomly for each transmission	• This scheme calculates avalanche effect	• The execution time of this scheme is slow

3. Background study

As discussed in Section 2 most of the researchers have used the central dogma of molecular biology to improve data security by proposing novel encryption algorithms. In central the dogma of molecular biology, DNA sequences are converted into RNA sequences in the transcription phase, and then, these RNA sequences are again converted into protein structure in the translation phase. Here, one important point to be noted about the transcription phase is that Uracil(U) is used instead of Thymine (T). During the formation of protein structure, the RNA sequences are coded as three bases as a unit, which is called a codon [51]. A total of 64 bases are formed by taking any three bases out of the four bases.

As shown in Fig. 1, the process of central dogma cannot be reversed that implies retrieving the DNA sequences from the protein structure is not possible. As per the rule of the codon table (i.e., 64 codons), protein structure is formed. Here, each codon is mapped to an amino acid. Out of 64 codons, one codon is used for the start signal, and three codons are used as stop codons and the remaining amino acids are used to form a protein structure. According to the structure of the codon table TTT and TTC can be replaced with Phenylalanine (P) that is represented. In the same manner, TTA and TTG can be replaced with Leucine (L)and many more. The letters of English alphabets can also be used to replace codons. Here, each codon is replaced with a corresponding alphabet letter. However, only 20 letters are used to represent these codons and it occurs an ambiguity in the representation of alphabets while performing the encryption and decryption process. In the first step, conversion of plaintext to DNA sequence is performed, and then, these DNA sequences are separated into three bases each. Later these three bases are represented with their corresponding alphabets. For example, a DNA sequence TTTTTATTG…..TTC is separated into three bases like TTT, TTA, TTG, …, TTC. Then, TTT, TTA, TTG and TTC are replaced with P, L, L and P, respectively. So, the codon sequence of the above DNA sequence is PLLP. When the encrypted information is sent to the receiver, it is not possible to decrypt the information because of ambiguity of putting the DNA sequence. In order to overcome this ambiguity, many researchers have introduced their own DNA codebooks and have assigned the alphabets and digits accordingly [52–54].

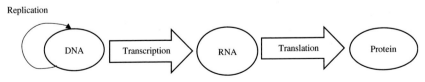

Fig. 1 Central dogma of molecular biology.

4. Proposed scheme

In this section, a DNA-based steganography algorithm is proposed to enhance the security of information using dual layer technique. The proposed scheme mainly focuses on hiding the information in the form of DNA bases in an image. The key feature of this proposed scheme is generating a random key of length 128-bit from the input image by using a pseudo random number generator. This proposed scheme uses a randomly generated DNA computing based ASCII table, i.e. Table 2. The different table is used during each data transmission process. The randomness of key and DNA computing based ASCII table support the proposed scheme to protect it against malicious users and hackers. The DNA computing based ASCII table is generated based on the four DNA bases, namely, A, T, G and C. These DNA bases are arranged in 4 * 4 * 4 * 4, i.e., 256 ways and mapped each DNA base codon to its equivalent decimal values. Before transmitting the information to the receiver, the sender and receiver must share this random DNA computing based ASCII table through the Secured Socket Layer (SSL). It also improves the proposed algorithm because if the intruder wants to hack the information, it is almost impossible due to the random behavior of this table. Moreover, 256!permutations must be performed by the intruder.

The entire process of the proposed scheme is executed in two layers. Fig. 2 shows the entire workflow of the proposed scheme. As mentioned in Section 4.1, the first layer is used to convert the secret message into DNA bases, and the second layer is used to embed the secret message into an input image, which is explained in Section 4.2. In order to hide or embed a secret message in a cover object or image, the cover image must contain some noises or redundant data, which is mainly used to conceal the secret message.

4.1 Encrypting secret message

The first layer is responsible for converting the secret message into a DNA sequence. In order to convert the secret message into a DNA sequence as shown in Table 3, there are the following steps:

Step 1: In the first step, the secret message M is converted into its equivalent ASCII values. Here M is considered as "Welcome to the new era of Steganography".

Table 2 DNA computing based ASCII table.

Decimal value	Binary	Character	Description	DNA codon
0	00000000	NUL	Null	CGCA
1	00000001	SOH	Start of header	TCTT
2	00000010	STX	Start of text	TCAT
3	00000011	ETX	End of text	TGTT
4	00000100	EOT	End of transmission	TGAT
5	00000101	ENQ	Enquiry	ACTT
6	00000110	ACK	Acknowledge	ACAT
7	00000111	BEL	Bell	CAGT
8	00001000	BS	Backspace	CTTC
9	00001001	HT	Horizontal tab	CTAC
10	00001010	LF	Line feed	AATC
11	00001011	VT	Vertical tab	AAAC
12	00001100	FF	Form feed	GTTC
13	00001101	CR	Enter/carriage return	GTAC
14	00001110	SO	Shift out	GATC
15	00001111	SI	Shift in	GAAC
.	.	.	.	.
.	.	.	.	.
.	.	.	.	.
253	11111101	ý		TCAG
254	11111110	þ		TCTG
255	11111111	ÿ		CGTG

Step 2: Here, each ASCII value is converted into an 8-bit binary digit.
Step 3: For generating a random key from the input image, the pseudo random generator selects 16 pixels randomly. Then, each pixel is converted into an 8-bit binary digit. The random key is generated from the input image Lenna of size 512×512. The key is given below:

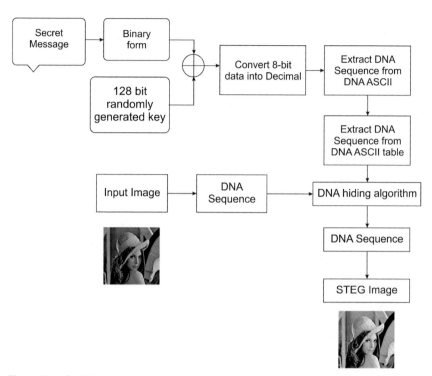

Fig. 2 Pseudo DNA steganography model.

01010111010100111001010101011110110101001111111101010
10011110101100010110100101100010011001001000011101110
00100011101010101010

Step 4: In the fourth step, XOR operation is applied between the randomly generated key of length 128 bit and the binary value of the plaintext. If the message length (*L*) is more than 128bit, then the key is repeated to make its length the same as the binary value of the plaintext.

Step 5: The resultant of the above step is divided into several parts in which each part contains 8-bit. These 8-bit numbers are converted into decimal values.

Step 6: In the sixth step, equivalent DNA sequences are taken from the DNA computing based ASCII table for the decimal values.

Step 7: In this step, complement DNA strand is calculated from the above DNA sequence.

Step 8: Finally, the complementary DNA strand is inserted into the image by using the data hiding algorithm.

Table 3 Conversion of secret message into a DNA sequence.

Sl. No.	Secret message	ASCII value	Binary value	Randomly generated key	XOR value	Decimal value	DNA sequence from DNA ASCII table	Complementary strand
1	W	127	01111111	01010111	00101000	40	CTCA	GAGT
2	e	101	01100101	10101001	11001100	204	CGGG	GCCC
3	l	108	01101100	11001010	10100110	166	TACG	ATGC
4	c	99	01100011	10101111	11001100	204	CGGG	GCCC
5	o	111	01101111	01101010	00000101	5	ACTT	TGAA
6	m	109	01101101	01111111	00010000	16	CACA	GTGT
7	e	101	01100101	10101010	11001111	207	AAGG	TTCC
8		32	00100000	01111010	01011010	90	AACT	TTGA
9	t	116	01110100	11000101	10110001	177	GCCA	CGGT
10	o	111	01101111	10100101	11,001,010	202	ACAA	TGTT
11		32	00100000	10001001	10101001	169	ACCG	TGGC
12	t	116	01110100	10010010	01100110	102	CATT	GTAA
13	h	104	01101000	00011101	01110101	117	ATAT	TATA
14	e	101	01100101	11100010	10000111	135	TGCA	ACGT
15		32	00100000	00111010	00011010	26	AATT	TTAA

16	n	110	01101110	10101010	11000100	196	TCTA	AGAT
17	e	101	01100101	01010111	00110010	50	TCGT	AGCA
18	w	119	01110111	10101001	11011110	222	CCGT	GGCA
19		32	00100000	11001010	11101010	234	CTAG	GATC
20	e	101	01100101	10101111	11001010	202	ACAA	TGTT
21	r	114	01110010	01101010	00011000	24	CTTT	GAAA
22	a	97	01100001	01111111	00011110	30	GATT	CTAA
23		32	00100000	10101010	10001010	138	CACG	GTGC
24	o	111	01101111	01111010	00010101	21	ATGC	TACG
25	f	102	01100110	11000101	10100011	163	GGAT	CCTA
26		32	00100000	10100101	10000101	133	TCCA	AGGT
27	S	83	01010011	10001001	11011010	218	ACAC	TGTG
28	t	116	01110100	10010010	11100110	230	ATGA	TACT
29	e	101	01100101	00011101	01111000	120	CCTC	GGAG
30	g	103	01100111	11100010	10000101	133	TCCA	AGGT
31	a	97	01100001	00111010	01011011	91	AGCT	TCGA

Continued

Table 3 Conversion of secret message into a DNA sequence.—cont'd

Sl. No.	Secret message	ASCII value	Binary value	Randomly generated key	XOR value	Decimal value	DNA sequence from DNA ASCII table	Complementary strand
32	n	110	01101110	10101010	11000100	196	TCTA	AGAT
33	o	111	01101111	01010111	00111000	56	CTCC	GAGG
34	g	103	01100111	10101001	11001110	206	CCGG	GGCC
35	r	114	01110010	11001010	10111000	184	ATCT	TAGA
36	a	97	01100001	10101111	11001110	206	CCGG	GGCC
37	p	112	01110000	01101010	00011010	26	AATT	TTAA
38	h	104	01101000	01111111	00010111	23	CAAC	GTTG
39	y	121	01111001	10101010	11010011	211	GAGG	CTCC

Algorithm 1. Encryption
   Input: Secret Message (M) and randomly generated 128-bit Key (K)
   Output: DNA based secret message
1. START
2. CONVERT M into decimal values
3. CONVERT into 8-bit binary number
4. If $L > 128$
5. ADD the key repeatedly
6. End If
7. PERFORM XOR operation between binary value and key value
8. DIVIDE the message into 8-bit each
9. APPLY decimal encoding rule
10. RETRIEVE the DNA sequence Table 2
11. APPLY complementary pair rule
12. END

The process of encrypting the secret message into DNA sequence is represented in Algorithm 1. After executing the above mentioned steps, the DNA sequence of the secret message is given below:

   GAGTGCCCATGCGCCCTGAAGTGTTTCCTTGACGGTTGT
   TTGGCGTAATATAACGTTTAAAGATAGCAGGCAGATCTGT
   TGAAACTAAGTGCTACGCCTAAGGTTGTGTACTGGAGAGG
   TTCGAAGATGAGGGGCCTAGAGGCCTTAAGTTGCTCC.

4.2 Hiding secret message in an input image

In this second layer, the resultant DNA sequence of the first layer is inserted into a randomly chosen position of the image. Here, the key value is also converted into the DNA sequence, which is also inserted into the secret position of the image. Here, the standard image of Lenna is considered as the input image, and as shown in Fig. 3, its size is 512×512 in the proposed scheme.

   Step 1: Here, the input image is converted into the grayscale image as shown in Fig. 4.

   Step 2: In the second step, the grayscale image is converted into a list of pixel values.

   Step 3: Each pixel is converted as an 8-bit binary value.

   Step 4: In the fourth step, the binary value is converted into a DNA sequence using the DNA encoding rule.

Fig. 3 Original image.

Fig. 4 Grayscale image.

Step 5: Here, the key value is converted into a DNA sequence using the DNA encoding rule.

Step 6: Now, the DNA sequence of the image is represented as an array and the secret DNA sequence is placed into the position of a prime number.

Step 7: In the seventh step, the DNA sequence of the key is placed in the even positions of the array.

Step 8: Here, the updated DNA sequence is converted into the binary value by using Table 2.

Step 9: In the ninth step, the binary value is transformed into the decimal value.

Step 10: At last, the decimal values are arranged in matrix form.

Once the embedding process is completed, the converted message, i.e., STEG image in which encrypted secret message and encrypted key, is ready to transmit over the network. Fig. 5 is the STEG image, which is transmitted over the network to the receiver. Algorithm 2 represents the hiding process of a secret message in an input image.

Fig. 5 STEG image.

4.3 Decryption process

In the proposed scheme, the decryption process contains two phases. In the first phase, the receiver retrieves the DNA sequences placed in the secret position of the STEG image and also retrieves the DNA sequence of the key placed in the even position. In order to retrieve the DNA sequences of the secret message and key, the receiver converts the STEG image sent by the sender into the grayscale image. In the second phase, the complementary strand of the secret message is retrieved from the prime position and find the equivalent decimal value from the DNA computing based ASCII Table. After that, an XOR operation is applied between the decimal value and key value. At last, the equivalent ASCII character is retrieved. There are the following steps in the decryption process:

Algorithm 2. DNA computing based data hiding
    Input: Image (IMG) and secret message's ciphertext (C)
    Output: STEG image
1. START
2. CONVERT IMG into grayscale
3. CONVERT grayscale into pixel matrix
4. CONVERT each pixel value into 8-bit binary number
5. CONVERT binary value into the DNA sequence
6. CONVERT the key into the DNA sequence
7. For $i = 1$ *to* L//Here, L is the length of the DNA sequence
8. If $i \bmod 2 = 0$
9. UPDATE the DNA sequence of the key of ith position
10. End If
11. If $i =$ a prime number
12. UPDATE the ciphertext of ith position

13. End If
14. End For
15. CONVERT the updated DNA sequence into binary
16. CONVERT binary value into decimal value
17. ARRANGE decimal data into pixel matrix
18. END

Step 1: At first, the STEG image is converted into the grayscale image.
Step 2: In the second step, the grayscale image is converted into its pixel values.
Step 3: Here, all pixel values are transformed into 8-bit binary values.
Step 4: In the fourth step, binary numbers are changes into the DNA sequence using the DNA encoding rule.
Step 5: In the fifth step, the DNA sequence is retrieved, which is stored in the even position and it is stored as the key value, i.e.:

> CCCTGGGCTAGGGGTTCGGGCTTTGGGGCTGGTACCG
> GCCGAGCGCAGACTCTGAGATGGGGGGCCCTGGGCTA
> GGGGTTCGGGCTTTGGGGCTGGTACCGGCCGAGCGCA
> GACTCTGAGATGGGGGGCCCTGGGCTAGGGGTTCGGG
> CTTTGGGG

Step 6: Here, the secret message's DNA sequence is retrieved, which is placed in the position of a prime number and it is stored as message *M*.

> GAGTGCCCATGCGCCCTGAAGTGTTTCCTTGACGGTT
> GTTTGGCGTAATATAACGTTTAAAGATAGCAGGCAGA
> TCTGTTGAAACTAAGTGCTACGCCTAAGGTTGTGTAC
> TGGAGAGGTTCGAAGATGAGGGGCCTAGAGGCCTTAA
> GTTGCTCC

Step 7: In the seventh step, the complementary strand of the above message *M*is calculated.

> CTCACGGGTACGCGGGACTTCACAAAGGAACTGCCAA
> CAAACCGCATTATATTGCAAATTTCTATCGTCCGTCTA
> GACAACTTTGATTCACGATGCGGATTCCAACACATGAC
> CTCTCCAAGCTTCTACTCCCCGGATCTCCGGAATTCAA
> CGAGG

Step 8: Here, the DNA sequence is divided into several parts containing four bases each, and then, the equivalent decimal values are taken from Table 2.

> 4020416620451620790177202169102117135
> 261965022223420224301382116313321823012013391196562061
> 842062623211

Step 9: In the ninth step, the decimal values are converted into their corresponding binary values.

00101000110011001010011011001100000001010001000011001111
10101101010110001110010101010100101100110011101011000010
11000110101100010000110010110111101110101011001010000011
00000011110100010100001010110100011100001011101101010111 0
01100111100010000101010110111100010000111000110011101010
11000110011100001101000010111110100111

Step 10: In this step, an XOR operation is performed between the resultant of the previous step and the binary value of the key

01111111011001010110110001100011011011110110110101100101 00
1001000001110100011011100100000001110100011010000110010
10010000000110111001100101011011011001000000011001010111 0
0100110000100100000001101111011001100010000000101001101111
01000110010101100111011000010110111001101111011011001110 11
10010011000010111000001101000011111001

Step 11: In the eleventh step, binary values are converted into ASCII values.

12710110899111109101321161113211610410132110101119321 01
11497321111023283
11610110397110111103114971121041 21

Step 12: Finally, the ASCII values are used to retrieve the secret message by using Table 2 as given below:

Welcome to the new era of Steganography.

Algorithm 3. Decryption

Input: STEG image

Output: Secret message (M)

1. START
2. CONVERT STEG into grayscale
3. CONVERT grayscale into pixel matrix
4. CONVERT each pixel value into 8-bit binary number
5. CONVERT binary data into the DNA sequence using DNA encoding rule
6. For $i = 1$ *to L*
7. If $i \bmod 2 = 0$
8. RETRIEVE DNA sequences placed in the even positions
9. STORE the sequences in K
10. CONVERT the DNA sequence of K into binary value
11. End If

12. If $i=$ a prime number

13. RETRIEVE DNA sequences placed in the prime positions

14. STORE the sequences in M

15. CONVERT the DNA sequences of M into binary number

16. End If

17. End For

18. APPLY XOR between the values of M and K

19. CONVERT binary to decimal values

20. RETRIEVE equivalent ASCII characters

21. STORE the characters in M

22. END

Algorithm 3 represents the decryption process. Using cryptography and steganography techniques, the proposed scheme improves data security.

5. Security analysis

Attackers and malicious users use many attacks to hack the data or key by using numerous attacks. For any cryptographic algorithm, it is very important to provide strong security of confidential data against security attacks. The proposed data encryption technique is secured against many attacks, namely masquerade attack, brute force attack, ciphertext only attack, known plaintext attack and man-in-the-middle attack. The security analysis of the proposed scheme has been discussed in this section.

5.1 Masquerade attack

Attackers create a fake identity to gain unauthorized access to any data, which is known as masquerade attack. In this attack, the attackers communicate several times with an authorized user by creating a fake identity. For example, sending an email to the victim as a banker and collect some secret information, execute some transactions like a victim, sending one time password and many more. A weak authentication system also allows the attacker to gain access. Sometimes, a coworker may also act as a masquerade to hack or steal the information. It is very hard to detect this attack in this situation. To prevent this attack, authorized users must not open any unknown email and must not share any confidential information with anyone. In the proposed scheme, the sender and receiver must agree on the DNA computing based ASCII table, i.e., Table 2 before transmission and this table is randomly generated. Table 2 acts a key role in encrypting the secret message as each time a different table is used for data encryption. Even though the attackers or hackers get the STEG image, they cannot retrieve the secret information from the STEG image because the secret key is encrypted by using the DNA computing based ASCII table and the encrypted information is embedded into some hidden positions. Without knowing the positions, as well as Table 2, no one can hack the information.

5.2 Brute force attack

In this attack, the attacker tries all possibilities to retrieve the ciphertext. This technique is also called the trial-and-error technique for guessing login information, keys and many more, to get the secret message. It is an exhaustive search by the intruder to retrieve the plaintext from the ciphertext by applying different permutations and combinations to a key value. It is one of the old attacks, but it is still effective and popular in many cases. A cryptosystem can use a long length key value, so that the permutation number increases. Thus, this attack can be prevented. If the length of a secret key is too small, retrieving the ciphertext can be easy for the hacker. The time taken to retrieve the information by using this attack is high. However, there is a maximum possibility of getting the original secret message. In the proposed scheme, the secret message in the form of a DNA sequence is embedded into a cover image known as STEG image. Before embedding the information, DNA computing based ASCII table is used to encrypt the information. In addition, the secret key is also randomly generated in the proposed scheme. Even though the key length is fixed 128bit, the attackers or malicious users must calculate 256! permutations to retrieve the original Table 2. Thus, the data security of the proposed scheme has been improved significantly, and it is not possible by the intruder to use brute force attack to compromise the proposed scheme.

5.3 Ciphertext only attack

The intruders or attackers know the ciphertext in this attack, and they try to get the plaintext. In a ciphertext only attack, the attackers is presumed to have access to the ciphertext or encrypted scrambled message and they have no clue about the plaintext and the key. While designing an encryption algorithm, it is critical to consider the ciphertext only attack as most of the algorithms consider that the communications are executed in a secured channel. In the proposed scheme, if the intruders know the secret information, which is embedded in the STEG image, it would be still difficult for them to get the original plaintext as data are encrypted before embedding it into the cover image. The attackers must know all the credentials, i.e., Table 2 and key, to get the original data. Most importantly, the ciphertext and the key value are randomly changed for each plaintext character because of the randomly generated Table 2. The proposed scheme generates different ciphertexts for the same data during each data communication process. Hence, the proposed scheme is secured form this attack.

5.4 Known plaintext attack

In the known plaintext attack, the hacker has access both the ciphertext and its corresponding plaintext. The intruders or hackers maintain a codebook that contains both plaintext and its corresponding ciphertext. By using the information of the codebook, the attackers try to decrypt the ciphertext. If the codebook contains a very large number of words, then only the attackers can decrypt the ciphertext. Here, attackers and malicious users guess the various keys by which they try to unscramble any further messages. Retrieving the key and knowing the working principles of the encryption algorithm are difficult tasks for malicious users even though they know the plaintext and ciphertext. In the proposed technique, for each transmission of the information, the receiver and the sender must agree on randomly arranged Table 2 and it is shared between them through a secured channel. Moreover, the encrypted message is embedded into a cover media in terms of a DNA sequence. So, if the attacker maintains a codebook for this proposed scheme, different code pairs must be formed for each transmission process because the key value, as well as Table 2 are changed during each data transmission process. So, the proposed scheme can resist this attack.

5.5 Man-in-the-middle attack

In this attack, the attackers or malicious users relay and alter the communications between two parties (i.e., sender and receiver), who believe that their communication is directly executed between them. The eavesdroppers secretly listen or get information about the communication between the sender and receiver, then, they act as an authorized user to get the original message or replace the original message with their fake message. This attack can be prevented only by implementing a strong encryption algorithm and by providing a strong encryption mechanism on the server's side, so that unauthorized users cannot join the network. The proposed technology supports two layers of security. At first, the information is encrypted by using the novel encryption technique, and then, this encrypted data is embedded into the cover image. Here, DNA computing and a random dictionary, i.e., Table 2 are used for improving data security. In the proposed scheme, if an eavesdropper acts as a sender to collect the encrypted information from the receiver, it is almost impossible to decrypt the hidden message of the STEG image because the encrypted message is embedded in terms of DNA bases. In addition, only the authorized parties (i.e., sender and receiver) rely on the DNA computing based ASCII Table before starting the communication process. The attackers or malicious users must have all the credentials to get the original data contents. As the credentials are only shared with the authorized parties through a secured channel, the proposed scheme can be considered secure against this attack.

6. Performance analysis

The performance of the proposed algorithm can be evaluated based on the three parameters PSNR, MSE and SSIM. These metrics are used to know the visual quality of an image after embedding the secret message in the original image.

The strength of a steganography technique mainly depends upon the image quality, when an image is considered as the cover media. There are several techniques to measure the quality of an image. They are classified as Full Reference (FR), Reduced Reference (RR) and No Reference (NR) methods [58]. Here, in this chapter, the FR method is considered, which contains the evaluation metrics as PSNR, MSE and SSIM. Embedding a secret message in a cover picture may change few pixel values that affect the visual nature of the STEG picture. These modifications must be correctly identified since they straightforwardly affect the indistinctness of the last unveiling of the STEG image.

6.1 Experimental setup

Many experiments are executed to evaluate the proposed scheme. Here, experimental environments are developed using Python on a Dell OptiPlex 5480 AIO Desktop. The system has core i7 processor (9^{th}generation), 225 SSD, 1 TB HDD and 8 GB RAM with Windows 10 as the operating system. The execution process includes OpenCV, NumPy, List and random method for generating STEG image, and Dictionary is one of the key factors for generating DNA computing based ASCII table.

6.2 Results and discussion

Four standard images, namely, Lenna, Pepper, Monalisa and Baboon, are considered for evaluation, and the size of all these images are 512×512 pixel. The results are shown in Table 4. Fig. 6 shows the grayscale images of test images, and Fig. 7 shows the STEG images in which the secret information, i.e., "Welcome to the new era of Steganography," is embedded. The values of MSE, PSNR and SSIM are calculated by using Eqs. (1)–(3), respectively, between the grayscale image and STEG image to identify the variation in the quality of an image for all the input images and compared with the three existing methods [55–57].

MSE is the cumulative squared error between the STEG image and the original image. A low value of MSE denotes less error. MSE and PSNR have the inverse relation, i.e., an image has better visual quality, when it has a low MSE and high PSNR value. In the proposed scheme, the test image Monalisa has the highest PSNR value and the lowest MSE value, when compared to all other test images. As shown in Fig. 8, Monalisa is better out of all test images to transmit the secret image. It can be easily understood from Table 4 that the proposed scheme is much better than the other existing schemes as it gives a low error rate. When the error rate is low, it gives high security.

$$MSE = \frac{1}{MN} \sum_{i-1}^{M} \sum_{j-1}^{N} \left(O(i,j) - D(i,j) \right)^2 \tag{1}$$

where.

$O(i,j) =$ Input image
$D(i,j) =$ STEG image
$M =$ Number of pixels in row
$N =$ Number of pixels in column
$i =$ Index of a row of the image
$j =$ Index of a column of the image

PSNR value is calculated for identifying the distortion of the cover images after embedding the secret data, and thus, the efficiency of the proposed scheme is evaluated. If the PSNR value is greater than $30\,dB$, the distortion of the STEG image is undetectable to human vision. Therefore, a high PSNR value implies less amount of distortion, which leads to high visual quality. From the observations from Table 4, the test image Monalisa has a high PSNR value as shown in Fig. 9 that means it has high visual quality, when compared to other images. The test image Baboon has a similar PSNR value as that of Monalisa. The other test images are also

Table 4 Comparisons of PSNR, MSE and SSIM values among different schemes.

Image/quality measure	Proposed scheme			Das and Kar [55]			Kumar et al. [56]			Sabry et al. [57]		
	PSNR	MSE	SSIM	PSNR	MSE	SSIM	PSNR	MSE	SSIM	PSNR	MSE	SSIM
Lenna	43.81	2.71	0.98156	40.81	3.91	0.72153	41.81	3.09	0.86582	42.81	2.91	0.92562
Pepper	44.45	2.32	0.98398	42.92	3.52	0.72893	42.15	2.52	0.85256	43.92	2.04	0.92514
Monalisa	54.41	0.23	0.99885	51.61	2.56	0.76525	50.61	2.48	0.92456	52.61	1.96	0.95262
Baboon	52.16	0.39	0.99915	49.26	1.89	0.72123	48.26	1.29	0.91256	50.26	1.94	0.95958

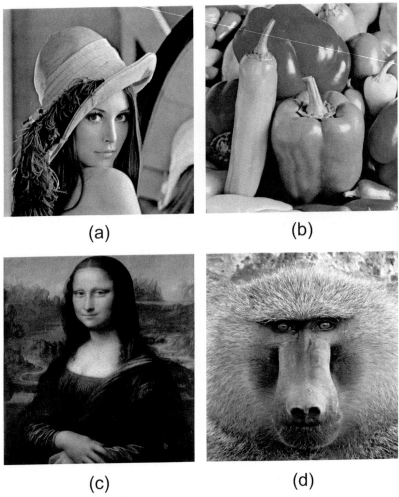

(a)

(b)

(c)

(d)

Fig. 6 Grayscale images. (A) Lenna, (B) Pepper, (C) Monalisa, (D) Baboon.

somewhat good visual quality because they have PSNR values greater than 30 dB. As the proposed scheme has higher PSNR values than the other existing models, it indicates lesser visual quality in the existing models. Thus, the data security of the proposed scheme is improved in the proposed scheme.

$$PSNR = 10 \log_{10} \frac{MAX^2}{MSE} \tag{2}$$

where
$MAX =$ Maximum possible pixel value of the image

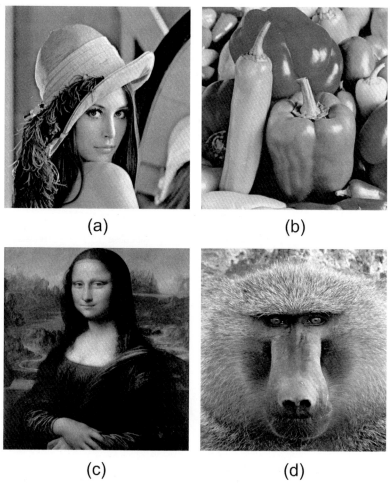

(a)   (b)

(c)   (d)

Fig. 7 STEG images. (A) Lenna, (B) Pepper, (C) Monalisa, (D) Baboon.

The last metric to evaluate the proposed scheme is SSIM, which is also called as Universal Quality Index(UQI). This metric is also calculated between the input image and STEG image to know the structural difference of the images. If the value of SSIM is 1, two images are structurally identical. The resultant SSIM index value is a decimal value between 0 and 1. Based on the conducted experiments, image Baboon as well as the other test images have the SSIM index value close to 1. The results of SSIM index values are shown in Fig. 10. When comparing to the existing models, the proposed scheme has an SSIM index value close to 1. Therefore, the malicious users and hackers cannot identify the difference between the original image and STEG image.

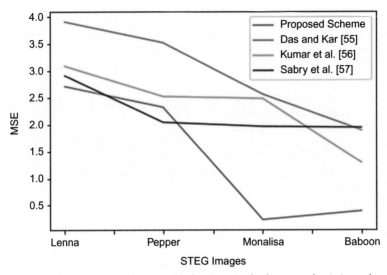

Fig. 8 Comparisons of MSE value among the proposed scheme and existing schemes.

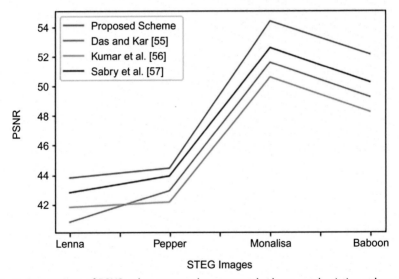

Fig. 9 Comparison of PSNR value among the proposed scheme and existing schemes.

$$SSIM(x, y) = \frac{\left(2\mu_x\mu_y + C_1\right)\left(2\sigma_{xy} + C_2\right)}{\left(\mu_x^2 + \mu_y^2 + C_1\right)\left(\sigma_x^2 + \sigma_y^2 + C_2\right)} \tag{3}$$

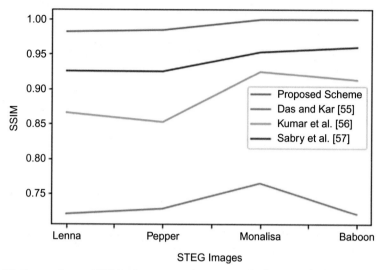

Fig. 10 Comparisons of SSIM value among the proposed scheme and existing schemes.

where

$x = $ Input image

$y = $ STEG image

$\mu_x = $ Luminance, i.e., average of all the pixel values of the input image

$\mu_y = $ Luminance, i.e., average of all the pixel values of the STEG image

$\sigma_x = $ Contrast, the standard deviation of all the pixel values of the input image

$\sigma_y = $ Contrast, the standard deviation of all the pixel values of the STEG image

$C_1 = $ Constant to ensure stability when the denominator is weak

$C_2 = $ Constant to ensure stability when the denominator is weak

In Figs. 8–10, there are some distortions in the STEG image due to embedding a huge amount of secret text. It is mainly occurred based on the randomly generated key values and changes in the DNA computing based ASCII table.

7. Conclusions

DNA cryptography and DNA steganography are emerging fields in supporting security of data transmission through an untrusted channel. In this chapter, a novel scheme has been proposed for data communication, which is secured against many security attacks, namely, masquerade attack, brute force attack, ciphertext only attack, known plaintext attack and man-in-the-middle attack. The proposed scheme has two stages or phases. In the first phase, data are encrypted by using a novel DNA computing based ASCII table, and in the second phase, the encrypted data is embedded into a cover media, i.e., image. Here, different DNA computing based ASCII table is used during each data accessing process. It is almost impossible to hack the proposed scheme due to the random behavior of this table. Many experiments are executed to evaluate the performance of the proposed scheme. Experimental results show that the distortion takes place in the STEG image is very low and the quality of the image is high. The rate of error of the proposed scheme is also low, which proves its high security, and the structure of the input image and STEG image is almost identical. There is a huge scope to apply this proposed scheme for big data. In the future, the proposed scheme can be improved by proposing a novel key generation algorithm.

References

[1] M.R.N. Torkaman, N.S. Kazazi, A. Rouddini, Innovative approach to improve hybrid cryptography by using DNA steganography, Int. J. New Comput. Archit. Appl. 2 (1) (2012) 224–235.
[2] D. Nixon, "DNA and DNA computing in security practices-is the future in our genes?". Available: https://www.giac.org/paper/gsec/1617/dna-dna-computing-security-practices-future-genes/102969 [Accessed on 30 November 2020].
[3] M. Yamuna, A. Elakkiya, Codons in datasafe transfer, Int. J. Eng. 2 (2015) 85–90.
[4] J. Ashok, Y. Raju, S. Munishankaraiah, K. Srinivas, Steganography: an overview, Int. J. Eng. Res. Technol. 2 (10) (2010) 5985–5992.
[5] E.S. Babu, C. Nagaraju, M.H.M.K. Prasad, Light-Weighted DNA Based Hybrid Cryptographic Mechanism against Chosen Ciphertext Attacks, in: Advanced Computing and Systems for Security, 2016, pp. 123–144, https://doi.org/10.1007/978-81-322-2650-5_9.
[6] M. Krenn, M. Malik, T. Scheidl, R. Ursin, A. Zeilinger, Quantum communication with photons, in: M.A. Amri, M.E. Gomati, M. Zubairy (Eds.), Optics in Our Time, Springer, Cham, 2016.
[7] Prentice Hall PTR, Modern cryptography: Theory and applications, 2020. Available: https://cs.stanford.edu/people/eroberts/courses/soco/projects/2004-05/cryptography/quantum.html[Accessed on 12 December 2020].
[8] S. Groblacher, T. Jennewein, A. Vazir, G. Weihs, A. Zeilinger, Experimental quantum cryptography with qutrits, New J. Phys. 8 (2006) 1–8.
[9] J. Mullins, Quantum cryptography's reach extended, 2003. Available:https://spectrum.ieee.org/telecom/security/quantum-cryptographys-reach-extended [Accessed on 17December 2020].
[10] M.M. Amin, M. Salleh, S. Ibrahim, M.R. Katmin, M.Z.I. Shamsuddin, Information hiding using steganography, in: Proceedings of the 4th National Conference of Telecommunication Technology, IEEE, Shah Alam, Malaysia, 2003, pp. 21–25.
[11] D. Heider, A. Barnekow, DNA watermarking: challenging perspectives for biotechnological applications, Curr. Bioinform. 6 (3) (2011) 375–382.

[12] R.P.K. Reddy, C. Nagaraju, N. Subramanyam, Text encryption through level based privacy using DNA steganography, Int. J. Emerg. Trends Technol. Comput. Sci. 3 (3) (2014) 168–172.

[13] G.M. Church, Y. Gao, S. Kosuri, Next-generation digital information storage in DNA, Science 337 (6102) (2012). https://doi.org/10.1126/science.1226355.

[14] R.M. Tank, H.D. Vasava, V. Agarwal, DNA based audio steganography, Orient. J. Comput. Sci. Technol. 8 (1) (2015) 43–48.

[15] M. Athira, R. Reshma, S.B. Sasidhar, N.V. Kalyankar, Audio-video using forensic technique for data security, Int. J. Comput. Eng. Technol. 5 (12) (2014) 154–157.

[16] K.U. Singh, Video steganography: Text hiding in video by LSB substitution, Int. J. Eng. Res. Appl. 4 (5) (2014) 105–108.

[17] B. Chandel, S. Jain, Video steganography: a survey, IOSR J. Comput. Eng. 18 (1) (2016) 11–17.

[18] F. Crick, Central dogma of molecular biology, Nature 227 (1970) 561–563.

[19] P.K. Gupta, Cell and molecular biology, 5th Edition, Rastogi Publications, 2005.

[20] X. Guozhen, L.U. Mingxin, Q. Lei, L. Xuejia, New field of cryptography: DNA cryptography, Chin. Sci. Bull. 51 (12) (1996) 1413–1420.

[21] ExtremeTech, Harvard cracks DNA storage, crams 700 terabytes of data into a single gram, 2012. Available: https://www.extremetech.com/extreme/134672-harvard-cracks-dna-storage-crams-700-terabytes-of-data-into-a-single-gram [Accessed on 17 February 2021].

[22] J.D. Watson, A.B. Tania, P.B. Stephen, G. Alexander, L. Michael, L. Richard, Molecular Biology of the Gene, 7th Ed, Pearson Education, 2017.

[23] S. Namasudra, R. Chakrabort, A. Majumder, N.R. Moparthi, Securing multimedia by using DNA based encryption in the cloud computing environment, ACM Trans. Multimedia Comput. Commun. Appl. 16 (3) (2020).

[24] S.E. El Khamy, N.O. Korany, A.G. Mohamed, A new fuzzy-DNA image encryption and steganography technique, IEEE Access 8 (2020) 148935–148951.

[25] Q.S. Alsaffar, H.N. Mohaisen, F.N. Almashhdini, An encryption based on DNA and AES algorithms for hiding a compressed text in colored image, in: Proceedings of the IOP Conference Series: Materials Science and Engineering, 2020. https://doi.org/10.1088/1757-899X/1058/1/012048.

[26] D.R.I.M. Setiadi, PSNR vs SSIM: imperceptibility quality assessment for image steganography, Multimed. Tools Appl. 80 (2021) 8423–8444.

[27] R.I. Abdelfattah, H. Mohamed, M.E. Nasr, Secure image encryption scheme based on DNA and new multi chaotic map, J. Phys. Conf. Ser. (2020), https://doi.org/10.1088/1742-6596/1447/1/012053.

[28] V. Verma, A. Kajal, I. Kajal, Enhancement of payload capacity for image steganography based on LSB, Int. J. Comput. Appl. Technol. Res. 5 (10) (2016) 678–682.

[29] G. Swain, S.K. Lenka, A novel steganography technique by mapping words with LSB array, Int. J. Signal Imaging Syst. Eng. 8 (1) (2015) 115–122.

[30] T. Anwar, S. Paul, S.K. Singh, Message transmission based on DNA cryptography: review, Int. J. BioSci. Biotechnol. 6 (5) (2014) 215–222.

[31] G. Xiao, M. Lu, L. Qin, X. Lai, New field of cryptography: DNA cryptography, Chin. Sci. Bull. 51 (2006) 1413–1420. https://doi.org/10.1007/s11434-006-2012-5.

[32] A. Gehani, T. LaBean, and J. Reif, "DNA-based cryptography," Lecture Notes in Computer Science, vol. 2950. Springer, Berlin, Heidelberg. DOI:https://doi.org/10.1007/978-3-540-24635-0_12.

[33] L.M. Adleman, P.W. Rothemund, S. Roweis, E. Winfree, On applying molecular computation to the data encryption standard, J. Comput. Biol. 6 (1) (1999) 53–63.

[34] L. Adleman, Molecular computation of solutions to combinatorial problems, Science 266 (5187) (1994) 1021–1024.

[35] L.M. Adleman, On constructing a molecular computer, DNA Based Comp. (1995).

[36] Stanford.edu, DNA computing, 2021. Available: https://cs.stanford.edu/people/eroberts/courses/soco/projects/2003-04/dna computing/slide1.html [Accessed on 10 January 2021].

[37] C.T. Clelland, R. Viviana, C. Bancroft, Hiding messages in DNA microdots, Nature 399 (1999) 533–534.

[38] S.T. Amin, M. Saeb, S.E. Gindi, A DNA-based implementation of YAEA encryptionalgorithm, Comput. Intell. (2006).

[39] M.E. Borda, O. Tornea, T. Hodrogea, Secret writing by DNA hybridization, Acta Tech. Napoc. 50 (2009) 21–24.

[40] Q. Gao, Biological alphabets and DNA-based cryptography, in: *Proceedings of the American Society for Engineering Education (ASEE)*, Easton, PA, 2010.

[41] K. Ning, A pseudo DNA cryptography method, arXiv (2009).

[42] G. Cui, L. Qin, Y. Wang, X. Zhang, An encryption scheme using DNA technology, in: *Proceedings of the 3rd International Conference on Bio-Inspired Computing: Theories and Applications*, Adelaide, SA, 2008, pp. 37–42.

[43] N.H.U. Rahman, C. Balamurugan, R. Mariappan, A novel DNA computing basedencryption and decryption algorithm, Procedia Comput. Sci. 46 (2015) 463–475.

[44] E.S. Babu, C. Nagaraju, M.H.M.K. Prasad, Inspired pseudo biotic DNA based cryptographic mechanism against adaptive cryptographic attacks, Int. J. Netw. Secur. 18 (2) (2016) 91–303.

[45] T. Tuncer, E. Avci, A reversible data hiding algorithm based on probabilistic DNA-XOR secret sharing scheme for color images, Displays 41 (2016) 1–8.

[46] P. Malathi, M. Manoaj, R. Manoj, R. Vaikunth, R.E. Vinodhini, Highly improved DNA based steganography, Procedia Comput. Sci. 115 (2017) 651–659.

[47] S. Namasudra, S. Sharma, G.C. Deka, P. Lorenz, DNA computing and table based data accessing in the cloud environment, J. Netw. Comput. Appl. 172 (2020).

[48] H. Singh, K. Chugh, H. Dhaka, A.K. Verma, DNA based cryptography: an approach to secure mobile networks, Int. J. Comput. Appl. 1 (19) (2010) 77–80.

[49] B.D. Patnalaand, R.K. Kumar, A novel level-based DNA security algorithm using DNA codons, in: Computational Intelligence and Big Data Analytics, Springer, Singapore, 2019. https://doi.org/10.1007/978-981-13-0544-3_1.

[50] R.K. Kumar, B.D. Patnala, A novel text encryption algorithm using DNA ASCII table with a spiral approach, Int. J. Recent Sci. Res. 9 (1) (2018) 23588–23595.

[51] H. Lodish, A. Berk, S.L. Zipursky, P. Matsudaira, D. Baltimore, J. Darnell, Molecular cell biology, fifth Edition, *W.H. Freeman & Company*, 2006.

[52] A. Cherian, S.R. Raj, A. Abraham, A survey on different DNA cryptographic methods, Int. J. Sci. Res. 2 (4) (2013) 167–169.

[53] A.A. Elhadad, A. Khalifa, S.Z. Rida, DNA-based data encryption and hiding using playfair and insertion techniques, J. Commun. Comput. Eng. 2 (3) (2011) 44–49.

[54] B. Anam, K. Sakib, M.A. Hossain, K. Dahal, Review on the advancements of DNA cryptography, arXiv (2010).

[55] P. Das, N. Kar, A highly secure DNA based image steganography, in: *Proceedings of the International Conference on Green Computing Communication and Electrical Engineering*, IEEE, Coimbatore, India, 2014.

[56] K.B.S. Kumar, K.B. Raja, R.K. Chhotaray, S. Pattnaik, Steganography based on payload transformation, IJCSI Int. J. Comput. Sci. 8 (2) (2011) 241–248.

[57] M. Sabry, T. Nazmy, M.E. Khalifa, Steganography in DNA sequence on the level of amino acids, in: *Proceedings of the 9th International Conference on Intelligent Computing and Information Systems*, IEEE, Cairo, Egypt, 2019, pp. 317–324.

[58] A. Pradhan, A.K. Sahu, G. Swain, K.R. Sekhar, Performance evaluation parameters of image steganography techniques, in: *Proceedings of the International Conference on Research Advances in Integrated Navigation Systems*, IEEE, Bangalore, India, 2016.

About the authors

Dr. P. Bharathi Devi presently working as Lecturer in department of Computer Science in SKBR Government Degree College, Macherla, Palnadu District, Andhra Pradesh. She obtained her Ph.D., degree in Computer Science & Engineering from Krishna University, Machilipatnam, Andhra Pradesh. She received her post graduate degree in Computer Applications in 2006 and M.Tech (CSE) in 2010. Earlier she worked as an Assistant Professor in K.B.N.College:: P.G.Centre since 2006 to 2021. She Qualified UGC-NET and AP-SET. Her research interests include Network Security, DNA Cryptography and Compression Techniques.

Mr. P. Ravindra presently working as a head, department of Computer Science & Applications, K.B.N.College(Autonomous), Vijayawada, Andhra Pradesh. He is techno-savvy person and enthusiastic towards new computer trends. He completed his post graduate degree in Computer Science & Applications in 2012. His research interests include Network Security, Data analytics and Image Processing.

Dr. R. Kiran Kumar is presently working as Assistant Professor in the department of Computer Science. He began his career as Lecturer from July, 1999. He obtained his Ph.D in Computer Science & Engineering form Acharya Nagarjuna University in 2009. He is currently serving as Head of the Department and Director, Directorate of Admissions. Dr. R. Kiran Kumar has previously held the positions Principal, University College of Engineering and Technology, Controller of Examinations, Director, Research Admissions, Coordinator, IQAC and Coordinator, Centre for Research Studies. Dr. R. Kiran Kumar as a Research mentor, has so far guided 28 (Twenty Eight) Ph.D's. He published more than 130 research papers, of which majority are in impact journals. He is one among the cited researchers in the field of Computer Science and Information Technology with h-index of 15 and i-10 index of 23 with 5 Published Patents.

CHAPTER SIX

The design of a S-box based on DNA computing and chaos theories

Jun Peng[a,b], Shangzhu Jin[b,c], Yingxu Wang[d], Xi Zheng[b], and Xiangren Wang[b]

[a]School of Mathematics, Physics and Data Science, Chongqing University of Science and Technology, Chongqing, China
[b]School of Intelligent Technology and Engineering, Chongqing University of Science and Technology, Chongqing, China
[c]Informatization Office, Chongqing University of Science and Technology, Chongqing, China
[d]International Institute of Cognitive Informatics and Cognitive Computing (ICICC), Department of Electrical and Software Engineering, Schulich School of Engineering and Hotchkiss Brain Institute, University of Calgary, Calgary, AB, Canada

Contents

Advances in Computers, Volume 129
ISSN 0065-2458
https://doi.org/10.1016/bs.adcom.2022.08.004

211

Abstract

Due to the continuous growth of systems operating on big data, the scope of cloud computing, and the applications of Internet of Things, data security becomes increasingly important. Substitution(S)-box is a vital non-linear component which challenges the design and implementation of cryptographic systems. This chapter presents a new method to design S-boxes synergizing deoxyribonucleic acid (DNA) computing and chaos theories. A 128-bit secret key is transformed to generate the initial conditions and parameters used in a hyperchaotic system with four-wing attractors to generate chaotic sequences. Subsequently, DNA complementary rules are randomly selected to encode the chaotic sequences to obtain three DNA codes, which dedrive two independent systems of piecewise linear chaotic maps (PWLCM) in order to generate required S-boxes. Numerical experiments for 1000 randomly generated S-boxes demonstrate that the results possess good bijection, ideal nonlinearity, strict avalanche criterion (SAC), valid bit independence criterion (BIC), differential approximation probability, and cryptographic key sensitivity. The proposed methodology for S-box generation provides better cryptographic properties and can be widely used in different security scenarios that overplay traditional technologies.

Abbreviations

AES	advanced encryption standard
BIC	bit independence criterion
CML	coupled map lattice
CNN	cellular neural network
DDEs	delay difference equations
DES	data encryption standard
DNA	deoxyribonucleic acid
DP	differential approximation probability
LFSR	linear feedback shift register
MAC	media access control
OTP	one time pad
PRNG	pseudorandom number generator
PSD	power spectral density
PSO	particle swarm optimization
PWLCM	piecewise linear chaotic maps
PWLM	piecewise linear map
SAC	strict avalanche criterion
S-box	substitution box
SPN	substitution–permutation network
TSP	traveling salesman problem
XOR	exclusive-or

1. Introduction

In the context of worldwide digitalization, mobile computing and Internet of Things, massive data are generated everyday that triggered increasingly high requirements for data security for various applications. For modern cryptographic systems with symmetric key cryptographic algorithms, the substitution box (S-box) is a nonlinear component that performs permutation calculations and its performance directly determines the quality of cryptographic algorithms [1]. In a block cipher similar to the Data Encryption Standard (DES), S-boxes are usually used to mask the relationship between the secret key and ciphertext to confuse Shannon properties. In many cases, S-boxes are carefully selected to make the cryptosystem resistant to cryptanalysis.

Because chaotic sequences have many excellent cryptographic properties, they have been widely used in the design of cryptosystems, including S-box construction algorithms [1–21]. In 2005, Tang et al. [12] proposed an encrypted dynamic S-box scheme based on iterated discrete 2D Baker chaotic map. Later, for this scheme, Chen et al. [13] employed 3D Baker map to overcome some errors, but the improved scheme is still weak against differential analysis attack. In 2010, Fatih et al. [14] proposed a methodology to design cryptographically S-boxes based on chaotic Lorenz system, but the nonlinearity of S-box generated is still not high. Then in 2013, Fatih et al. [15] studied an S-box design algorithm based on time-delay chaotic systems. However, due to the lack of key sensitivity analysis, the security is slightly insufficient. Wang et al. [16] studied a method for S-box construction based on chaos, genetic algorithm and Traveling Salesman Problem (TSP). However, the method takes too much time. Recently, the S-box scheme based on chaos has been proposed by Khan et al. [17], Çavuşoğlu et al. [18], Wang et al. [19], and Tian et al. [20] in their respective perspectives. However, these S-boxes have low nonlinearity, which leads to poor performance against differential analysis. Further details of these schemes [12–20] will be discussed in Section 2.1 and summarized in Table 1.

Table 1 Key findings of Chaos-based S-box schemes.

Schemes	Key findings	Disadvantages
Tang et al. [12]	2D Baker mapping and Logistic map are used to generate S-box	The nonlinearity of S-box is not very high
Chen et al. [13]	3D Baker mapping is used and the errors in Tang et al. [12] are corrected	Weak performance against differential cryptanalysis
Fatih et al. [14]	Lorenz chaotic system is used and algorithm is enhanced by shifting columns and rows of S-box	The nonlinearity of S-box is still not high
Fatih et al. [15]	Delay difference equations (DDEs) with rich dynamics is employed	Lack of key sensitivity analysis and insufficient security
Wang et al. [16]	The S-box performance optimization problem is transformed into a TSP problem solved by genetic algorithm	S-box optimization takes too much time
Khan et al. [17]	Chaotic Boolean functions are used	Low nonlinearity
Çavuşoğlu et al. [18]	Scaled Zhongtang chaotic system is used to produce PRNG for S-box generation	Low nonlinearity
Wang et al. [19]	3D chaotic systems without equilibrium are adopted	Low nonlinearity
Tian et al. [20]	A system called FLDSOP is used	Low nonlinearity

Adleman [22] first applied the concept of DNA to computing in 1994, and cryptography based on DNA computing has gradually become a branch of informatics. The benefits of DNA computing are high speed, minimum storage requirements and minimum power requirements. When it comes to security, Tornea et al. [23] pointed out that DNA encryption is a recent trend, which is developed on the basis of DNA Computing. In 2018, Namasudra et al. [24] discussed the security, privacy, trust and anonymity in DNA computing. Many DNA-based encryption algorithms have been proposed [25–33]. In order to exploit the advantages of chaos and DNA computing in cryptography, joint cryptography based on DNA computing and chaotic system has become an exciting research direction [34–39]. In 2010, Zhang et al. [35] studied a scheme for image encryption using DNA sequence operation and chaos. Subsequently, in 2013, Hermassi et al. [36] pointed out that this scheme is irreversible and cannot resist the chosen plaintext attack. Enayatifar et al. [37] presented a hybrid image encryption model based on genetic algorithm and chaotic system. However, this algorithm does not support strong security. In 2017, Mondal et al. [38] proposed a light-weight image secure encryption scheme based on DNA computing and cross-coupled chaotic Logistic map. Low dimensional Logistic map leads to the complexity of chaotic sequences and the security of the scheme is not enough. In 2018, Wang et al. [39] and Rehman et al. [40] proposed an image encryption scheme based on chaotic system and DNA sequence. In order to resist the chosen plaintext attack, both schemes use SHA-256 hash function to disturb the initial value and control parameters of chaotic system, but the efficiency of image encryption is still not high. Further details of these schemes [35,37–40] will be discussed in Section 2.2 and summarized in Table 2.

To overcome the drawbacks of chaos-based S-box and DNA-based encryption schemes mentioned above, a construction scheme for designing cryptographically strong S-box based on the combination of chaos and DNA

Table 2 Key findings of the DNA-based encryption schemes.

Schemes	Key findings	Disadvantages
Zhang et al. [35]	Image encryption scheme based on DNA operation and chaotic Logistic map	Cryptosystem is non-invertible and may not resist the chosen plaintext attack
Enayatifar et al. [37]	Genetic algorithm and chaotic function are used to determine the best DNA mask for image encryption	Proposed scheme does not support strong security
Mondal et al. [38]	A light-weight secure cryptographic scheme based on cross-coupled chaotic Logistic map and DNA computation	Complexity of chaotic sequences and security of the scheme are not enough
Wang et al. [39]	Based on CML and DNA sequences, SHA-256 is used to enhance the ability to resist chosen plaintext attacks	Efficiency of proposed scheme for image encryption is still not high
Rehman et al. [40]	The substitution for each pixel of a channel is achieved by XOR operation with the DNA complementary rule	Proposed algorithm consumes time to encrypt color image

computing is presented in this chapter. First, the 128 bit key is transformed to obtain the initial conditions and parameters, which are applied to a hyperchaotic system with four wing attractors to generate chaotic sequences with complex dynamic characteristics. Second, DNA complementary rules are randomly selected to encode these chaotic sequences to obtain three DNA codes, which will be used to drive two independent PWLCM systems and generate the required S-boxes through exclusive OR operations. Although chaos-based S-box has made some achievements, as far as we know, none of the chaos-based S-box can achieve the high performance which is used in Advanced Encryption Standard (AES). The performance gap between chaos-based S-box and classical S-box still exists. Therefore, the performance of chaos-based S-box needs to be further improved. In this chapter, one of the main motivations of proposed scheme is to use hyperchaotic system to realize more complex random sequences, and fully combine the characteristics of DNA complementary rules to generate strong S-boxes, which is expected to have better security performance, and can be applied to data encryption, user authentication and system access control, etc. The main contributions of this chapter are summarized as follows:

(1) A scheme of S-box generation based on the combination of DNA computing and chaotic system is introduced in this chapter.

(2) A hyperchaotic system with four wing attractors is adopted, and the 128-bit secret key is applied to generate the initial values and iterative parameters needed to drive the hyperchaotic system, which enhances the security performance of the scheme and improves the high sensitivity to the secret keys.

(3) DNA complementary rules are randomly selected to encode hyperchaotic sequences, which greatly increases the dynamic performance of the scheme.

(4) Two DNA coding values are used to drive two independent PWLCM, and then the required S-box elements can be obtained by exclusive-OR (XOR) operation between the output chaotic sequence and the remaining third DNA coding value, which not only increases the complexity of the scheme, but also improves the efficiency of the algorithm.

(5) Experimental results and performance analysis of the proposed scheme are presented to prove that the S-boxes generated by the scheme has ideal cryptographic properties and can be applied in different security requirement scenarios.

The remainder of this chapter is organized as follows. In Section 2, some related literature about the chaos-based S-box schemes and DNA-based encryption schemes are reviewed first. Then, in Section 3, the chaotic systems and DNA computing which will be adopted in the proposed scheme are introduced. The details of S-box generation scheme combining chaotic system and DNA computing will be described in Section 4. Furthermore, in Section 5, experiments are carried out on the proposed S-boxes, and the cryptography characteristics such as bijection, nonlinearity, strict avalanche criterion, output bit independence criterion and differential approximation probability are analyzed, followed by performance comparison of proposed S-box with those chaos-based one in the literature. Finally, the conclusion and future work are summarized in Section 6.

2. Literature reviews

In this section, several encryption schemes based on chaotic system, DNA computing, and the combination of them are briefly described.

2.1 Chaos and cryptographic system

It is well known that chaos has good cryptographic characteristics, which is widely used in the design of information security system. The literature in this field contains a large number of studies based on chaotic cipher systems. Wang et al. [2], Kadir et al. [3], Yavuz et al. [4], Murillo-Escobar et al. [5] and Peng et al. [6] proposed a new chaotic encryption system, pseudo-random number generator (PRNG) or Hash function in their studies. Recently, quantum chaos has attracted much attention for cryptosystem design due to its excellent cryptographic properties [7–11].

In 2005, Tang and Liao et al. [12] proposed a scheme to obtain cryptographically strong dynamic S-boxes based on iterating discretized chaotic map. The method consists of two stages: First, by iterating Logistic chaotic map, the output real number is converted into 8 bits of decimal integer in the range of 0–255, and then a real number table is generated. Second, the nonlinear permutation of the table is carried out through a 2D Baker map, and finally the desirable S-box is obtained. Although the scheme is efficient, due to the use of low dimensional chaotic map, the nonlinearity of S-box is not very high. Furthermore, in 2007, Chen et al. [13] pointed out some implementation errors in Tang's scheme [12] and employed a 3D Baker map to overcome these errors. Although 3D map can generate more random sequences than 2D map, the S-box proposed by the improved scheme shows the weak performance against differential cryptanalysis attack.

In 2010, Fatih and Ahmet [14] proposed a methodology to design cryptographically S-boxes based on continuous-time chaotic Lorenz system instead of chaotic maps. The algorithm is enhanced by shifting columns and rows of S-box, which makes the performance of S-box no longer depend on chaotic system alone. The results show that proposed cryptosystem using the designed S-boxes is very suitable for secure communication. The use of three-dimensional Lorenz chaotic system improves the randomness and complexity of the sequences, but the nonlinearity of S-box generated by the scheme is still not high, which is similar to Tang's scheme. Subsequently, in 2013, Fatih et al. [15] also studied an S-box design algorithm based on time-delay chaotic systems. Compared with other algorithms in literature, the proposed algorithm is considered to be more useful according to the criteria such as simple and efficient implementation. However, due to the lack of key sensitivity analysis, the security is slightly insufficient.

In 2012, Wang et al. [16] represented a method to design S-box based on chaos and genetic algorithm by making full use of the traits of chaotic map and evolution process. One of the highlights is to transform the construction problem of S-box into a TSP, aiming at the subsequent optimization of S-box performance. The problem is that optimization can take too much time. In 2016, Khan et al. [17] studied a construction method for designing S-box by using chaotic boolean functions and applied the obtained S-box to encrypt image. Measurable analysis of the proposed framework shows that the encryption quality and security of the framework are improved. Unfortunately, the nonlinearity of the proposed S-box is very low, and its performance against differential analysis is weak.

In 2017, Çavuşoğlu et al. [18] represented a novel method for designing strong S-box generation algorithm using PRNG produced by a chaotic scaled Zhongyang system. Performance tests show the proposed S-box is stronger and more effective. In 2019, Wang et al. [19] used a new 3D chaotic systems without equilibrium to construct S-boxes, and the experiment results indicate that S-box based encryption algorithm can be used safely in image encryption operations. Recently, in 2020, Tian et al. [20] presented a system called FLDSOP to design S-box. First, a preliminary binary S-box is constructed using six dimensional fractional Lorenz Duffing chaotic system. Second, an O-path scrambling scheme is designed to scramble the elements in the S-box. Experimental results have shown that the proposed S-box can effectively resist to multiple types of cryptanalysis attacks. However, there is still a problem that the nonlinearity of the above proposed S-boxes in [18–20] is relatively low, which leads to the poor performance of against differential analysis. Table 1 summarizes the key findings of the Chaos-based S-box schemes.

For S-box performance optimization, recently in 2020, Hematpour et al. [21] studied an optimization method for generating S-box based on chaotic map and particle swarm optimization (PSO). First, the performance of PSO was improved using ergodic chaotic maps. Second, the improved PSO was used for optimization to obtain the best S-boxes. With the help of this method, the performance of S-box is greatly improved.

2.2 DNA computing and cryptographic system

With the rapid popularity of mobile applications, a large number of personal sensitive information is spread on the network, resulting in the need for secure and reliable encryption algorithm for privacy protection. With the gradual development of big data and cloud computing technology, the computing power of computer is constantly enhanced, and the traditional encryption algorithm has been greatly challenged. In addition to the application of chaotic system introduced in Section 2.1 to information security, the application of DNA computing to the design of cryptographic system has become a new and promising field in recent years.

DNA was first observed by a German biochemist named Frederich Miescher in 1869. Watson, Crick and Wilkins were awarded the Nobel Prize in Medicine in 1962 "for their discoveries concerning the molecular structure of nucleic acids and its significance for information transfer in living material." Since the discovery of DNA double helix structure in the last century, Adleman [22] first applied the concept of DNA to computation in 1994, and cryptography based on DNA computing has gradually become an important branch of informatics in the future. DNA cryptography is a new technology based on traditional cryptography, which is more secure and reliable than traditional encryption methods. It will inject new vitality into modern encryption system.

Xiao et al. [25] gives a comprehensive introduction on the biological background of DNA cryptography and the principle of DNA computing, summarizes the research progress and several key problems of DNA cryptography, and discusses the security and application fields of DNA cryptography with those of traditional cryptography and quantum cryptography. Soni et al. [26] studies a new image encryption algorithm based on DNA sequence addition operation. In this algorithm, the image is divided into several blocks, and each block is encoded by a DNA sequence addition operation. Finally, DES is used to encrypt the encoding result, so as to get the final encrypted image. In literature [27], Wang et al. realized a reversible data hiding algorithm based on the combination of histogram correction algorithm and DNA computing. The experimental results show that the embedding rate of this algorithm has been significantly improved and is very suitable for image copyright protection. Chen et al. [28] used rand function to generate random DNA coding rules. Random DNA coding may obtain a balanced DNA distribution in the encoded DNA matrix, while hackers may not be able to obtain any useful information about the statistical characteristics of original images. However, in Chen's scheme, different original images may have the same DNA coding rules, which makes the encryption algorithm vulnerable to statistical attacks.

Shujaa and Hussein [29] presented an encryption-decryption scheme based on stream ciphers. In this scheme, the linear feedback shift register (LFSR) is used as a pseudo digital key generator to generate the required key sequence, and the messages transmitted between nodes of the Internet of Things are encrypted using One Time PAD (OTP) and DNA computing. In 2020, Pavithran et al. [30] proposed a new cryptosystem based on DNA cryptography and finite automata theory, which is characterized by using randomly generated mealy machine to encode DNA sequences, improving the security of ciphertext. In the same year, Namasudra et al. [31–34] presented a DNA based data encryption scheme for cloud computing environment. Based on DNA computing, user attributes and media access control (MAC) address, a 1024-bit key is generated. DNA base and complementary rules are used to prevent many security attacks. Experimental results and theoretical analysis show that the proposed scheme is more efficient and effective than some well-known existing schemes.

Due to the high sensitivity of chaotic system to initial conditions and system parameters, this will bring potential high security performance to the encryption system designed based on chaos theory. At the same time, because of its strong storage capacity, low energy consumption and high parallelism of DNA molecules, joint cryptography based on DNA computing and chaotic system has become a new research direction.

In 2010, Zhang et al. [35] presented a scheme for image encryption using DNA sequence operation and chaos theory. The main idea is to encode the original image by DNA sequence first, then divide the encoded image into equal blocks for DNA addition operation. Then, chaotic Logistic mapping and DNA complementary operation are carried out to obtain the final encrypted image through DNA decoding. However, unfortunately, Hermassi et al. [36] pointed out that this scheme is non-invertible by demonstrating a failure of the decryption process. In addition, a chosen plaintext attack on the invertible part of this cryptosystem is performed.

In 2014, Enayatifar et al. [37] studied a hybrid model based on genetic algorithm and chaotic function to determine the best DNA mask for image encryption. In the first stage of the algorithm, multiple DNA masks are generated by using logistic map and DNA sequence. In the second stage, the genetic algorithm determines the best DNA mask in the evolution process. This algorithm does not support strong security.

In 2017, Mondal et al. [38] proposed a lightweight image secure encryption scheme based on DNA computing and chaotic map. In this scheme, two pseudo-random number sequences are generated based on cross-coupled chaotic Logistic map, and then random DNA sequences are generated to scramble and encrypt the image. This algorithm can also be extended to encrypt text and color images. Low dimensional Logistic map leads to the complexity of chaotic sequences and the security of the scheme is not enough.

In 2018, Wang et al. [39] studied an image encryption scheme based on Coupled Map Lattice (CML) and DNA sequence. First, the original image is encoded into a DNA matrix, then some specific lines of the matrix are circularly shifted, and the matrix is further diffused and scrambled by DNA sequences generated by CML and DNA calculation rules. Finally, the encrypted image is obtained by DNA decoding. Here, the parameters and initial values of CML are generated by SHA-256 algorithm combined with the given key. DNA matrix scrambling operation can be parallelized by multithreading technology, but the efficiency of this scheme for image encryption is still not high.

In 2018, Rehman and Liao et al. [40] proposed a novel color image encryption algorithm by employing XOR with DNA complementary rules based on chaotic system and SHA-256. In order to resist the chosen plaintext attack, SHA-256 hash function is used to disturb the initial value and control parameters of the chaotic system, and then XOR operation is carried out through the DNA complementary rules and pixels, and each color image pixel is encoded into the DNA base independently, in which the selection of DNA rules is chaotic. However, due to three chaotic systems and a large number of DNA sequence operations, the efficiency of the algorithm is relatively low.

Table 2 summarizes the key findings of the DNA-based encryption schemes. The above results provide a good reference value for designing S-box generation algorithm based on chaos and DNA computing.

3. Background studies
3.1 Chaotic systems

There are many chaotic systems that can be used in cryptography, such as Logistic map [42], Chebyshev map [43], Baker map [44], Henon map [45], Rossler system [46,47], cellular neural network (CNN) [48], Lorenz system [10,41,49], time-delay chaotic system [15,50] and spatiotemporal chaotic system [51,52].

It has been found that most low dimensional chaotic maps are simple in structure and have two main problems. First, the chaotic sequence generated by low dimensional simple chaotic map has a short period, and second, the key space of low dimensional chaotic map is small. However, high dimensional hyperchaotic systems are more suitable for cryptographic applications due to their multiple positive Lyapunov exponents, strong spatiotemporal complexity and mixing. Hence, in this chapter, a new hyperchaotic system with four-wing attractors [53,54] is employed. The system is defined as follows:

$$
\begin{cases}
\dot{x} = 10(y - x) + u \\
\dot{y} = 28x - y - xw^2 - v \\
\dot{w} = k_1 xyw - k_2 w + k_3 x \\
\dot{u} = -xw^2 + 2u \\
\dot{v} = 8y
\end{cases}
\tag{1}
$$

where k_1, k_2 and k_3 are control parameters, $\dot{x}$ means the derivative of x with respect to the independent variable. For the convenience of description, the system defined according to Eq. (1) is hereinafter referred to as system (1). Hu et al. has proved in Refs. [53,54] that this system is in hyperchaotic state with four wing attractors when $k_1 = 1$, $k_2 = 4$ and $k_3 > 0.4$. Fig. 1 shows the hyperchaotic attractors of system (1) for parameter $k_1 = 1$, $k_2 = 4$ and $k_3 = 1.2$ and initial condition $(x_0, y_0, w_0, u_0, v_0) = (-1, -1, -1, -1, -1)$. The trajectories of the attractors are always entangled repeatedly in a limited phase space and never intersect. The inherent properties of hyperchaotic system are very useful for the purpose of cryptographic application.

Furthermore, Fig. 2A describes a time series diagram of state variable x, y, and w of system (1), and Fig. 2B describes the corresponding power spectral

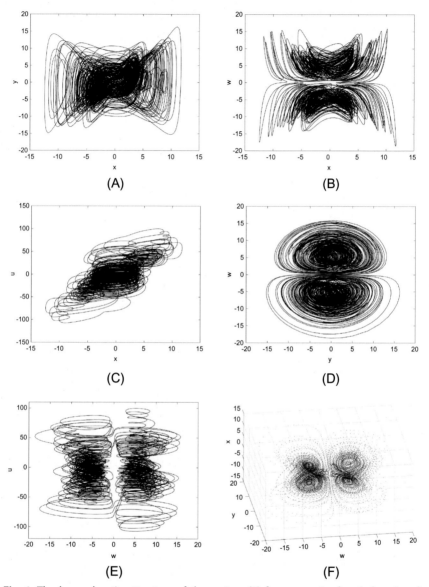

Fig. 1 The hyperchaotic attractors of the system (1) for parameter $k_1 = 1$, $k_2 = 4$ and $k_3 = 1.2$ and initial condition $(x_0, y_0, w_0, u_0, v_0) = (-1, -1, -1, -1, -1)$. (A) x-y plane; (B) x-w plane; (C) x-u plane; (D) y-w plane; (E) w-u plane; (F) w-y-x plane.

density (PSD) gained by the Fourier transform of the correlation function, and the difference is that the power spectra of these variables are uninterrupted, and there are no obvious troughs or peaks, which just proves the characteristics of chaos.

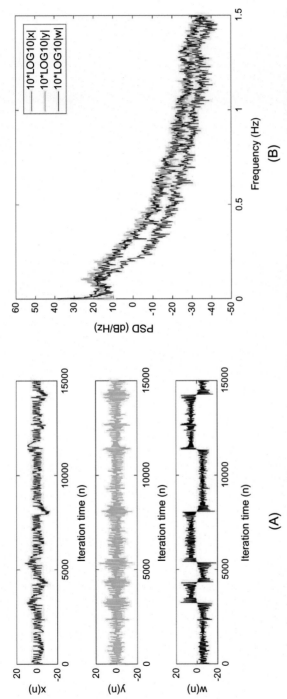

Fig. 2 Time series of the system (1) for parameter $k_1 = 1$, $k_2 = 4$ and $k_3 = 1.2$ and initial condition $(x_0, y_0, w_0, u_0, v_0) = (-1, -1, -1, -1, -1)$. (A) time series; (B) power spectral density.

Another chaotic system used by this chapter is PWLCM which is a chaotic map composed of multiple linear segments. Given an interval $X = [\alpha, \beta] \subset \mathbb{R}$, consider the following Piecewise Linear Map (PWLM) [55]:

$$F : X \to X, \quad i = 1 \sim m, F(x)|C_i = F_i(x) = a_i x + b_i \tag{2}$$

where $\{C_i\}_{i=1}^m$ is a partition of X, which satisfies $\cup_{i=1}^m C_i = X$ and $\forall i \neq j$, $C_i \cap C_j = \varnothing$. If each linear segment is mapped onto X by F_i: $\forall i = 1 \tilde{m}$, $F_i(C_i) = X$, then the above Eq. (2) satisfies piecewise *onto* property.

The Studies by Li and Chen et al. [55] show that PWLM with piecewise *onto* property has many excellent dynamic properties, including ergodicity, random-like behavior, large positive Lyapunov exponent, uniform invariant density function and exponential decay auto-correlation function on its defining interval X:

(a) its Lyapunov exponent $\lambda = -\sum_{i=1}^m \|C_i\| \cdot \ln \|C_i\|$ and satisfies $0 < \lambda < \ln m$;

(b) it is exact, mixing and ergodic;

(c) it has a uniform invariant density function, $f(x) = 1/\|X\| = 1/(\beta - \alpha)$;

(d) its auto-correlation function $\tau(n) = \frac{1}{\sigma^2} \lim_{N \to \infty} \frac{1}{N} \sum_{i=0}^{N-1} (x_i - \bar{x})(x_{i+n} - \bar{x})$

tends to zero as $n \to \infty$, where $\bar{x}$, σ are the mean value and the variance of x, respectively; especially, if $\sum_{i=1}^m \text{sign}(a_i) \cdot \|C_i\|^2 = 0$, then $\tau(n) = \delta(n)$.

These properties are very useful for applications in chaotic cryptography and chaotic PRNG. In the following, without loss of generality, the term PWLCM is used to represent the above chaotic PWLM. In this chapter, especially, the PWLCM is described as follows:

$$z_{n+1} = f_\mu(z_n) = \begin{cases} z_n \cdot \dfrac{1}{\mu}, & z_n \in [0, \mu) \\[2mm] (z_n - \mu) \cdot \dfrac{1}{0.5 - \mu}, & z_n \in [\mu, 0.5) \\[2mm] f_\mu(1 - z_n), & z_n \in [0.5, 1] \end{cases} \tag{3}$$

where $z_n \in [0, 1]$, μ is a control parameter. For the convenience of description, the system defined according to Eq. (3) is hereinafter referred to as system (3). It's found that the system (3) is in chaotic state when $\mu \in (0, 0.5)$. The Fig. 3A and B are the PWLCM chaotic sequences of z_n when $z_0 = 0.251$ and $\mu = 0.286$, and $z_0 = 0.252$ and $\mu = 0.285$, respectively, and from Fig. 3C it is found that these two sequences are quite difference with a tiny change to initial condition and control parameter.

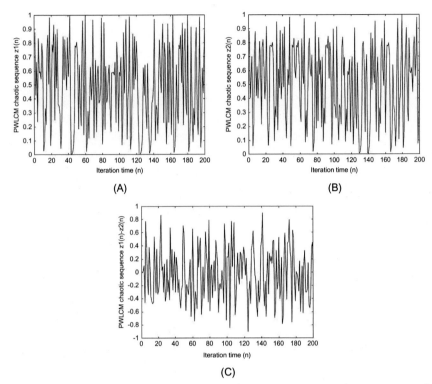

Fig. 3 The PWLCM chaotic sequences with (A) $z_0 = 0.251$ and $\mu = 0.286$, and (B) $z_0 = 0.252$ and $\mu = 0.285$, respectively. (C) is the difference of (A) and (B).

3.2 DNA computing

A DNA sequence contains four nucleic acid bases, A (adenine), C (cytosine), G (guanine) and T (thymine). In 1953, Watson and Crick [56] published an article in Nature defining the principle of complementary base pairing, also known as the complementary rule. According to this complementary rule, the purine adenine (A) always pairs with the pyrimidine thymine (T), and the pyrimidine cytosine (C) always pairs with the purine guanine (G). It can be concluded that A and T are complementary, and C and G are complementary.

It is well known that in the binary system, 0 and 1 are complementary. So it follows that 00 and 11 are complementary, and 01 and 10 are complementary. There are 24 mapping rules for DNA coding and decoding in total, but only eight of them satisfy the Watson-Crick complementary rule [56]. Table 3 shows these eight kinds of mapping rules.

Table 3 Eight kinds of DNA sequences mapping rules.

DNA base	Mapping rule number							
	1	2	3	4	5	6	7	8
A	00	00	11	11	10	01	10	01
T	11	11	00	00	01	10	01	10
C	10	01	10	01	00	00	11	11
G	01	10	01	10	11	11	00	00

Table 4 DNA complementary rule.

Rule number	Complementary rule
1	(AT) (TC) (CG) (GA)
2	(AT) (TG) (GC) (CA)
3	(AC) (CT) (TG) (GA)
4	(AC) (CG) (GT) (TA)
5	(AG) (GT) (TC) (CA)
6	(AG) (GC) (CT) (TA)

In this chapter, the sixth mapping rule is adopted, that is, C, A, T, and G represent the binary strings 00, 01, 10, and 11, respectively. Therefore, any 8-bit character can be encoded as a nucleotide string. For example, the decimal value 156 can be expressed in binary as 10011100, and its corresponding DNA sequence is TAGC.

Suppose n_i represents each nucleotide in the nucleotide string, then the following complementary rule described in Eq. (4) must be satisfied:

$$n_i \neq B(n_i) \neq B(B(n_i)) \neq B(B(B(n_i))), n_i = B(B(B(B(n_i)))) \qquad (4)$$

where $B(n_i)$ is the base pair of n_i. According to Eq. (4), there are six groups complementary rules, which are shown in Table 4 [57]:

In the subsequent process of DNA coding, different rules according to the specific algorithm design can be dynamically selected. For instant, the binary string of the number 201 is 11001001, which is first mapped to the DNA sequence GCTA according to the sixth mapping rule in Table 3, and then TGAC is obtained by executing the fourth DNA complementary rule randomly selected in Table 4, whose binary string is 10110100.

This means that the number 201 is eventually converted to the number 180 through a DNA coding transformation.

4. Proposed scheme

In this section, the design details of 8×8 S-box construction algorithm is given. Mathematically, an $n \times m$ S-box is a nonlinear mapping (or substitution) from V_n to V_m, where V_n and V_m represent the vector spaces of n and m tuples of elements from $GF(2)$, respectively. On the whole, the input of the algorithm is 128 bits secret key, and the output is 8×8 S-box. The generation algorithm will make full use of the inherent characteristics of chaotic sequences and combine with DNA computing to produce S-boxes that can meet the requirements of cryptography.

The schematic diagram of S-box generation system is shown in Fig. 4.

4.1 Parameters generation

Assume a given 128-bit random secret key $K = k_1 k_2 \cdots k_{16}$. Calculate the values of the following parameters, which will be used for subsequent S-box generation.

Calculate parameters $h_i (i = 1, 2, \cdots, 8)$, Gk according to the following Eqs. (5) and (6), respectively:

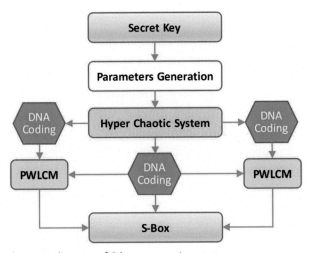

Fig. 4 The schematic diagram of S-box generation system.

$$h_i = \begin{cases} \sum_{j=1}^{4} k_{i+4 \times (j-1)} \bmod 8, & 1 \le i \le 4 \\ (k_{i-4} \times k_{i+4} + k_i \times k_{i+8}) \bmod 8, & 5 \le i \le 8 \end{cases} \tag{5}$$

$$Gk = (k_1 \oplus k_2 \oplus \cdots \oplus k_{16})^2 \bmod 256 \tag{6}$$

where $x \bmod y$ represents the remainder of x after dividing it by y, and $\oplus$ denotes bitwise XOR. The basic rule for generating the above parameters is to keep the correlation with the secret key as much as possible. In particular, the value of Gk is related to every character of the secret key. The main purpose of this is to make the final generated S-box highly sensitive to the secret key.

4.2 The Scheme of S-box generations

After parameters are generated, the S-box will be generated as follows.

Step 1: Initialization settings. Let S_{out} represent the output S-box of the scheme, and let $m = 1$, $S_{out} = \varnothing$.

Step 2: Hyperchaotic system iteration. The initial values and iteration number of the chaotic system (1) are determined based on the following condition and Eqs. (7)–(12) or Eqs. (13)–(18):

(a) If $m = 1$, let

$$x_0 = (k_1 \oplus k_3 \oplus k_5 \oplus k_7 \oplus Gk)^{\ll h_4}/256 \tag{7}$$

$$y_0 = (k_2 \oplus k_4 \oplus k_6 \oplus k_8 \oplus Gk)^{\ll h_3}/256 \tag{8}$$

$$w_0 = (k_9 \oplus k_{11} \oplus k_{13} \oplus k_{15} \oplus Gk)^{\ll h_2}/256 \tag{9}$$

$$u_0 = (k_{10} \oplus k_{12} \oplus k_{14} \oplus k_{16} \oplus Gk)^{\ll h_1}/256 \tag{10}$$

$$v_0 = Gk/256 \tag{11}$$

$$n = 10 + ((k_1 + k_4 + k_7) \times (k_2 + k_6 + k_8) \bmod 256)^{\ll h_5 \oplus h_6 \oplus h_7 \oplus h_8} \bmod 32 \tag{12}$$

(b) If $m > 1$, use the previous output to disturb the system (1) as follows.

$$x_0 \leftarrow x_0 + \tilde{s}_{out,m} \times (C_{x,n-1})^{\ll h_1}/256^2 \tag{13}$$

$$y_0 \leftarrow y_0 + \tilde{s}_{out,m} \times (C_{y,n-1})^{\ll h_2}/256^2 \tag{14}$$

$$w_0 \leftarrow w_0 + \tilde{s}_{out,m} \times (C_{w,n-1})^{\ll h_3}/256^2 \tag{15}$$

$$u_0 \leftarrow u_0 + \tilde{s}_{out,m} \times (C_{u,n-1})^{\ll h_4}/256^2 \tag{16}$$

$$v_0 \leftarrow v_0 + \tilde{s}_{out,m} \times (C_{v,n-1})^{\ll h_5 \oplus h_6}/256^2 \tag{17}$$

$$n \leftarrow 10 + \left[n \times \left(C_{x,n-1} \oplus C_{y,n-1} \right)^{\ll h_7 \oplus h_8} \mod 32 \right] \qquad (18)$$

where $R^{\ll h}$ means cyclic left-shift by h bits of R. According to the above parameters, n iterative operations are performed on the hyperchaotic system (1), and the obtained value is set as $C_{x,\,n}$, $C_{y,\,n}$, $C_{w,\,n}$, $C_{u,\,n}$, and $C_{v,\,n}$.

Step 3: DNA coding. The following DNA coding transformation is carried out by Eqs. (19)–(21):

$$DC_1 = DNA\_encode\left(C_{y,n} \oplus C_{x,n}, \mod\left(C_{x,n}, 6 \right) \right) \qquad (19)$$

$$DC_2 = DNA\_encode\left(C_{y,n} \oplus C_{w,n}, \mod\left(C_{w,n}, 6 \right) \right) \qquad (20)$$

$$DC_3 = DNA\_encode\left(C_{y,n} \oplus C_{u,n} \oplus C_{v,n}, \mod\left(C_{u,n} \times C_{v,n}, 6 \right) \right) \qquad (21)$$

where DNA_*encode* (p, q) represents DNA coding transformation function, which encodes binary string p through the q-th DNA complementary rule (see Table 4).

Step 4: PWLCM system iteration. The initial values, control parameters and iteration number of two PWLCM systems are determined based on the following Eqs. (22)–(24) (for left PWLCM system) or Eqs. (25)–(27) (for right PWLCM system):

$$z_{0,left} = (DC_1 \times DC_2)^{\ll h_7} / 256 \qquad (22)$$

$$\tilde{n}_{left} = 30 + (DC_1 + DC_2)^{\ll h_5} \mod 64 \qquad (23)$$

$$\mu_{left} = (DC_1 \oplus DC_2)^{\ll h_1 \oplus h_3} / 512 \qquad (24)$$

$$z_{0,right} = (DC_1 \times DC_2)^{\ll h_8} / 256 \qquad (25)$$

$$\tilde{n}_{right} = 30 + (DC_1 + DC_2)^{\ll h_6} \mod 64 \qquad (26)$$

$$\mu_{right} = (DC_1 \oplus DC_2)^{\ll h_2 \oplus h_4} / 512 \qquad (27)$$

The above initial condition, control parameter and iteration number are used to iterate the two PWLCM systems (left part and right part) in Fig. 4, and the values $\tilde{C}_{z,\tilde{n}}^{left}$ and $\tilde{C}_{z,\tilde{n}}^{right}$ are obtained, respectively.

Step 5: S-box element generation. The S-box elements are generated according to the following rules:

(a) Let $\tilde{s}_{out,m} = \tilde{C}_{z,\tilde{n}}^{left} \oplus \tilde{C}_{z,\tilde{n}}^{right} \oplus DC_3$;

(b) If $\#S_{out} < 256$ and $\tilde{s}_{out,n} \notin S_{out}$, then $S_{out} \leftarrow S_{out} \cup \tilde{s}_{out,m}$;

(c) If $\#S_{out} = 256$ then stop the algorithm. The desinged S-box, i.e., S_{out}, is obtained. Otherwise, let $m = m + 1$, go to **Step 2** to continue the algorithm.

Remark 1: In order to accelerate the generation of S-box, multiple chaotic sequence values can be taken simultaneously during the iteration of the PWLCM system (3).

Remark 2: When iterating system (3), if $DC_1 = 0$ or $DC_2 = 0$, then let $z_{0,\ left} = 0.5$ and $z_{0,\ right} = 0.5$.

In order to show the main idea of S-box generation scheme more clearly, the flow chart of the algorithm is shown in Fig. 5, and the detail is shown in Algorithm 1.

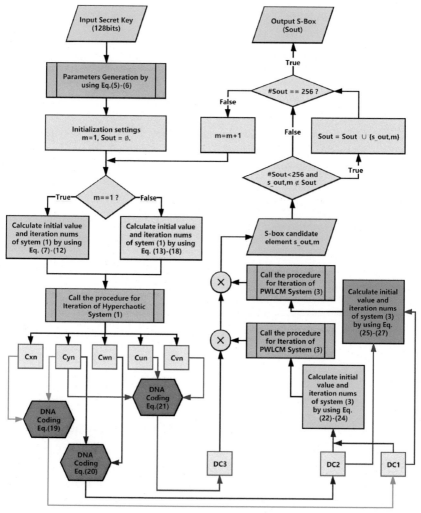

Fig. 5 The flowchart of the S-box generation algorithm.

Algorithm 1 S-box Generation Algorithm

Input: 128-bit secret key $\mathbf{K} = \mathbf{k_1 k_2 ... k_{16}}$

Output: 8×8 S-box $\mathbf{S_{out}}$

1. Parameters generation by Eqs.(5) and (6)
2. Initialization settings, let $m = 1$, $S_{out} = \varnothing$
3. **While** $\#S_{out} < 256$ **Do**
4.   **If** $m = 1$ **Then**
5.   Calculate initial values and iteration number of chaotic system (1) by Eqs. (7)–(12)
6.   **Else**
7.   Calculate initial values and iteration number of chaotic system (1) by Eqs. (13)–(18)
8.   **End if**
9.   Call the procedure for iteration of hyperchaotic system (1) to obtain the chaotic seuqunces C_{xn}, C_{yn}, C_{wn}, C_{un}, C_{vn}
10.   Perform DNA coding transformation by Eqs. (19)–(21) to obtain DC_1, DC_2, DC_3
11.   Calculate initial values and iteration number of PWLCM system (3) by Eqs. (22)–(24) (for left part) or Eqs. (25)–(27) (for right part)
12.   Call the procedure for iteration of PWLCM system (3) to obtain C_{left} and C_{right}
13.   Let $S_{out}^* = C_{left} \oplus C_{right} \oplus DC_3$
14.   **If** $S_{out}^* \notin S_{out}$ **Then**
15.   Let $S_{out} = S_{out} \cup \{S_{out}^*\}$
16.   **End if**
17.   Let $m = m + 1$
18. **End while**

---

5. Security analysis

How to measure the strength of S-box is always a difficult problem in block cipher design and analysis. S-boxes must satisfy certain cryptographic characteristics before they can be used in practical systems [58]. The following properties are chosen as the evaluation criteria for general cryptographically strong S-boxes. They include bijection property, nonlinearity property, differential approximation probability, strict avalanche criterion, output bit independence criterion, and sensitivity to secret key.

5.1 Experiment environment and result

Experiments have been implemented on ASUSPRO D540MA desktop computer with 3.20 GHz Intel Core i7-8700 CPU, 8GB RAM and 128GB SSD + 1 TB HDD with MS Windows 10 operating system, the programming platform is MATLAB R2016a (64bit). According to the algorithm flow chart shown in Fig. 5, the corresponding MATLAB script program is programmed and implemented. This program takes 128-bit secret key as input parameter and outputs corresponding S-box. Table 5 shows the obtained S-box corresponding to a 128-bit secret key "Kd8bT2bW#A5PsG." In addition, 1000 different S-boxes are randomly generated for the purpose of security analysis.

The usage of S-box is explained as follows: suppose the input of the S-box is a byte A, which can be represented in hexadecimal as xy, then the element in row x and column y of the S-box is the output of the S-box. For example, the hexadecimal representation of the decimal number 155 is 9b, from Table 5 it can found that the corresponding output is 59.

5.2 Bijection property

S-boxes are usually required to be reversible, especially the S-box used in substitution–permutation network (SPN) must be bijective. The following methods for checking bijective properties of S-box are given [59]. A Boolean function $f(x) = (f_1, f_2, \ldots, f_n)$ is bijective if it satisfies the following conditions:

$$wt(a_1 f_1 \oplus a_2 f_2 \oplus \cdots \oplus a_n f_n) = 2^{n-1} \tag{28}$$

where the $a_i \in \{0, 1\}$, $(a_1, a_2, \ldots, a_n) \neq (0, 0, \ldots, 0)$ and $wt(\cdot)$ is the Hamming weight. The above condition in Eq. (28) guarantees that any linear combination of boolean function f_i has Hamming weights $2^{n-1} (i = 1, 2, \ldots, n)$. In other words, the bijective property ensures that all possible 2^n n-bit input vectors map to different output vectors.

According to Eq. (28), it is found that the Hamming weight of all the obtained S-boxes are 128, which means that the bijectivity property is satisfied.

Table 5 8 × 8 S-box generated by the proposed scheme.

165	24	100	255	35	200	101	49	166	222	9	163	98	211	71	207
179	93	229	23	209	127	131	96	138	25	67	55	230	227	178	108
46	236	20	148	215	78	250	70	95	242	120	62	151	73	47	212
65	153	51	243	223	54	195	170	89	40	201	112	193	124	214	141
183	185	123	167	134	64	146	104	171	6	4	181	92	128	37	192
226	176	41	90	60	220	116	38	208	252	198	19	17	149	233	11
85	118	105	244	205	232	94	239	191	107	82	111	48	143	187	159
221	174	119	248	87	106	253	241	10	177	188	125	194	29	160	32
219	122	238	22	162	50	245	81	144	240	59	97	13	246	182	39
27	175	129	137	52	224	0	206	91	30	180	202	139	197	190	83
53	173	210	150	168	34	140	77	80	1	152	66	8	117	69	154
115	172	14	57	216	204	136	33	2	45	43	5	44	164	254	26
84	58	156	225	28	109	247	18	157	102	235	234	110	169	196	158
135	217	218	103	121	113	132	75	203	145	251	68	184	21	42	3
12	237	74	186	79	161	147	36	228	142	231	56	189	16	126	99
63	31	86	72	213	199	155	7	114	76	133	15	249	88	130	61

5.3 Nonlinearity property

In fact, the cryptographic properties of S-box directly affect the security of cryptographic algorithm. A "good" S-box requires a high degree of non-linearity, which means that there is no linear equation to generate S-box. The nonlinear criteria for boolean functions are classified according to their applicability to cipher design [60]. In general, the nonlinearity of boolean functions $f(x)$ can be expressed by the Walsh spectrum:

$$N_f = 2^{n-1}\left(1 - 2^{-n}\max_{\omega \in GF(2^n)}|S_{\langle f \rangle}(\omega)|\right) \tag{29}$$

The Walsh spectrum of $f(x)$ is defined by

$$S_{\langle f \rangle}(\omega) = \sum_{x \in GF(2^n)}(-1)^{f(x) \oplus x \cdot \omega} \tag{30}$$

where $\omega \in GF(2^n)$ and $x \cdot \omega$ denotes the dot-product of x and ω over $GF(2)$. The nonlinear property ensures that the S-box is not a linear mapping from the input vector to the output vector.

For the S-box in Table 5, according to Eqs. (29) and (30), the nonlinear value is calculated to be 109. In addition, the 1000 secret keys are randomly selected for the generation of 1000 S-boxes, and each secret key is composed of 16 visible characters except Spaces. The nonlinearity of these S-boxes is shown in Fig. 6. The maximum, minimum and average nonlinearity of these S-boxes are 109, 94 and 103.7520, respectively. In particular, 92.30% of the S-boxes have a nonlinearity between [100, 109], and only 0.016% of the S-boxes have a nonlinearity between [94, 96], indicating that most of S-boxes have desired nonlinearity property and are difficult to be approximated linearly by the cryptoanalysts.

5.4 Differential approximation probability

For an S-box, it should ideally have differential uniformity to resist the differential cryptanalysis, which means that an input differential Δx should uniquely map to an output differential Δy, thereby ensuring a uniform mapping probability for each x. The differential approximation probability is a measure for differential uniformity and is defined by Biham and Shamir [61] as:

$$DP_f = \max_{\Delta x \neq 0, \Delta y}\left(\frac{\#\{x \in X | f(x) \oplus f(x \oplus \Delta x) = \Delta y\}}{2^n}\right) \tag{31}$$

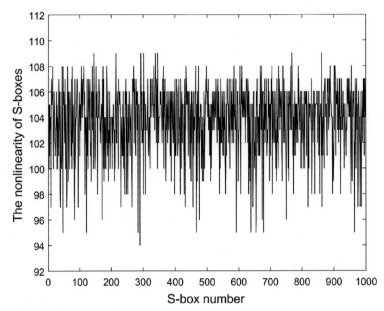

Fig. 6 The nonlinearity of 1000 S-boxes.

where X is the set of all possible input values, 2^n is the number of its elements, and # is the number of elements in X that satisfy specific condition. In fact, DP_f means the maximum probability of output differential Δy corresponding to the input differential Δx. The smaller the value of DP_f, the better the performance against differential cryptanalysis. For the S-box in Table 5, according to Eq. (31), the differential approximation probability (DP) matrix is calculated and listed in Table 6.

It can be found from Table 6 that the number 6 and number 8 are the most, 168 times and 71 times, respectively, and the maximum number is 10, indicating that DP_f value of the S-box in Table 5 is 0.03906 (i.e., 10/256). For all the 1000 S-boxes, the DP_f value of each S-box is calculated, and the results are shown in Fig. 7 and Table 7.

From the results in Table 7, it is found that 94.3% of DP_f have the values of 0.03906 and 0.04688 (<0.05), which indicates that these S-boxes have good resistance to differential cryptanalysis to a certain extent.

5.5 Strict avalanche criterion

SAC is introduced by Webster and Tavares [62] and is a very important criterion for the cryptographic properties of S-boxes. This criterion embodies the completeness and avalanche effect. Avalanche effect means

Table 6 Differential approximation probability matrix of the S-box in Table 5.

−	6	6	6	8	8	6	6	6	6	8	6	8	6	6	8
6	6	6	6	8	6	6	8	8	10	4	10	6	8	6	8
10	8	8	6	4	10	6	8	6	6	6	6	8	8	6	8
6	6	6	6	8	6	6	6	6	8	6	6	6	6	6	8
8	6	6	8	6	6	8	6	6	6	6	8	6	6	6	6
8	6	8	8	6	6	8	8	8	6	6	6	6	6	6	8
8	6	8	6	6	8	6	6	6	6	4	6	6	6	6	6
8	6	6	6	6	8	6	6	6	6	6	6	4	6	6	8
6	6	8	8	6	6	6	6	6	6	6	6	10	6	8	6
8	8	6	6	6	6	6	6	6	6	6	6	6	6	6	6
6	8	6	6	6	8	6	6	10	8	6	8	6	8	6	6
4	4	6	6	6	6	6	8	8	8	8	8	10	6	8	6
8	8	6	6	8	6	6	8	6	6	8	6	6	8	6	6
10	8	6	6	6	6	6	6	6	6	8	6	6	8	6	6
8	6	6	10	8	6	6	6	6	8	6	6	6	6	8	6
6	6	8	8	6	6	6	4	6	6	8	8	6	8	6	8

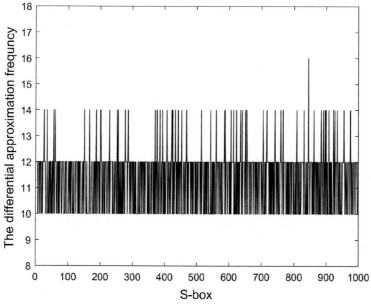

Fig. 7 The differential approximation frequency of 1000 S-boxes.

Table 7 The differential approximation probability.

Frequency	DP_f	Numbers (%)
10	0.03906	408 (40.8%)
12	0.04688	535 (53.5%)
14	0.05469	56 (5.6%)
16	0.06250	1 (0.1%)

Table 8 The dependence matrix of the S-box in Table 5.

0.4531	0.4219	0.5156	0.5313	0.4219	0.5156	0.4688	0.6250
0.5469	0.5000	0.5625	0.5313	0.5156	0.5938	0.5156	0.5156
0.5938	0.4531	0.4531	0.4531	0.4844	0.3750	0.5000	0.5313
0.4375	0.5156	0.4844	0.5313	0.4844	0.5000	0.5313	0.5313
0.5156	0.5000	0.5781	0.4531	0.4844	0.6406	0.5469	0.5781
0.5000	0.5313	0.5000	0.5469	0.5313	0.5156	0.5469	0.5156
0.5313	0.5000	0.5313	0.5781	0.4688	0.5000	0.5469	0.4375
0.5781	0.5625	0.5000	0.4063	0.5781	0.5156	0.4844	0.5000

that when one input bit of plaintext is complemented, each output bit of ciphertext should change with a half probability. Completeness means that every bit of plaintext must contribute to every ciphertext bit. If some bits of the ciphertext change only some bits of the plaintext, the cryptoanalyst can detect this relationship between the input and output, and use this relationship to search for the secret key with the chosen plaintext attack.

The dependence matrix is constructed to determine whether a given S-box meets the SAC. If the S-box satisfies SAC, then the mean of the dependent matrix is close to 0.5, that is, the value of each element in the dependent matrix must be close to half.

The dependence matrix of S-box in Table 5 is listed in Table 8 using the method in [62], and the minimum, maximum and average value are 0.3750, 0.6406 and 0.5125, respectively.

Besides, it is found that the mean value of the dependence matrix of 1000 S-boxes is in [0.4839, 0.5208] (see Fig. 8), which is close to 0.5, and the mean value of standard deviation is 0.0110, which indicates that all S-boxes have excellent SAC performance.

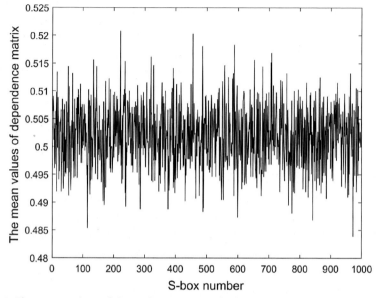

Fig. 8 The mean values of dependence matrix of 1000 S-boxes.

5.6 Output bit independence criterion

Another desirable feature of S-boxes is that they should meet the output bit independence criterion. This means that all avalanche variable pairs must be independent of the avalanche vector set generated by the inverse of a single plaintext bit [13]. The boolean function in the 8×8 S-box supposed as $f_1, f_2, ..., f_8$. If $F_i = f_j \oplus f_k$ is highly nonlinear and very close to the SAC function, and this ensures that each pair of output bits has a correlation as close to zero as possible when input bit is reversed. If BIC of f_j and f_k is satisfied, nonlinearity and SAC should also be satisfied.

The computational nonlinearity of S-boxes is shown in Fig. 6. The nonlinearities mean value of $f_j \oplus f_k$ is >100, and the mean value of $f_j \oplus f_k$ dependence matrix is close to 0.5 (see Table 8 and Fig. 8), indicating that all S-boxes basically meet the BIC performance requirements.

5.7 Sensitivity to secret key

In order to study the sensitivity of S-boxes to secret key, the correlation coefficient between S-boxes generated by two different keys can be observed. Suppose the key has a value of "Kd8bT2bW#A5PsG" and only one character in the key is modified at a time. For example, subtract 1 from the ASCII value of each character, and then 16 new keys that are slightly

Table 9 S-boxes correlation coefficient analysis results.

S-box number	Correlation coefficient	S-box number	Correlation coefficient
1	−0.0501	9	−0.0045
2	0.0144	10	−0.0045
3	−0.0718	11	0.0539
4	0.0470	12	−0.1002
5	0.0384	13	−0.0009
6	0.0720	14	0.0641
7	−0.0752	15	0.0299
8	−0.0866	16	0.0334

different from the original key are obtained. Then these 16 keys are used to generate 16 new different S-boxes. The correlation coefficient between each new S-box and the original S-box is calculated, and the results are shown in Table 9. The experimental data show that these correlation coefficients are very small, and the average value is only −0.0025, indicating that the proposed S-box scheme is very sensitive to secret keys.

This is due to one of the design concepts, such as parameter Gk in Eq. (6), which causes each character of secret key to participate in the generation of S-box.

6. Performance comparison

In this section, the performance of the S-boxes generated by the proposed algorithm with the chaotic S-boxes in the literature is compared. The comparison mainly focuses on three properties, namely, nonlinearity, dependence matrix and differential approximation probability. The results are shown in Table 10.

It is found that the S-box proposed in this chapter has the largest nonlinearity, followed by the S-box proposed in literature [16,21]. From the point of view of dependence matrix, the mean value of the S-box proposed is close to 0.5, and the performance is almost the same as other S-boxes. In terms of differential approximation probability, all of them have the same performance with smaller probability (3.906%) except the S-box generated in [13,17,47] with a larger probability (5.469%, 6.250% and 4.688%, respectively).

Table 10 The performance comparison of chaos-based S-box.

S-box	Nonlinearity	Dependence matrix			DP
		Average	Min	Max	
Tang et al. [12]	104	0.4971	0.4063	0.5781	10 (3.906%)
Chen et al. [13]	106	0.5000	0.4219	0.6094	14 (5.469%)
Fatih et al. [14]	104	0.5049	0.4219	0.5938	10 (3.906%)
Fatih et al. [15]	107	0.5061	0.4141	0.6094	10 (3.906%)
Wang et al. [16]	108	0.5068	0.4063	0.5781	10 (3.906%)
Khan et al. [17]	102	0.4812	0.1250	0.6250	16 (6.250%)
Çavuşoğlu et al. [18]	104	0.5039	0.4219	0.5938	10 (3.906%)
Wang et al. (S-box3) [19]	106	0.4917	0.3594	0.5781	10 (3.906%)
Tian et al. [20]	106	0.4976	0.4063	0.6250	10 (3.906%)
Hematpour et al. [21]	108	0.5056	0.4063	0.6406	10 (3.906%)
Belazi et al. [47]	102	0.4929	0.4063	0.5938	12 (4.688%)
Proposed S-box	109	0.5125	0.3750	0.6406	10 (3.906%)

Based on the above comparison results, the S-box proposed in this chapter has better security performance. It should be noted that the algorithm in this chapter can generate a large number of S-boxes using different keys. In practice, in order to meet the requirements of high security, it is best to choose the S-box with good performance.

In order to make full use of the randomness of chaotic sequence and increase the complexity of the scheme, two chaotic systems are used when designing the S-box generation algorithm in this chapter. In addition, it is suggested to use at least one hyperchaotic system in the algorithm to increase the resist cryptanalysis ability of the system.

Furthermore, the sensitivity of the algorithm to the secret key may be enhanced by carefully selecting a variable through which all characters of the secret key participate in the operation, as shown in Eq. (6). In this way, the tiny change of each key character will significantly affect the result.

7. Conclusions and future work

In this chapter, a scheme for S-box generation is proposed based on the combination of chaos and DNA computing. In this scheme, a 128-bit secret key is first applied to generate the initial values and iterative parameters needed to drive the hyperchaotic system with four-wing attractors. Then the DNA sequences are generated according to the randomly selected DNA complementary rules and hyperchaotic sequences. Finally, two independent PWLCM systems are driven by DNA sequences, and the S-box is generated by XOR operation. The numerical results of S-boxes show that the characteristics of bijection, nonlinearity, SAC, BIC, differential approximation probability and key sensitivity are satisfied well. Compared with the S-boxes in the existing literature, the S-box proposed in this chapter has better security features. The scheme can generate S-boxes quickly in cryptographic systems to meet the security requirements of different application scenarios.

The novelty of the methodology and algorithm are that each character of the secret key participates in the generation of S-box, which greatly enhances the sensitivity of S-box to the secret key. In addition, the adoption of hyperchaotic system and the combined application of DNA computing increase the dynamic, complexity and randomness of S-box generation. Furthermore, the dynamic performance of the scheme is greatly improved by randomly selecting DNA complementarity rules to encode hyperchaotic sequences.

The limitation of the scheme lies in the periodicity of chaotic sequence in the process of digitization, and the degradation of random performance in the process of S-box generation is inevitable, which is an aspect that needs to be paid attention to and solved in the future work. A feasible way to improve the random performance is to apply XOR, addition and subtraction operations of DNA sequences in the scheme, rather than just using DNA complementary rules. In a word, there are many ways to explore the application of DNA computing in cryptography design and there are still many contents worthy of further research and experiment.

Acknowledgments

The presented work is partially funded by the National Science and Technology Major Project (No. 2016ZX05060), the National Natural Science Foundation of China (No. 61873043), the Science and Technology Research Program of Chongqing Municipal Education Commission (No. KJ1713329), the Natural Science Foundation of Chongqing (No. cstc2019jcyjmsxmX0355 and No. cstc2018jcyjAX0048), the Scientific Research Fund of Chongqing University of Science and Technology (No. ckzg201914), the Cooperation Project between Chongqing Municipal Undergraduate Universities and Institutes Affiliated to the Chinese Academy of Sciences in 2021 (No. HZ2021015), the Postgraduate Innovation Project of CQUST (No. 2020833 and No. 2020838). The authors would like to thank the anonymous reviewers for their valuable suggestions and comments.

References

[1] B. Schneier, Applied Cryptography: Protocols, Algorithms, and Source Code in C (Second Edition), Wiley Inc., New York, NY, 1996, p. 758.

[2] Y. Wang, K. Wong, X. Liao, G. Chen, A new chaos-based fast image encryption algorithm, Appl. Soft Comput. J. 11 (1) (2009) 514–522.

[3] A. Kadir, A. Hamdulla, W. Guo, Color image encryption using skew tent map and hyper chaotic system of 6th-order CNN, Optik 125 (5) (2014) 1671–1675.

[4] E. Yavuz, R. Yazici, M.C. Kasapbaşi, E. Yamaç, A chaos-based image encryption algorithm with simple logical functions, Comput. Electr. Eng. 54 (2016) 471–483.

[5] M.A. Murillo-Escobar, C. Cruz-Hernández, L. Cardoza-Avendaño, R. Méndez-Ramírez, A novel pseudorandom number generator based on pseudorandomly enhanced logistic map, Nonlinear Dyn. 87 (1) (2017) 407–425.

[6] J. Peng, D. Zhang, Y. Liu, X. Liao, A Double-Piped iterated hash function based on a hybrid of chaotic maps, in: *Proceedings of the 7th IEEE International Conference on Cognitive Informatics*, IEEE, Stanford, CA, USA, 2008, pp. 358–365.

[7] A.A. Abd El-Latif, L. Li, N. Wang, Q. Han, X. Niu, A new approach to chaotic image encryption based on quantum chaotic system, exploiting color spaces, Signal Process. 93 (11) (2013) 2986–3000.

[8] A. Akhshani, A. Akhavan, A. Mobaraki, S.C. Lim, Z. Hassan, Pseudorandom number generator based on quantum chaotic map, Commun. Nonlinear Sci. Numer. Simul. 19 (1) (2014) 101–111.

[9] D. Lambić, Security analysis and improvement of the pseudo-random number generator based on quantum chaotic map, Nonlinear Dyn. 94 (2) (2018) 1117–1126.

[10] U. Arshad, S.I. Batool, M. Amin, A novel image encryption scheme based on Walsh compressed quantum spinning chaotic Lorenz system, Int. J. Theor. Phys. 58 (10) (2019) 3565–3588.

[11] K.K. Butt, G. Li, F. Masood, S. Khan, A digital image confidentiality scheme based on pseudo-quantum chaos and lucas sequence, Entropy 22 (11) (2020).

[12] G. Tang, X. Liao, Y. Chen, A novel method for designing S-boxes based on chaotic maps, Chaos Solit. Fractals 23 (2) (2004) 413–419.

[13] G. Chen, Y. Chen, X. Liao, An extended method for obtaining S-boxes based on three-dimensional chaotic baker maps, Chaos Solit. Fractals 31 (3) (2007) 571–579.

[14] Ö. Fatih, B.Ö. Ahmet, A method for designing strong S-boxes based on chaotic Lorenz system, Phys. Lett. A. 374 (36) (2010) 3733–3738.

[15] Ö. Fatih, Y. Sırma, Designing chaotic S-boxes based on time-delay chaotic system, Nonlinear Dyn. 74 (3) (2013) 551–557.

[16] Y. Wang, K.W. Wong, C. Li, Y. Li, A novel method to design S-box based on chaotic map and genetic algorithm, Phys. Lett. A. 376 (6) (2012) 827–833.

[17] M. Khan, T. Shah, S.T. Batool, Construction of S-box based on chaotic Boolean functions and its application in image encryption, Neural Comput. Applic. 27 (3) (2016) 677–685.

[18] Ü. Çavuşoğlu, A. Zengin, I. Pehlivan, S. Kaçar, A novel approach for strong S-box generation algorithm design based on chaotic scaled Zhongtang system, Nonlinear Dyn. 87 (2) (2017) 1081–1094.

[19] X. Wang, Ü. Çavuşoğlu, S. Kacar, A. Akgul, V.T. Pham, S. Jafari, F.E. Alsaadi, X.Q. Nguyen, S-box based image encryption application using a chaotic system without equilibrium, Appl. Sci. 9 (2019) 781–798.

[20] Y. Tian, Z. Lu, Chaotic S-box: six-dimensional fractional Lorenz-duffing chaotic system and O-shaped path scrambling, Nonlinear Dyn. 94 (3) (2018) 2115–2126.

[21] N. Hematpour, S. Ahadpour, Execution examination of chaotic S-box dependent on improved PSO algorithm, Neural Comput. Applic. (2020) 1–23.

[22] L.M. Adleman, Molecular computation of solutions to combinatorial problems, Science 266 (5187) (1994) 1021–1024.

[23] O. Tornea, M.E. Borda, DNA cryptographic algorithms, in: *IFMBE Proceedings of International Conference on Advancements of Medicine and Health Care through Technology 2009*, 26, Springer, 2009, pp. 223–226.

[24] S. Namasudra, D. Devi, S. Choudhary, R. Patan, S. Kallam, Security, privacy, trust, and anonymity, in: S. Namasudra, G.C. Deka (Eds.), Advances of DNA Computing in Cryptography, Taylor & Francis, 2018, pp. 138–150.

[25] G. Xiao, M. Lu, L. Qin, X. Lai, New field of cryptography: DNA cryptography, Chin. Sci. Bull. 51 (12) (2006) 1413–1420.

[26] R. Soni, A. Johar, V. Soni, in: An Encryption and Decryption Algorithm for Image Based on DNA, *Proceedings of the IEEE 2013 International Conference on Communication Systems & Network Technologies*, IEEE, Gwalior, India, 2013, pp. 478–481.

[27] B. Wang, Y. Xie, S. Zhou, C. Zhou, X. Zheng, Reversible data hiding based on DNA computing, Comput. Intell. Neurosci. 2017 (2017) 1–9.

[28] J. Chen, Z. Zhu, L. Zhang, Y. Zhang, B. Yang, Exploiting self-adaptive permutation-diffusion and DNA random encoding for secure and efficient image encryption, Signal Process. 142 (2018) 340–353.

[29] M.I. Shujaa, N.A. Hussein, DNA computing based stream cipher for internet of things using MQTT protocol, Int. J. Electr. Comput. Eng. 10 (1) (2020) 1035–1042.

[30] P. Pavithran, S. Mathew, S. Namasudra, P. Lorenz, A novel cryptosystem based on DNA cryptography and randomly generated mealy machine, Comput. Secur. 104 (2021) 102160.

[31] S. Namasudra, D. Devi, S. Kadry, R. Sundarasekar, A. Shanthini, Towards DNA based data security in the cloud computing environment, Comput. Commun. 151 (2020) 539–547.

[32] S. Namasudra, R. Chakraborty, A. Majumder, N.R. Moparthi, Securing multimedia by using DNA based encryption in the cloud computing environment, ACM Trans. Multimedia Comput. Commun. Appl. 16 (3s) (2020) 1–19.

[33] S. Namasudra, S. Sharma, G.C. Deka, P. Lorenz, DNA computing and table based data accessing in the cloud environment, J. Netw. Comput. Appl. 172 (15) (2020), 102835.

[34] S. Namasudra, Fast and secure data accessing by using DNA computing for the cloud environment, IEEE Trans. Serv. Comput. (2020) 1.

[35] Q. Zhang, L. Guo, X. Wei, Image encryption using DNA addition combining with chaotic maps, Math. Comput. Model. 52 (11–12) (2010) 2028–2035.

[36] H. Hermassi, A. Belazi, R. Rhouma, S.M. Belghith, Security analysis of an image encryption algorithm based on a DNA addition combining with chaotic maps, Multimed. Tools Appl. 72 (3) (2013) 2211–2224.

[37] R. Enayatifar, A.H. Abdullah, I.F. Isnin, Chaos-based image encryption using a hybrid genetic algorithm and a DNA sequence, Opt. Lasers Eng. 56 (2014) 83–93.

[38] B. Mondal, T. Mandal, A light weight secure image encryption scheme based on chaos & DNA computing, J. King Saud Univ. - Comput. Inf. Sci. 29 (4) (2017) 499–504.

[39] X. Wang, Y. Hou, S. Wang, R. Li, A new image encryption algorithm based on CML and DNA sequence, IEEE Access 6 (2018) 62272–62285.

[40] A.U. Rehman, X. Liao, R. Ashraf, S. Ullah, H. Wang, A color image encryption technique using exclusive-OR with DNA complementary rules based on chaos theory and SHA-256, Optik 159 (2018) 348–367.

[41] M.A. Ben Farah, R. Guesmi, A. Kachouri, M. Samet, A novel chaos based optical image encryption using fractional Fourier transform and DNA sequence operation, Opt. Laser Technol. 121 (2020).

[42] L. Kocarev, G. Jakimoski, Logistic map as a block encryption algorithm, Phys. Lett. A. 289 (4–5) (2001) 199–206.

[43] X. Huang, Image encryption algorithm using chaotic Chebyshev generator, Nonlinear Dyn. 67 (4) (2012) 2411–2417.

[44] J. Fridrich, Symmetric ciphers based on two-dimensional chaotic maps, Int. J. Bifurc. Chaos 8 (6) (1998) 1259–1284.

[45] Y. Zheng, J. Jin, A novel image encryption scheme based on Hénon map and compound spatiotemporal chaos, Multimed. Tools Appl. 74 (2015) 7803–7820.

[46] M.K. Mandal, M. Kar, S.K. Singh, V.K. Barnwal, Symmetric key image encryption using chaotic Rossler system, Secur. Commun. Netw. 7 (11) (2014) 2145–2152.

[47] A. Belazi, R. Rhouma, S. Belghith, A Novel Approach to Construct S-Box Based on Rossler System, in: *Proceedings of the International Wireless Communications and Mobile Computing Conference*, IEEE, Dubrovnik, Croatia, 2015, pp. 611–615.

[48] Q. Li, X.S. Yang, F. Yang, Hyperchaos in a simple CNN, Int. J. Bifurc. Chaos 16 (8) (2006) 2453–2457.

[49] J. Chen, L. Chen, L. Zhang, Z. Zhu, Medical image cipher using hierarchical diffusion and non-sequential encryption, Nonlinear Dyn. 96 (1) (2019) 301–322.

[50] J. Peng, X. Liao, Z. Yang, A novel feedback block cipher based on the chaotic time-delay neuron system and feistel network, in: *Proceedings of Intentional Conference on Communications Circuits and Systems*, III, IEEE, Guilin, China, 2006, pp. 1618–1622.

[51] K. Kaneko, Pattern dynamics in spatiotemporal chaos: pattern selection, diffusion of defect and pattern competition intermittency, Phys. D: Nonlinear Phenom. 34 (1–2) (1989) 1–41.

[52] P. Li, Z. Li, W.A. Halanga, G. Chen, A stream cipher based on a spatiotemporal chaotic system, Chaos Solit. Fractals 32 (5) (2007) 1867–1876.

[53] G. Hu, A family of hyperchaotic systems with four-wing attractors, Acta. Phys. Sin. 58 (6) (2009) 3734–3738.

[54] G. Hu, B. Yu, A hyperchaotic system with a four-wing attractor, Int. J. Mod. Phys. C 20 (2) (2009) 323–335.

[55] S. Li, G. Chen, X. Mou, On the dynamical degradation of digital piecewise linear chaotic maps, Int. J. Bifurc. Chaos 15 (10) (2005) 3119–3151.

[56] J.D. Watson, F.H.C. Crick, A structure for deoxyribose nucleic acid, Nature 171 (4356) (1953) 737–738.

[57] H. Liu, X. Wang, A. Kadir, Image encryption using DNA complementary rule and chaotic maps, Appl. Soft Comput. J. 12 (5) (2012) 1457–1466.

[58] C. Adams, S. Tavares, The structured design of cryptographically good S-boxes, J. Cryptol. 3 (1) (1990) 27–41.

[59] G. Jakimoski, L. Kocarev, Chaos and cryptography: block encryption ciphers based on chaotic maps, IEEE Trans. Circuits Syst. I. Fundam. Theory Appl. 48 (2) (2001) 163–169.

[60] W. Meier, Nonlinearity criteria for crytographic functions, in: Advances in Cryptology: Proceeding of Eurocrypt'89, Springer-Verlag, 1990, pp. 549–562.

[61] E. Biham, A. Shamir, Differential cryptanalysis of DES-like cryptosystems, in: Advances in Cryptology: Proceeding of Crypto'90, Springer-Verlag, 1991, pp. 2–21.

[62] A.F. Webster, S.E. Tavares, On the design of S-boxes, in: Advances in Cryptology: Proceedings of Crypto'85, Springer-Verlag, 1985, pp. 523–534.

About the authors

Jun Peng received a Ph.D. in Computer Software and Theory from Chongqing University in 2003, a MA in Computer System Architecture from Chongqing University in 2000, and a BSc in Applied Mathematics from the Northeast University in 1992. From 1992 to present he works at Chongqing University of Science and Technology, where he is currently a Professor and Dean in School of Mathematics, Physics and Data Science. He was a visiting scholar in the Laboratory of Cryptography and Information Security at Tsukuba University, Japan in 2004, and Department of Computer Science at California State University, Sacramento in 2007, respectively. He has authored or coauthored over 150 peer reviewed journal or conference papers. He has served as the program committee member or session co-chair for over 20 international conferences including IEEE ICCI*CC'11-22. His current research interests are on cryptography, chaos and network security, image processing and big data analysis.

Shangzhu Jin received the B.Sc. degree in computer science from Beijing Technology and Business University, China, and the M.Sc. degree in control theory and control engineering from Yanshan University, China, and the Ph.D. degree from Aberystwyth University, UK. He is currently an associate professor at the School of Intelligent Technology and Engineering, Chongqing University of Science and Technology. His research interests include fuzzy systems, approximate reasoning, and network security. His paper, entitled "Backward Fuzzy Interpolation and Extrapolation with Multiple Multi-antecedent Rules" has won the best student paper award at the 21th IEEE International Conference on Fuzzy Systems.

Dr. Yingxu Wang is professor of cognitive systems, brain science, software science, and intelligent mathematics. He is the founding President of International Institute of Cognitive Informatics and Cognitive Computing (I2CICC). He is FIEEE, FBCS, FI2CICC, FAAIA, and FWIF. He has held visiting professor positions at Univ. of Oxford (1995, 2018–2022), Stanford Univ. (2008, 2016), UC Berkeley (2008), MIT (2012), and distinguished visiting professor at Tsinghua Univ. (2019–2022). He received a PhD in Computer Science from the Nottingham Trent University, UK, in 1998 and has been a full professor since 1994. He is the founder and steering committee chair of IEEE Int'l Conference Series on Cognitive Informatics and Cognitive Computing (ICCI*CC) since 2002. He is founding Editor-in-Chiefs and Associate Editors of 10+ Int'l Journals and IEEE Transactions. He is Chair of IEEE SMCS TC-BCS on Brain-inspired Cognitive Systems, and Co-Chair of IEEE CS TC-CLS on Computational Life Science. His basic research has been across contemporary science disciplines of intelligence, mathematics, knowledge, robotics, computer, information, brain, cognition, software, data, systems, cybernetics, neurology, and linguistics. He has published 600+ peer reviewed papers and 38 books/proceedings. He has presented 63 invited keynote speeches in international conferences. He has served as honorary, general, and program chairs for 40 international conferences. He has led 10+ international, European, and Canadian research projects as PI. He is recognized by Google Scholar as world top 1 in Software Science, top 1 in Cognitive Robots, top 8 in Autonomous Systems, top 2 in Cognitive Computing, and top 1 in Knowledge Science with an h-index 60. He is recognized by ResearchGate as among the world's top 2.5% scholars with a remarkable readership record of 498,000+.

Xi Zheng, born in February 1997 in Sichuan, China, received her bachelor's degree in software engineering from Chongqing University of Technology in 2020. Since 2020, she has studied for a master's degree in resources and environment from Chongqing University of Science and Technology, studied the direction of artificial intelligence, and published an IEEE conference paper. Her current research interests are cryptography, chaos theory, deep learning and image processing.

Xiangren Wang was born in Henan, China. He obtained a bachelor's degree from Henan University of Science and Technology in 2020, and is currently studying for a master's degree at Chongqing University of Science and Technology. In 2021, he published an article on image encryption in IEEE ICESIT'21. His research interests include network security and image encryption.

CHAPTER SEVEN

DNA computing-based Big Data storage

Deepak Sharma and Manojkumar Ramteke
Department of Chemical Engineering, Indian Institute of Technology Delhi, New Delhi, India

Contents

Abstract

In the current digital age, the rate of digital data generation is growing exponentially. Though the capacity of conventional storage devices is continuously increasing, it is far from matching the current exponential growth rate of digital data generation. Further, there is an urgent need for a high-density and high-capacity medium to store the information for a prolonged period. Deoxyribonucleic acid (DNA) seems to be a favorable alternative for storing such exponentially growing digital information for a prolonged period with high density and capacity as it keeps the information at a molecular level using nucleotides. DNA stores genetic information of all living things and the information is transferred from one generation to another accurately due to its precise Watson–Crick base pairing. Researchers have successfully used DNA for storing digital data, which opened the possibility of storing Big Data using DNA-based systems. In this chapter, different conventional tools and challenges associated with Big Data storage are reviewed. Further, the various encoding and encryption methods used for DNA-based data storage are critically analyzed. In addition, the challenges for DNA-based Big Data storage are reviewed, and the capabilities of different approaches to overcome these shortcomings are discussed.

Advances in Computers, Volume 129
ISSN 0065-2458
https://doi.org/10.1016/bs.adcom.2022.08.005

Abbreviations

A	adenine
bp	base pair
C	cytosine
DCN	dynamic circuit network
DNA	deoxyribonucleic acid
G	guanine
GIS	geographic information systems
GPS	global positioning system
HDFS	Hadoop distributed file system
HPCC	high-performance computing cluster
IoT	Internet of Things
MB	megabyte
NoSQL	nonstructured query language
PCR	polymerase chain reaction
RNA	ribonucleic acid
SAT	satisfiability
SQL	structured query language
T	thymine

1. Introduction

In the current digital age, the amount of data is growing at an exponential rate [1–3]. This is referred to as Big Data. The analysis of Big Data is becoming increasingly important primarily due to its widespread applications. It is used for the Internet of Things (IoT) [4–6], for global positioning system (GPS) tracking [7], wearable sensors [8], and smart grids [5]. In healthcare, a quite significant amount of data is collected daily from the laboratory, monitoring sensors, and patients' medical history [8–10]. The healthcare-generated data is used for making personalized medicine [3]. This gives appropriate treatment at a reduced cost. Big Data analysis can be used for identifying the affected area from some diseases [8–10]. This can help in timely planning for diagnosis, and vaccination. In production and manufacturing, zero downtime and transparency can be gained by utilizing Big Data analysis. The government uses Big Data to increase the use of limited resources and services. Data collected from various sensors planned at different public infrastructure locations, such as water supplies, power supplies, etc., is used to identify the utilization rate in different areas [11–13]. Big Data analysis not only helps in proper distribution but also helps in improving government facilities. The data extracted from different government departments play an important role in e-governance in their respective areas [11–13]. Various areas such as cyber cell, cybersecurity, tax, traffic supervision, weather collect the Big Data and help improve the facilities. Another important area of the Big Data application is transportation and logistics [5–7]. Transporters equipped with radio frequency identification generate the data used for planning and managing the routes and maintaining the staff's track record [9–13].

Today, humans live in a world filled with information all around. The information transfer from one generation to another has helped humanity evolve to a current level of development. Documentation of essential data allows the next generation to use that knowledge to make it better. For instance, up to 50,000 years ago, the Paleolithic inhabitants' circumstances and situations were understood by the cave paintings present [14]. A naturally preserved man died in the Neolithic age (7000–4000 years ago) discovered in 1991 in Austrian–Italian Alps named Ötzi [15]. Ötzi opened many ways to understand the anatomic and pathogenic discoveries stored in genetic material over 5000 years. Likewise, today people are generating a considerable amount of digital data that needs to be stored and archived, if possible, for all time [16]. Even though a wide range of data storage technologies is present, current data storage methods will turn outdated due to the limited prolonged existence of the hardware used for storing digital data and the high associated cost [17,18].

The rate of digital data production is increasing the demand for data storage [17] exponentially. As of February 20, 2021, the World Wide Web indexed webs comprised of at least 55 Billion web pages [19]. Although only a small fraction of the generated digital data is needed to be reserved, fulfilling the growth in archiving information is becoming difficult using conventional storage technologies [20,21]. Moreover, the attributes such as data density, data access speed, data retention time, resting, and accessing energy cost of data are becoming extremely crucial [20,22]. This inspired the researchers to work on new data storage technologies that can preserve the data efficiently over an extended period of time.

Researchers are now looking at deoxyribonucleic acid (DNA) as a material for data storage to meet the increasing demand for digital data storage. Neiman [23] proposed the idea of storing digital data on DNA. The reason for choosing DNA for storing the data is primarily because of its proven ability to store biological information efficiently. The genetic information of organisms is transferred from one generation to another through DNA or

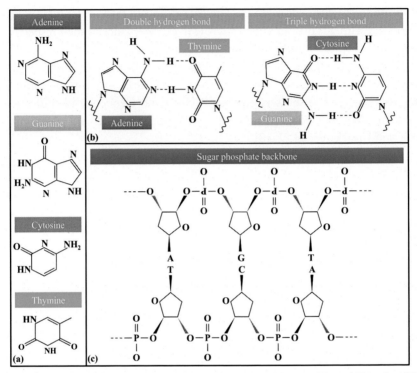

Fig. 1 (a) Nucleotide bases Adenine (A), Guanine (G), Cytosine (C), and Thymine (T), (b) base pairing of A & T, and G & C of two opposite strands by hydrogen bonds (shown by dotted lines), (c) hybridization of two antiparallel and complementary DNA strands.

ribonucleic acid (RNA). As shown in Fig. 1, Adenine (A), Guanine (G), Cytosine (C), and Thymine (T) are the four basic building blocks of the DNA which are called nucleotides [24–26]. Like the digital computers that store the information in binaries (0 and 1), DNA stores the biological information in the quaternary system of four nucleotides (A, T, G, and C). Further, specific DNA sequences used for storing the data can be easily synthesized chemically at a reasonably low cost up to a few hundred base pairs in nanomolar quantities. Moreover, the durability of the DNA is very high as compared to the other biomolecule. The absence of the 2′-hydroxyl group gives a more stable phosphodiester bond in DNA. Under physiological conditions, the half-life of hydrolysis of phosphodiester bond is around 30,000,000 years [27]. The above advantages make DNA an attractive choice for data storage.

The hybridization of individual nucleotides (i.e., A joins only with T and G joins only with C) of two single-stranded DNA is selective (see Fig. 1). This property of selective hybridization serves in identifying the molecules accurately and is referred to as "Watson–Crick complementarity base pairing." The precise Watson–Crick base pairing in DNA is the central property for different types of manipulations such as addition, amplification, cutting, and separation of the DNA. The process of joining two DNA strands is ligation. Two DNA stands join with each other by forming the phosphodiester bond using the ligase enzyme. The method of amplifying the DNA is polymerase chain reaction (PCR), where the formation of multiple copies of the DNA takes place in the presence of the enzyme Taq polymerase. Cutting of the DNA at the precise location is performed using the restriction enzyme digestion. Further, the desired size of DNA is separated using gel electrophoresis, whereas the affinity separation separates the DNA molecules if a specific sequence is present in the DNA molecule. Also, there are numerous high-throughput sequencing methods for sequencing the DNA. In these methods, the DNA strand of interest works as a template for the DNA polymerase reaction. It generates a complementary sequence of the strand comprising fluorescent nucleotides, which emit different colors to read the complementary sequence optically.

The data storage density of DNA is up to 10^{18} bytes per mm^3. This density is six times higher than today's densest media present [21]. The data storage using DNA also enables the protection of the data in molecules at a low cost for an extended period [28]. It has been verified and tested by the time that it is possible to read DNA sequences from thousands of years old fossils [28]. Further, DNA can remains for millennia compared to decades, a usual storage time for media archives when maintained under specific light humidity and temperature [12,18].

Rapid progress in synthesizing and reading DNA establishes DNA as viable storage material, particularly for archival storage, which incurs a high maintenance cost using conventional technology [2,29]. It is useful when duplication of the data is required as the DNA-based storage devices can make millions of copies of the DNA at a low cost and time using PCR [21,30–32], unlike the proportionate time required in conventional technologies. With these advantages, DNA can provide a solution for Big Data storage.

Researchers have successfully used DNA to store digital data up to a megabyte scale [21,33–35]. Further, the encapsulation in silica may help in protecting the data for millennia-long [28]. Also, DNA is a highly dense material that can work as a data archive [17,30,36]. Recently, Erlich and Zielinski [37] practically demonstrated high-density data storage. These methods are then scaled up to store ~ 200 megabytes (MB) data on DNA by Organick et al. [38]. Microsoft and the University of Washington [32] developed and demonstrated a fully automated machine that can encode digital data on DNA and can decode it back with high accuracy. The researchers [39–41] have generated DNA computing-based passwords for confidential multimedia files or messages to enhance security. This model's advantage is that it provides useful and robust access control to the user to minimize the data access time with high confidence in the cloud-based environment [40–42]. The model also helps in data management and improvement in the works related to healthcare like telemedicine and virtual medication [8]. Several of the above studies and their limitations are listed in Table 1. The practical demonstration of data storage on DNA in the above studies opened the possibility of storing Big Data on DNA. Though several conventional tools and technologies are available for analyzing and storing Big Data, the use of molecular computing methods such as DNA computing and DNA-based storage has the potential to address critical issues such as the exponential growth of Data and its prolonged storage. This motivated the authors to review the existing technologies for Big Data storage, the challenges associated with conventional Big Data storage methodologies, and possible solutions using DNA-based storage methods and their associated challenges.

The significant contributions of this chapter are listed below.

1. It provides a review on DNA-based computing and data storage
2. It offers a review of challenges and tools used for Big Data storage and analytics
3. It gives a detailed experimental procedure of DNA-based data storage
4. It provides a critical review on different DNA-based storage methods, their advantages, and limitations for Big Data storage

This chapter focuses on the analysis of various encoding and encryption methods used for DNA-based data storage. Section 2 reviews the background studies on DNA-based computing, types, and characteristics of Big Data, and technologies and tools used to manage Big Data and their associated challenges. The encoding, decoding methods, and experimental procedure to store the digital data on DNA are presented in Section 3. Challenges for DNA-based Big Data storage are discussed in Section 4, followed by conclusions and future work in Section 5.

Table 1 A literature review of DNA-based data storage.

Study	Total data (MB)	DNA synthesis by	DNA sequencing by	Coverage	Reassembly	Length of DNA strands (base pair)	Bits per base with primer	Bits per base without primer	Random access	Limitations
Church et al. [36]	0.65	Deposition	Illumina	3000×	Index	115	0.6	0.83	No	PCR amplification is required multiple times for sequencing to reduce reading error
Goldman et al. [35]	0.63	Deposition	Illumina	51×	Overlap	117	0.19	0.29	No	PCR amplification is required multiple times, and the cost of synthesis and sequencing is high
Grass et al. [28]	0.08	Electrochemistry	Illumina	372×	Index	158	0.86	1.16	No	DNA synthesis and the sequencing error rates are estimated to be ~1% per nucleotide
Bornholt et al. [30]	0.15	Electrochemistry	Illumina	40×	Index	117	0.57	0.85	Yes	DNA synthesis and sequencing are restricting the feasibility of using it in the existing state
Erlich and Zielinski [37]	2	Deposition	Illumina	10.5×	Luby seed	152	1.18	1.55	No	DNA sequences created by synthesis and amplification are unequal and uneven. Indexing limits the size of the DNA

Continued

Table 1 A literature review of DNA-based data storage.—cont'd

Study	Total data (MB)	DNA synthesis by	DNA sequencing by	Coverage	Reassembly	Length of DNA strands (base pair)	Bits per base with primer	Bits per base without primer	Random access	Limitations
Blawat et al. [22]	22	Deposition	Illumina	160×	Index	230	0.89	1.08	No	The length of the DNA is limited
Organick et al. [38]	200	Deposition	Illumina	5×	Index	150–200	0.81	1.1	Yes	Cost is high, and throughput is low
Anavy et al. [43]	8.5	Deposition	Illumina	164×	Index	194	1.94	2.64	No	Prone to errors in DNA synthesis and sequencing
Choi et al. [44]	0.000854	Column	Illumina	250×	Index	85	1.78	3.37	No	Prone to errors in DNA synthesis and sequencing
Yazdi et al. [45]	0.003	Column	Nanopore	200×	Index	880–1000	1.71	1.74	Yes	Prone to errors in DNA sequencing
Organick et al. [38]	0.033	Deposition	Nanopore	36×	Index	150	0.81	1.1	Yes	Cost is high, and throughput is low
Lee et al. [46]	1.80E−05	Column (Enzymatic)	Nanopore	175×	NA	150–200	1.57	1.57	No	It acquires a sixfold loss in the volumetric density of data

2. Background studies

The DNA storage workflow involves encoding digital information into DNA sequences. For this purpose, suitable DNA sequences are synthesized. These DNA sequences are substantially acclimatized into a library for long-term data storage, retrieving, and random access. The DNA sequencer reads these DNA sequences. The reading (output in fasta format) of the DNA sequencer is in the form of quaternary data ("A," "T," "G," and "C") which is converted back to digital binary data ("0" and "1"). To illustrate the workflow for DNA-based Big Data storage, the fundamentals of DNA computing and Big Data analytics are described next.

2.1 DNA-based computing

Adleman [47] is the first researcher to start using DNA for computing. He solved the Hamiltonian path problem for a supergraph with 7 vertices and 14 edges. The objective was to obtain a path in a supergraph such that all vertices are visited, and each one is visited only once. For this purpose, he represented the vertices and the edges of a given supergraph using single-stranded DNA sequences. The sequences of edges are selected as complementary to half of the predecessor vertex and the remaining half to the successor vertex. These single-stranded DNA sequences ligate selectively as only specific complementary sequences are used in edges to form double-stranded DNA in such a way that these represent only the possible paths present in the given supergraph. From this initial pool of double-stranded DNA, the correct length DNA molecules that satisfy the Hamiltonian path's length are separated using the gel electrophoresis separation. These separated DNA molecules are amplified using polymerase chain reaction (PCR) for better accuracy. Though these separated DNA molecules may have a correct length, some of the vertices may be missing, and some may be repeated multiple times. Therefore, the sequences are further subjected to an affinity separation process that extracts only those paths representing the DNA sequences corresponding to each vertex of the supergraph only once. Thus, after ligation, gel electrophoresis, PCR, and affinity separation, if a DNA molecule is present in the final solution, it represents the Hamiltonian path of a given supergraph. Otherwise, it proves that no Hamiltonian path is possible for the given supergraph.

After the classic study by Adleman [47], the development of several exciting adaptations occurred for solving the computationally intractable problems using DNA. Lipton [48] developed a DNA computer to solve the satisfiability (SAT) problem. Subsequently, Smith et al. [49] and Liu et al. [50] created a new surface-based DNA computing model to solve the SAT problem. The researchers [51–53] have developed the modified approaches for solving the other nondeterministic polynomial-time complete (NP-complete) problems such as 3-SAT, chess, and maximal clique. Sakamoto et al. [54] developed an interesting method to solve the SAT problem using DNA hairpin formation. Sharma and Ramteke [55] solved the practical example of polymer grade scheduling using a DNA computer. Recently, Chao et al. [56] solved a maze problem using a DNA navigator. Most of these studies are illustrated on small-scale problems. To improve the scalability, Sharma and Ramteke [57] developed a DNA computer based on circular DNA structure formation to solve a Hamiltonian cycle problem of the 18-vertex supergraph. In this model, the single-stranded DNA represents the vertices and the supergraph edges. The formation of a circular double-stranded DNA represents a possible path after ligation and hybridization. The restriction enzyme corresponding to each vertex cuts the circular DNA and makes it a linear double-stranded DNA. The obtained linear DNA is then separated using gel electrophoresis. The correct circular DNA cuts only at one location, and the incorrect DNA cuts at multiple locations or remain circular. The iterative circularization (using ligation), cutting by respective restriction enzyme for each vertex, and separation by gel electrophoresis for correct length yields the right solution if it exists after N such steps for N-vertex problem (see Fig. 2). Most of the above studies are reviewed extensively by Sharma and Ramteke [58].

2.2 Big Data analytics

2.2.1 Types of Big Data

Big Data analysis is performed on various data collected from multiple sources such as network, event, time series, natural language, and real-time data. The collected data is in the structured or unstructured form, which is continuously growing. The distribution of this data for analysis turns out to be a crucial step in today's era. The data is noisy and unevenly distributed, which requires some preprocessing steps for analysis; otherwise, data can be misinterpreted. In this section, we are discussing the different data types.

(i) Structured data

Data stored in a proper format that can be readily utilized by a person or computer program is referred to as structured data. A typical example is numerical data stored in the form of a table with labeled rows and columns.

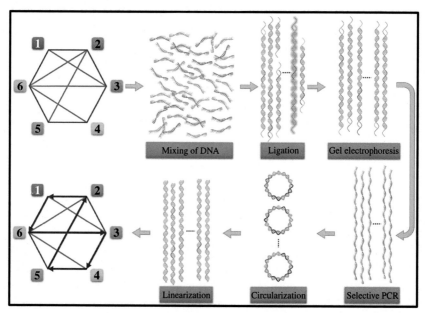

Fig. 2 Circular structure-based DNA computing for solving the Hamiltonian cycle problem.

Database management system (DBMS) facilitates the access of structured data. Only 10% of the total data fit into the structured data generated from various government departments, real estate, different organizations, and transactions.

(ii) Unstructured data

Any data which is not arranged in a predefined form, or a predefined data model is referred to as unstructured data. About 90% of the total data belongs to unstructured data. Unstructured data is in the form of text, audio, video, or image. Growth in social media resulted in the explosion of unstructured data in the past decades. Conventional storage devices and techniques are not sufficient for handling unstructured data. A suitable database [for example, nonstructured query language (NoSQL)] is required for storing and handling this type of data.

Both structured and unstructured data can be further classified based on its source as follows.

(a) Real-time data

Various sources such as images, audio, and video generate real-time data by streaming live media data. Due to the live streaming, storing, and processing, real-time media turn out to be a big challenge. YouTube, flicker, and videoconference are some of the examples of real-time media data generating resources.

(b) Geographic data

Geographic information systems (GISs) generate data such as an address, building, route, workplace, etc. fits into the geographic data. This information is collected from the GIS and sensors installed. The collected raw data is processed to provide valuable information. Environmental conditions are also analyzed by using geographic data.

(c) Natural language

Verbal conversation generates data. This data is known as natural language data. The major sources of natural language data generation are the Internet of Things, speech capturing devices, landlines, and mobile phones.

(d) Network data

Network data is generated from large networks such as Facebook, Twitter, and YouTube. Along with this biological network, information networks are also a few major sources of network data generation. In network data, the connection among the nodes can be established in two forms, first, one-to-one, second, one-to-many. A major challenge in network data is maintaining the connection among the nodes and network architecture.

2.2.2 Characteristics of Big Data

Big Data refers to exponentially growing data sets containing the heterogeneous formats generated from internet sources, social media, social networks, sharing data, sensors, and clickstreams [7,59]. For Big Data applications, conventional business intelligence tools are not efficient. Robust technology and advanced algorithms are required for processing Big Data due to their complex nature. Due to this, researchers and Big Data scientists have defined Big Data using the 3V characteristics [5,12,59]. The next section discusses these 3V characteristics of Big Data.

The first characteristic is *volume*. Today millions of devices (social networks, smartphones, sensors, bar codes, etc.) use the large number of applications that are generating a massive volume of digital data continuously [5]. McAfee et al. [60] forecasted about the generation of digital data on each day in 2012 is approximately 2.5 EB. International Data Corporation published a study estimating the production, replication, and usage of 4.4 zettabytes (ZB) of digital data in 2013. The volume of the data is doubling every 2 years. In 2015, it was 8 ZB [61], and was estimated that it would reach 40 ZB in 2020.

The second characteristic is *velocity*. Digital data is generated at a high rate and needs to be processed instantly to extricate useful information and appropriate insights. YouTube is one of the good examples of the fast speed of Big Data. Another example is Walmart which generates more than 2.5 petabytes (PB) of digital data every hour from its customer's transactions [5].

The third characteristic is *variety*. Audio, videos, documents, etc., are the different formats of the digital data generated by the various resources. This data may be public or private, organized or unorganized, local or distant, shared or confidential global, complete or incomplete, shared or secret, etc. [5,12].

For a better definition of Big Data, some more Vs with extended characteristics are added in the above list by Emani et al. [4] and Gandomi and Haider [62]. These include *vision* (purpose of data), *verification* (confirmation of the data to some specification), *validation* (fulfillment of the purpose of data), *value* (extracting the information from the data for other sectors), and *complexity* (difficulty in organizing and analyzing the data due to evolving relationships).

2.2.3 Technologies and tools to manage Big Data

Big Data management is becoming increasingly important. However, the raw data generated from various sources are messy, noisy, and unduplicable. This type of data management issue is resolved by using appropriate tools and technologies. There are three layers in the usual Big Data project: infrastructure, computing, and application layer. The infrastructure layer operates the interaction between the systems, data storage, and devices in the network. It also organizes the data retrieval requests received from the other layers in the architecture. The computing layer offers an abstraction on the infrastructure layer and the access, organization, and retrieval of data. The data is indexed in this layer and organized over the allocated storage devices. The data is managed over the available repositories after dividing them into blocks. The application layer consists of tools, techniques, and logic for the development of area-specific analytics. The prominent tools and techniques used are presented next.

(i) Apache Hadoop

Apache Hadoop is one of the most famous and used Big Data tools which stores petabytes of structured and unstructured data [63,64]. MapReduce model-based structure of Hadoop works in a distributed environment. Especially, Hadoop is an appropriate choice for the storage of huge amounts of datasets. Data storage and computation are the two fundamental tasks in

Hadoop. Hadoop distributed file system (HDFS) has data storage account-ability, and the MapReduce framework usually performs data computations.

(ii) Apache Spark

Apache Spark is an open-source, fast cluster computing framework written in Scala [63–65]. It offers an interface for whole cluster programming, inherent data parallelism, high scalability, and fault tolerance. Though MapReduce-based platforms are suitable for managing large-scale applications, iterative tasks across parallel processes cannot be handled efficiently if an acyclic flow model is applied. Apache Spark enables the implementation of iterative as well as interactive data analysis. It is important to locate the HDFS in query processing using a structured query language (SQL) in Big Data analytics.

(iii) Hadoop's SQL Tools

Hadoop's SQL tools are effective for managing interactive and complex inquiries on big datasets [64]. Therefore, in recent years multiple SQL tools querying on Hadoop tools have been developed. Stream processing systems are service offerings that provide important information to the consumers and empower businesses to make suitable and quick judgments. Tools of stream computing process the live streaming and produce low potential outputs. The applications are denoted in directed acyclic graphs with operators as vertices and flow channels between the operators as edges.

(iv) Apache Storm

Apache Storm is developed for better benchmark processing [65]. It empowers developers to create real-time distributed processing systems. It is also called Hadoop for real-time data. It is highly scalable, user-friendly and can process the unbounded streams of data extremely fast with low latency and guaranteed data processing.

(v) High-Performance Computing Cluster System

High-Performance Computing Cluster (HPCC) System is an open-source computing platform [66]. HPCC helps in managing Big Data workflow. This system is a single platform with a single architecture. Data simulation is performed on the HPCC system in a single programming language.

(vi) Qubole

Qubole is a cloud-based Big Data platform developer that permits customers to formulate, incorporate, and analyze Big Data [67]. Customers of Qubole can connect to Microsoft Excel, Micro Strategy, Tableau, and Qlik through Open Database Connectivity (ODBC) tools. It also offers massively parallel processing.

(vii) Cassandra

Cassandra is a distributed, and open-source database storage system used to manage a huge amount of structured data [68,69]. The structured data is distributed among many commodity servers. The major advantage of Cassandra is that it provides high service with no single point of failure. Cassandra intends to operate over an arrangement of hundreds of nodes possibly distributed around different data centers.

The key features of the above technologies and the tools are listed in Table 2.

Table 2 Technologies and tools to manage Big Data [3,5,7,12].

Sr. no.	Names of the tool	Key features
1.	Apache Hadoop	• Open-source, free license tool • A high-end server is not required • Manages Volume, Variety, Velocity, and Value characteristics • Cost-effective • Scalable with data size • Flexible for different types of data • Involves faster data processing • High fault tolerance resulting nearly no data loss • It offers a robust ecosystem for data • Hadoop Distributed File System (HDFS) can store up to tens of Petabytes of data
2.	Apache Spark	• Open-source tool • Reusable for batch processing • In-memory processing • Cost-efficient • Dynamic in nature • A faster and real-time stream processing tool • High fault tolerance resulting nearly no data loss
3.	Hadoop's SQL tool	• Open-source tool • Compatible with various data sources such as Hive, Avro, Parquet, Optimized Row Columnar (ORC), Java Script Object Notation (JSON), Java Database Connectivity (JDBC)

Continued

Table 2 Technologies and tools to manage Big Data [3,5,7,12].—cont'd

Sr. no.	Names of the tool	Key features
4.	Apache Strom	• Better benchmark processing • Can process up to one million 100-byte messages per second per node • Performs massive parallel calculations • In case a node dies, it automatically restarts on an alternative node • Each unit of data processed at least once • One of the easiest tools for Big Data analysis
5.	High-Performance Computing Cluster System	• Open-source tool • Used for rigorous data computing • It is built on one platform, one architecture, and one programming language • Enhances Scalability, Performance, and Accessibility of the data • Automatically performs the optimization of parallel processing
6.	Qubole	• Cloud-based open-source tool • Autonomous Big Data management platform • Optimizes performance inherently
7.	Cassandra	• NoSQL, distributed, and open-source database • Used to handle large data within multiple servers • Assures high availability with no failure • Replicates data in multiple locations or clouds to ensure the fault tolerance

2.2.4 Challenges of Big Data

The efficient utilization of Big Data may lead to a new trend of valuable progress [3]. Businesses are now shaping to take advantage of Big Data analysis in an extremely competitive environment. New employers focus on attracting employees with proficiencies in dealing with Big Data. It improves operating efficiency, consumer services, and obtaining the updated planned direction, new consumers, and new products and marketplaces. It seems to be a promising approach for dealing with various appealing opportunities. However, the users face multiple challenges as the extent of information exceeds the channeling capabilities. The data is increasing exponentially, whereas improvement in the data processing speed is bounded. There are limited tools available that mainly focus on Big Data analysis. The real-life problem of data storage, sharing, searching, and real-time analysis is challenging for state-of-the-art methods and technologies such as Hadoop.

The challenges associated with Big Data analysis are described below.

(i) Data understanding

Understanding the data is one of the major challenges in Big Data storage. Most of the users do not understand the details of the data. Handling Big Data leads to some open questions such as what is the size of the data? What is the source of the data? How much important the present data is? What type of hardware is required for handling the data? Therefore, an organization needs to find the answers to the above questions for a better understanding of the data.

(ii) Big Data tool selection

After a clear understanding of Big Data, the user should choose an appropriate tool for handling the data. For example, which tool can perform better for analysis of given data (i.e., Hadoop MapReduce, or Apache Spark)? Which tool is better for storage of given data [i.e., Cassandra or Hadoop database (HBase)]? A bad decision on selecting the tools results in loss of data, time, effort, and money.

(iii) Data acquisition

Data acquisition is the process of collecting, sorting, and cleaning the data before transferring it to the storage device. The data acquisition is governed by characteristics such as volume, velocity, variety, and value. Data is acquired from various sources such as media, sensors, social networks, etc., at a considerable rate in structured and unstructured forms (e.g., video, text, pictures, etc.). However, the analysis often requires the data in a specific format. Although the available data can be converted to a particular format, there is always a risk of losing some data in the conversion process.

(iv) Data preprocessing

Before analyzing the data, it should be constructed correctly and represented efficiently. Preprocessing is essential to enhance data quality. It results in improved analysis. Data generated from different sources is extensive and usually, it is unreliable, incomplete, noisy, and inconsistent. The noise can be removed, and inconsistencies can be corrected by applying various data preprocessing methods. Data preprocessing methods include data integration, cleaning, reduction, and transformation.

(v) Data storage

The selection of an appropriate Big Data storage is a critical decision. Attaining consistency in data accessing is also an important factor for selecting data storage. Data storage architecture should give information about the storage facility with the available and reliable storage space, dynamically. Also, access to the data should be provided while analyzing the Big Data. Currently, data storage tools are categorized as direct-attached storage (DAS) and network storage. Network storage can be further classified as storage area network (SAN) and network-attached storage (NAS).

(vi) Data sharing and data transfer

Processing and analysis of Big Data require reliable data sharing and transfer. Intra-dynamic circuit network (DCN) transmission and Inter-DCN transmissions are two types used for data transmission. Intra-DCN transmits the data within the data center, whereas inter-DCN transmits the data from the source to the data center. The data grows continuously, but the capacity of wired and wireless hardware does not increase accordingly. The limited growth of hardware ultimately restricts the complicated data flow across the network and its storing and sharing. This problem is referred to as the internet plumping problem.

(vii) Data analysis

Analyzing the massive data and obtaining beneficial information is a challenging task. Often, the customers desire the outcome of the analysis in high-quality formats that are simple to understand and operate. However, meeting such requirements is difficult when a high amount of heterogeneity and variety is present in the data. Data generated from different fields can be analyzed by using various tools. Choice of the tool and the analysis is very important for extracting meaningful information. Different data analysis methods used are correlation analysis, factor analysis, cluster analysis, regression analysis, memory-level analysis, massive analysis, and business intelligence analysis.

(viii) Upscaling hurdles

Data is increasing at an exponential rate. Therefore, the storage capacities of the current hardware need upscaling according to the present and future requirements. An appropriate architecture based on the forecast can be used for upscaling the hardware and avoiding difficulties in the future.

(ix) Data security

Big Data generally comprises some personal information of the users. To avoid the misuse of personal information, the data is typically anonymized before storage. For this purpose, all identifiers are removed from the data. Even with this encryption, sometimes the data is not completely anonymized. In such cases, fully homomorphic encryption (FHE) is used to anonymize the data completely. However, anonymization reduces the efficiency of data usage as it removes some part of the data. Therefore, the right balance between the efficiency of the data usage and privacy of the data is required. An organization's security model involves achieving confidentiality, integrity, and availability as the key objectives [70].

Further research is required to enhance the efficiency, analysis, and storage of Big Data. Since storage is one of the biggest bottlenecks with Big Data analysis, it inspired several researchers to look beyond the conventional data storage method to DNA-based data storage.

3. DNA computing-based Big Data storage

Although the DNA-based storage devices are entirely different from conventional data storage devices, both run parallelly at the lowest level. The traditional storage device stores the information in binary bits. Unlike this, a quaternary system comprising four nucleotide bases (A, T, G, or C) joined by a phosphodiester bond in a DNA molecule is used to store the data in DNA-based storage.

3.1 Write down to DNA

The steps performed for writing the digital data on the DNA are discussed in this section.

Step 1: In the first step, n binary (0 and 1) digits are converted to $n/2$ quaternary digits (0, 1, 2, and 3) (i.e., $11010001 \rightarrow 3101$).

Step 2: In the second step, quaternary digits (0, 1, 2, and 3) are mapped to DNA nucleotides (A, C, G, and T) (i.e., $3101 \rightarrow TCAC$). For instance, the binary string 11010001 represents the base four string as 3101. This base four string further means TCAC. Additionally, the probability of several experimental errors is reduced by encoding binary data in base three instead of base four. A binary string 01100001 can be mapped to the base three string as 01112 by using the Huffman code. The mapping of each ternary digit to a DNA nucleotide is based on rotating code. This string 01112 maps into DNA sequence CTCTG by rotating nucleotide encoding. It prevents the repetition of the same nucleotide twice and is extremely important as the repetition of the same nucleotide considerably enhances the chance of error in the sequencing.

Step 3: In the third step, the corresponding single-stranded DNA sequences are synthesized by using a DNA synthesizer. Using the existing DNA synthesis technology, it is not possible to synthesize large DNA sequences, and only the sequences with hundreds of bits can be synthesized with a very low error rate. Therefore, the encoding of a complete solution is difficult as it requires large-size DNA sequences to be synthesized. This challenge is handled to a large extent by segmenting the large information in multiple small DNA sequences arranged in blocks. In these blocks, each DNA strand is synthesized separately, and therefore it allows storing large values with a very low error rate.

Step 4: In step 4, the strands are assigned with some indexes so that the DNA strands tag to corresponding primers that segregate molecules of interest and thus execute random access. Since DNA sequences synthesized in a DNA pool are randomly collected to the decoder, indexing becomes very important. Amplification of each data block occurs by addressing the information to distinguish its location in the input data sharing. There are two parts to the address space. The one part of the address recognizes the key. The second part is the address index of the block which contains the value linked with the key. The combined address is represented by a fixed length of nucleotides. Typically, 24 bits are used for indexing as it gives $2^{24} = 16,777,216$ distinct indexed sequences.

Further, the encoding process is associated with errors occurring in DNA writing. Errors such as missing the nucleotide (deletion error), the addition of nucleotide when it is not required (insertion error), and the addition of wrong nucleotide (substitution error) occur at the time of synthesis. Also, a few DNA sequences are lost, or some redundant ones are generated at the time of writing, affecting the recovery of information. Some additional information is encoded in the DNA sequences to eliminate the above errors to ensure that the data recovery is perfect.

Step 5: In step 5, all the strands generated are amplified by performing the PCR. This improves the accuracy.

3.2 Read out from DNA

The steps performed for reading the DNA sequences are discussed in this section.

Step 1: In the first step, the DNA sample is drawn from the DNA pool obtained in the write-down procedure.

Step 2: In this step, the DNA sample is amplified using PCR with primers specific to intended information, and DNA of the desired length is extracted by using gel electrophoresis.

Step 3: In this step, the unused DNA templates, primer-dimers, unbound primers, and unused Taq DNA polymerase are removed from the pool of the DNA as these can decrease sequencing efficiency.

Step 4: After extraction of the DNA, two adaptors are added to the two ends of the DNA. These adaptors are the sequencing adaptors added to the DNA sequence one by one in two consecutive PCR. After each step of the PCR, agarose gel electrophoresis is performed for extracting the DNA of the desired length. Again, it removes unused DNA templates, primer–dimers, unbound primers, and unused Taq DNA polymerase from the DNA pool.

Step 5: In this step, the sample is subjected to the DNA sequencing process. Here, each DNA strand is randomly chosen from the DNA pool, and nucleotide sequences of this stand are identified. Thus, the output file generated by the sequencer has DNA sequences of millions of molecules. Since the process is stochastic, there are chances that some strands are more often read than the others. Also, there are high chances that several DNA strands remain unread, so a large quantity of samples is used to avoid this. Since more individual DNA strands are naturally readout than available DNA variants in the pool, the sequencing coverage, which is a ratio of the variants present to the total number of DNA readout, is an important factor. Typically, the required content is directly proportional to the natural logarithm of the number of diverse DNA variants present in the pool. This gives the estimation of required sequencing coverage for reading the complete information. The output of the DNA Sequencing machine is generated in the fasta format.

Step 6: In this step, the DNA sequences present in the fasta file are converted back to the binary format. Each sequence is translated back to the bits by mapping and protecting the individual sequence. There is a high possibility of obtaining multiple sequences for each index due to numerous errors. Therefore, sequences that present most frequently are chosen. This methodology operates excellently in the low-error regime for multiple reads. Organick et al. [38] proposed another method that works well for the high error regime. In this method, candidate DNA sequences for decoding are identified from the reads by first clustering the reads. This method also shows a slight advantage in the low-error regime. However, it additionally requires clustering. Unlike this, correction can be directly used in the encoding by using the additional redundant sequences. Since the errors occurring during the synthesis, storage, and sequencing yield the original sequence subset, reading of the redundant sequences is used to correct the missing sequences of real digital information. These sequences are then transformed back to the actual binary sequences. For this conversion, the same encoding used for converting the binaries to quaternary and then to A, T, G, and C sequences is reversed.

3.3 Experimental details

3.3.1 Experimental environment

A specific experimental environment is required for practically storing the information using DNA. Since all the digital data is present in binary format, it is first converted to a quaternary format using the appropriate encoding scheme. Based on this, the appropriate DNA sequences are designed. A high-performance computing device is essential for this task. The DNA sequences designed are synthesized by using the DNA synthesizer. A clean room with a controlled temperature requires the experimental process to avoid contamination and degradation. Further, the DNA sequences can be stored at 4 °C for 2–4 weeks, at −20 °C, or −80 °C for a longer storage period. DNA sequencing is used to retrieve and read the information, which generates the output in fasta format. These DNA sequences are then decoded back to binary form using the computer program.

3.3.2 Experimental steps

Fig. 3 shows the experimental procedure of DNA-based data storage. Four major experimental steps are typically involved in DNA data storage. These steps are writing, storage, retrieval, and reading.

(i) Writing

First, the digital data is converted to DNA sequences by using the appropriate coding. This coding scheme may vary as per the user. The coding and decoding should be the same to get an accurate result. Few properties and parameters such as termination level, bits encoded per base, and possible length of the sequences are calculated using the appropriate code. This code takes the digital data file as the input and gives a text file containing the DNA sequences as the output file.

(ii) Storage

The DNA sequences generated are synthesized in the writing step using a DNA synthesizer. These DNA sequences are highly concentrated. Therefore, DNA sequences are diluted to a concentration of 5 ng/μL using PCR-grade water. This stock solution is stored at 4 °C for 2–4 weeks, at −20 °C or −80 °C for a longer storage period. Next, the PCR is performed on the synthesized DNA sequences three times. In the first PCR, the desired DNA sequences generated are amplified for data storage. Then the PCR product is separated using gel electrophoresis for 120–170 nucleotides based on the payload. There are some possibilities of loss of the sample due to the removal of short oligonucleotides. Therefore, it is required to check the

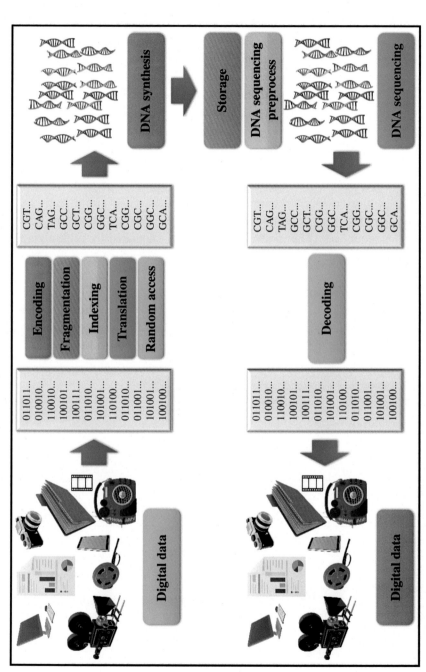

Fig. 3 Process flow of the DNA-based storage system.

quality of output using fluorescence-based concentration measurement methods. Then, the second PCR is performed for adding the sequencing adaptors at one end of the DNA. The PCR product is then separated using gel electrophoresis to extract an oligonucleotide band of 150–200 nucleotides. The third PCR is performed on these separated DNA sequences to add the sequencing adaptors at the remaining end of the DNA. Finally, the DNA sequences of 200–250 nucleotides are separated using gel electrophoresis. Tables 3 and 4 show the detailed parameters used in PCR.

(iii) Retrieve

The sequencing coverage is selected based on the quality of DNA synthesis and PCR cycles for retrieving the information. Conventionally, the number of initially written DNA sequences becomes almost 200 times after the reading. Illumina iSeq system or any other DNA sequencer is used to read the DNA sequences. The sequencing reads are calculated by multiplying sequencing coverage with the number of sequences encoded.

(iv) Reading

DNA sequences are obtained as a fasta file with all information of the reads and the quality control data. This file works as input data to the computer code, which decodes the DNA sequences to binary. The accuracy is calculated by comparing the original binary file with the decoded binary file.

Table 3 Typical quantities used in PCR.

Reagent	Amount (µL in 1 well)	Final concentration
PCR master mix	10	$1\times$
PCR-grade water	7	—
Primer 1F ($10\,\mu M$)	1	$0.5\,\mu M$
Primer 1R ($10\,\mu M$)	1	$0.5\,\mu M$
DNA stock solution	1	$\sim 0.005\,ng$ per μL
Total	20	—

Table 4 Typical cycling parameters used in PCR.

Cycle number	Denaturation (temperature, duration)	Annealing (temperature, duration)	Extension (temperature, duration)
1	95 °C, 5 min	—	—
2–15	95 °C, 15 s	54 °C, 30 s	72 °C, 30 s

4. Challenges for DNA-based data storage

There are several challenges in using DNA for Big Data storage and analysis. The discussion on the significant challenges is given next.

(i) Spatial segregation

The size of the data that can be stored is limited mainly by the expense of DNA synthesis and DNA sequencing. In recent years, researchers are putting their efforts into improving the extent of DNA synthesis and encoding strategies. The increase in file index labeling significantly reduces the number of nucleotide bits available for data storage as the length of the DNA synthesized is limited. For instance, for a data set of exabyte-size, at least three barcodes are required for indexing each file [30]. Three barcodes acquire up to 60 nucleotides in each DNA sequence, ultimately reducing the number of nucleotides available for encoding the data [71].

(ii) Errors in DNA synthesis

The DNA synthesis process is carried out by attaching the DNA molecule to a solid surface, and each nucleotide is added one after another through a chemical process. Current DNA synthesis methods generate millions of DNA strands per chip. Secondly, various errors occur at the time of the extension (i.e., new nucleotide addition) of the DNA strand. These errors are (a) *Deletion* (i.e., the attachment of the desired nucleotide is not at its position), (b) *Insertion* (i.e., a wrong nucleotide is attached at a position somewhere it should not be), and (c) *Substitution* (i.e., the addition of a nucleotide other than the expected). In some cases, it might be possible that the distribution of the DNA generated at each spot on a solid surface is uneven [17,38].

(iii) Errors in DNA storage

The DNA processing and DNA data storage steps such as removing DNA from the solid surface and heating cycles of PCR may result in DNA decay. The significant consequences of damage are depurination which ultimately results in DNA strand break down. The PCR amplification of these inadequate DNA cannot be performed as primers are absent on both ends of DNA strands. This may lead to substitution errors and considerable DNA loss at the time of storage. Error-correcting codes can be employed for the correction of storage-related errors [28,33,72].

(iv) Errors in DNA sequencing

DNA sequencing errors depend on the methodology used for the sequencing. Illumina is the most widely used DNA sequencing method. The error rates of Illumina are higher at the end of a read as these errors are strand-specific. The substitution reading error rates depend on the dataset. An exact number is about 0.0015–0.0004 errors per base for substitution and 10^{-6} for insertions and deletion errors [17,32,38].

(v) Data redundancy
Duplication of the data arises if two or more data samples represent the same object. Data inconsistency and duplication severely affect storage. Although many techniques are available to detect duplicate files, they are not feasible for Big Data storage using DNA [30,31].

(vi) Data noise
In DNA-based data storage systems, data noise is created at different stages of the process (e.g., substitution, deletion, and insertion errors on the DNA sequences create unwanted DNA sequences). In some cases, homopolymers and hairpin formation also generate noisy data. These unwanted DNA sequences generate noisy reads at the time of DNA sequencing.

(vii) The time required for retrieving the information
The DNA synthesis and DNA sequencing are mostly automated, but some intermediate steps are performed manually. A fully automated DNA-based storage device is developed and demonstrated by Takahashi et al. [32]. Each process step requires some time. Also, preparation of the experiment takes some time. The overall time required for automatic DNA data storage system is approximately 21 h.

(viii) Extending to Big Data analytics
Though several researchers have illustrated DNA-based data storage, the processing and analytics of Big Data using DNA are still awaited. Unlike the conventional tools developed for Big Data analytics, such DNA-based processing tools are yet to develop.

5. Conclusions and future work

High endurance and massive information density make DNA a suitable medium for data storage. Although DNA is known for storing biological information, storing digital information is challenging. Recent progress of storing data up to 200 MB using DNA opens new prospects for handling Big Data storage challenges. Particularly, low energy requirements and long storage times offered by DNA systems can be extremely useful in storing important information for a much-extended period than conventional storage devices. Also, DNA-based computers' progress for solving a variety of combinatorial and computing problems opens the possibility of developing

DNA-based tools for Big Data analytics. Developing DNA-based data security, making DNA-based storage more convenient for the nonexpert user, and applying DNA-based storage for real-time Big Data storage are some of the open problems in this area.

Acknowledgment

The partial financial support from the Science and Engineering Research Board, Government of India, New Delhi (through grant EEQ/2020/000460, dated December 18, 2020) is gratefully acknowledged.

References

[1] Y. Cevallos, et al., On the efficient digital code representation in DNA-based data storage, in: *Proceedings of the 7th ACM International Conference on Nanoscale Computing and Communication, NanoCom 2020*, 2020, https://doi.org/10.1145/3411295.3411314.

[2] A. Agrahari, D.T.V.D. Rao, A review paper on big data: technologies, tools and trends, Int. Res. J. Eng. Technol. (2017) 640–649.

[3] J. Manyika, et al., Big Data: The Next Frontier for Innovation, Competition, and Productivity, McKinsey Global Institute, 2011.

[4] C.K. Emani, N. Cullot, C. Nicolle, Understandable big data: a survey, Comput. Sci. Rev. 17 (2015) 70–81, https://doi.org/10.1016/j.cosrev.2015.05.002.

[5] A. Oussous, F.Z. Benjelloun, A. Ait Lahcen, S. Belfkih, Big data technologies: a survey, J. King Saud Univ. Comput. Inf. Sci. 30 (4) (2018) 431–448, https://doi.org/10.1016/j.jksuci.2017.06.001.

[6] J. Wang, Y. Yang, T. Wang, R. Simon Sherratt, J. Zhang, Big data service architecture: a survey, J. Internet Technol. 21 (2) (2020) 393–405, https://doi.org/10.3966/160792642020032102008.

[7] T.R. Rao, P. Mitra, R. Bhatt, A. Goswami, The big data system, components, tools, and technologies: a survey, Knowl. Inf. Syst. 60 (3) (2019). Springer London.

[8] R. Urwongse, K. Culver, Applications of Blockchain in Healthcare, Springer Nature, 2021.

[9] L. Zhou, S. Pan, J. Wang, A.v. Vasilakos, Machine learning on big data: opportunities and challenges, Neurocomputing 237 (January) (2017) 350–361, https://doi.org/10.1016/j.neucom.2017.01.026.

[10] M. Chen, S. Mao, Y. Zhang, V.C.M. Leung, Big data related technologies, challenges and future prospects, Inf. Technol. Tour. 15 (3) (2015) 283–285, https://doi.org/10.1007/s40558-015-0027-y.

[11] S. Ullah, M.D. Awan, M. Sikander Hayat Khiyal, Big data in cloud computing: a resource management perspective, Sci. Program. 2018 (2018), https://doi.org/10.1155/2018/5418679.

[12] I. Anagnostopoulos, S. Zeadally, E. Exposito, Handling big data: research challenges and future directions, J. Supercomput. 72 (4) (2016) 1494–1516, https://doi.org/10.1007/s11227-016-1677-z.

[13] M. Wada, N. Tanaka, Big data: survey, technologies, opportunities, and challenges Nawsher, Jpn. J. Appl. Phys. 29 (8) (1990) L1497–L1499, https://doi.org/10.1143/JJAP.29.L1497.

[14] H. Valladas, et al., Radiocarbon AMS dates for Paleolithic cave paintings, Radiocarbon 43 (2B) (2001) 977–986.

[15] W. Kutschera, W. Rom, Ötzi, the prehistoric Iceman, Nucl. Instrum. Methods Phys. Res. Sect. B Beam Interact. Mater. Atoms 164 (2000) 12–22, https://doi.org/10.1016/S0168-583X(99)01196-9.

[16] A. Keller, et al., New insights into the Tyrolean Iceman's origin and phenotype as inferred by whole-genome sequencing, Nat. Commun. 3 (2012), https://doi.org/10.1038/ncomms1701.

[17] L. Ceze, J. Nivala, K. Strauss, Molecular digital data storage using DNA, Nat. Rev. Genet. 20 (8) (2019) 456–466, https://doi.org/10.1038/s41576-019-0125-3.

[18] M.G.T.A. Rutten, F.W. Vaandrager, J.A.A.W. Elemans, R.J.M. Nolte, Encoding information into polymers, Nat. Rev. Chem. 2 (11) (2018) 365–381, https://doi.org/10.1038/s41570-018-0051-5.

[19] M. Kunder, Daily estimated size of the world wide web. https://www.worldwidewebsize.com/.

[20] P.Y. De Silva, G.U. Ganegoda, New trends of digital data storage in DNA, Biomed. Res. Int. 2016 (2016) 1–14, https://doi.org/10.1155/2016/8072463.

[21] S. Shrivastava, R. Badlani, Data storage in DNA, Int. J. Electr. Energy 2 (2) (2014) 119–124, https://doi.org/10.12720/ijoee.2.2.119-124.

[22] M. Blawat, et al., Forward error correction for DNA data storage, Procedia Comput. Sci. 80 (2016) 1011–1022, https://doi.org/10.1016/j.procs.2016.05.398.

[23] M.S. Neiman, Some fundamental issues of microminiaturization, Radiotekhnika 1 (1) (1964) 3–12.

[24] H. Lodish, et al., Molecular Cell Biology, seventh ed., W. H. Freeman, 2012.

[25] D.L. Nelson, A.L. Lehninger, M.M. Cox, Lehninger Principles of Biochemistry, Macmillan, 2008.

[26] J.D. Watson, F.H.C. Crick, Molecular structure of nucleic acids, Nature 171 (4356) (1953) 737–738.

[27] K.S. Gates, An overview of chemical processes that damage cellular DNA: spontaneous hydrolysis, alkylation, and reactions with radicals, Chem. Res. Toxicol. 22 (11) (2009) 1747–1760.

[28] R.N. Grass, R. Heckel, M. Puddu, D. Paunescu, W.J. Stark, Robust chemical preservation of digital information on DNA in silica with error-correcting codes, Angew. Chem. Int. Ed. 54 (8) (2015) 2552–2555, https://doi.org/10.1002/anie.201411378.

[29] A.K.-Y. Yim, et al., The essential component in DNA-based information storage system: robust error-tolerating module, Front. Bioeng. Biotechnol. 2 (2) (2014) 1–5, https://doi.org/10.3389/fbioe.2014.00049.

[30] J. Bornholt, et al., A DNA-based archival storage system, IEEE Micro 44 (2) (2017) 637–649, https://doi.org/10.1109/MM.2017.264163456.

[31] R. Laddha, K. Honwadkar, Digital data storage on DNA, Int. J. Comput. Appl. 142 (2) (2016) 43–46.

[32] C.N. Takahashi, B.H. Nguyen, K. Strauss, L. Ceze, Demonstration of end-to-end automation of DNA data storage, Sci. Rep. 9 (1) (2019) 1–5, https://doi.org/10.1038/s41598-019-41228-8.

[33] R. Lopez, et al., DNA assembly for nanopore data storage readout, Nat. Commun. 10 (1) (2019) 2933, https://doi.org/10.1038/s41467-019-10978-4.

[34] L.C. Meiser, et al., Reading and writing digital data in DNA, Nat. Protoc. 15 (1) (2020) 86–101, https://doi.org/10.1038/s41596-019-0244-5.

[35] N. Goldman, et al., Toward practical high-capacity low-maintenance storage of digital information in synthesised DNA, Nature 494 (7435) (2013) 77–80, https://doi.org/10.1038/nature11875.

[36] G.M. Church, Y. Gao, S. Kosuri, Next-generation digital information storage in DNA, Science 337 (6102) (2012) 1628, https://doi.org/10.1038/21092.

[37] Y. Erlich, D. Zielinski, DNA Fountain enables a robust and efficient storage architecture, Science 355 (6328) (2017) 950–954, https://doi.org/10.1126/science.aaj2038.

[38] L. Organick, et al., Random access in large-scale DNA data storage, Nat. Biotechnol. 36 (3) (2018) 242–248, https://doi.org/10.1038/nbt.4079.

[39] S. Namasudra, R. Chakraborty, A. Majumder, N.R. Moparthi, Securing multimedia by using DNA-based encryption in the cloud computing environment, ACM Trans. Multimed. Comput. Commun. Appl. 16 (3s) (2021) 1–19, https://doi.org/10.1145/3392665.

[40] S. Namasudra, Fast and secure data accessing by using DNA computing for the cloud environment, IEEE Trans. Serv. Comput. (2020), https://doi.org/10.1109/TSC.2020.3046471.

[41] S. Namasudra, S. Sharma, G.C. Deka, P. Lorenz, DNA computing and table based data accessing in the cloud environment, J. Netw. Comput. Appl. 172 (April) (2020), 102835, https://doi.org/10.1016/j.jnca.2020.102835.

[42] S. Namasudra, An improved attribute-based encryption technique towards the data security in cloud computing, Concurr. Comput. Pract. Exp. 31 (3) (2019) e4364, https://doi.org/10.1002/cpe.4364.

[43] L. Anavy, I. Vaknin, O. Atar, R. Amit, Z. Yakhini, Improved DNA based storage capacity and fidelity using composite DNA letters, bioRxiv (2018) 433524, https://doi.org/10.1101/433524.

[44] Y. Choi, et al., Addition of degenerate bases to DNA-based data storage for increased information capacity, bioRxiv (2018) 367052, https://doi.org/10.1101/367052.

[45] S.M.H.T. Yazdi, R. Gabrys, O. Milenkovic, Portable and error-free DNA-based data storage, Sci. Rep. 7 (1) (2017) 1–6, https://doi.org/10.1038/s41598-017-05188-1.

[46] H.H. Lee, R. Kalhor, N. Goela, J. Bolot, G.M. Church, Enzymatic DNA synthesis for digital information storage, bioRxiv (2018) 348987, https://doi.org/10.1101/348987.

[47] L.M. Adleman, Molecular computation of solutions to combinatorial problems, Science 266 (5187) (1994) 1021–1024.

[48] R.J. Lipton, DNA solution for hard computational problems, Science 268 (5210) (1995) 542–545.

[49] L.M. Smith, et al., A surface-based approach to DNA computation, J. Comput. Biol. 5 (2) (1998) 255–267.

[50] Q. Liu, L. Wang, A.G. Frutos, A.E. Condon, R.M. Corn, L.M. Smith, DNA computing on surfaces, Nature 403 (6766) (2000) 175–179.

[51] R.S. Braich, N. Chelyapov, C. Johnson, P.W.K.K. Rothemund, L.M. Adleman, Solution of a 20-variable 3-SAT problem on a DNA computer, Science 296 (5567) (2002) 499–502.

[52] D. Faulhammer, A.R. Cukras, R.J. Lipton, L.F. Landweber, Molecular computation: RNA solutions to chess problems, Proc. Natl. Acad. Sci. U. S. A. 97 (4) (2000) 1385–1389.

[53] Q. Ouyang, P.D. Kaplan, S. Liu, A. Libchaber, DNA solution of the maximal clique problem, Science 278 (5337) (1997) 446–449.

[54] K. Sakamoto, et al., Molecular computation by DNA hairpin formation, Science 288 (5469) (2000) 1223–1226.

[55] D. Sharma, M. Ramteke, A note on short-term scheduling of multi-grade polymer plant using DNA computing, Chem. Eng. Res. Des. 135 (2000) (2018) 78–93.

[56] J. Chao, et al., Solving mazes with single-molecule DNA navigators, Nat. Mater. 18 (3) (2019) 273–279.

[57] D. Sharma, M. Ramteke, In vitro identification of the Hamiltonian cycle using a circular structure assisted DNA computer, ACS Comb. Sci. 22 (5) (2020) 225–231, https://doi.org/10.1021/acscombsci.9b00150.

[58] D. Sharma, M. Ramteke, DNA computing: methodologies and challenges, in: DNA- and RNA-Based Computing Systems, Wiley Online Library, 2021, pp. 15–29.

[59] W. Fan, A. Bifet, Mining big data: current status, and forecast to the future, ACM SIGKDD Explor. Newslett. 1 (1) (2013) 1–5, https://doi.org/10.21742/ijpccem.2014.1.1.01.

[60] A. McAfee, E. Brynjolfsson, T.H. Davenport, D. Patil, D. Barton, Big data: the management revolution, Harv. Bus. Rev. 90 (10) (2012) 60–68.

[61] V. Rajaraman, Big data analytics, Resonance 21 (8) (2016) 695–716.

[62] A. Gandomi, M. Haider, Beyond the hype: big data concepts, methods, and analytics, Int. J. Inf. Manage. 35 (2) (2015) 137–144, https://doi.org/10.1016/j.ijinfomgt.2014.10.007.

[63] R. Ratra, P. Gulia, Big data tools and techniques: a roadmap for predictive analytics, Int. J. Eng. Adv. Technol. 9 (2) (2019) 4986–4992, https://doi.org/10.35940/ijeat.B2360.129219.

[64] A. Mohan, Big data analytics: recent achievements and new challenges, Int. J. Comp. Appl. Technol. Res. 5 (7) (2016) 460–464.

[65] M. Hussain Iqbal, T. Rahim Soomro, Big data analysis: Apache Storm perspective, Int. J. Comput. Trends Technol. 19 (1) (2015) 9–14.

[66] B. Thillaieswari, Comparative study on tools and techniques of big data analysis, Int. J. Adv. Netw. Appl. 8 (5) (2017) 61–66.

[67] M.S. Al-Hakeem, A proposed big data as a service (BDaaS) model, Int. J. Comput. Sci. Eng. 4 (11) (2016) 15–21.

[68] A. Lakshman, M. Prashant, Cassandra: a decentralized structured storage system, ACM SIGOPS Oper. Syst. Rev. 44 (2) (2010) 35–40.

[69] D. Featherston, Cassandra: Principles and Application, 2010.

[70] S. Namasudra, D. Devi, S. Choudhary, R. Patan, S. Kallam, Security, privacy, trust, and anonymity, in: Advances of DNA Computing in Cryptography, Taylor & Francis, 2018, pp. 138–150.

[71] J.L. Banal, T.R. Shepherd, J. Berleant, H. Huang, M. Reyes, Random access DNA memory using Boolean search in an archival file storage system, Nat. Mater. 20 (9) (2021) 1272–1280, https://doi.org/10.1038/s41563-021-01021-3.

[72] K. Chen, E. Winfree, Error correction in DNA computing: misclassification and strand loss, Computing (2000) 49–63.

About the authors

Manojkumar Ramteke received a B. Tech degree in Chemical Engineering from Dr. Babasaheb Ambedkar Technological University, Lonere (MH), M. Tech, and Ph. D. in Chemical Engineering from the Indian Institute of Technology, Kanpur. Dr. Ramteke is an Associate Professor at the Department of Chemical Engineering, Indian Institute of Technology Delhi. His current research interests are industrial process modeling and multi-objective optimization, scheduling, planning and control of process operations, metaheuristic algorithms, and novel computing methods.

Deepak Sharma received a B. Tech from Maharshi Dayanand University, Rohtak, and M. Tech. from the Indian Institute of Technology Roorkee. He acquired a Ph.D. from the Indian Institute of Technology Delhi, New Delhi, India. He also worked as a Research Associate at the Department of Chemical Engineering, Indian Institute of Technology Delhi, on Deep Learning algorithms for early diagnosis of pancreatic cancer and cardiovascular disease.

CHAPTER EIGHT

Securing confidential data in the cloud environment by using DNA computing

Divyansh Agrawal and Sachin Minocha
School of Computing Science and Engineering, Galgotias University, Greater Noida, India

Contents

Abstract

With the immense growth of the data generated every day, cloud computing has gained attention due to its flexibility, on-demand services, scalability, etc. Nevertheless, the internet-based services through cloud computing platforms have provided ample

Advances in Computers, Volume 129
ISSN 0065-2458
https://doi.org/10.1016/bs.adcom.2022.08.006

opportunity for hackers and malicious users to violate and compromise data security and integrity. Thus, ensuring confidentiality and security of data became a major research challenge. As reported in the literature, chaos-based cryptography and DNA base encryption have gained popularity to meet the aforementioned challenge. This chapter proposes a two-layer encryption scheme that uses chaos-based cryptography and DNA computing to achieve a solid encryption scheme for the cloud environment. The proposed model is further robust by using the SHA-512 technique in selecting the control parameters. DNA computing is used to generate a secret key along with a decimal encoding and complementary rule. Theoretical analysis and experimental results show the efficiency of the proposed technique over the other well-known existing encryption methods for the cloud computing environment.

Abbreviations

AES	Advanced Encryption Standard
ASCII	American Standard Code For Information Interchange
CBIE	chaos-based image encryption
CIS	Cloud Information Service
CTAC	Cross Tenant Access Control
DNA	deoxyribonucleic acid
DNA-ChSHA	deoxyribonucleic acid-Chaotic Secure Hash Algorithm
MIPS	million instructions per second
mRNA	messenger ribonucleic acid
ODM	one-dimensional map
OTP	one-time pad
PCR	polymerase chain reaction
RDHS	reversible data hiding scheme
RSA	Rivest-Shamir-Adleman
SHA	Secure Hash Algorithm
SSL	Secure Socket Layer
TDM	two-dimensional map
TTP	trusted third party
VMM	Virtual Machine Manager

1. Introduction

Since the inception of cloud technology, data storage and retrieval are becoming more efficient without worrying about maintenance overhead and costs. Cloud computing has enabled businesses to only pay for what they consume. By using this technology, businesses can outsource the management of their resources and focus on their niche work without concerning about managing the computing resources. Nowadays, all the information technology companies are using cloud computing technology directly or indirectly and can manage their data for business insights in terms of time series performance using data analysis and machine learning. The popularity of cloud computing comes at the expense of facilitating unauthorized data access by malicious users. The traditional approaches for data accessing and data encryption in cloud computing are going to become obsolete because of the ease of attack. Therefore, developing more secure and faster data access and encryption techniques have become a significant concern.

In DNA computing, a sequence of nucleotides is combined to form a DNA strand, and DNA strands are intertwined with each other to form a DNA molecule. DNA strands have the capability of storing a larger amount of data with improved security. It can be well suited to encrypt any data before storing it on the cloud server. While the traditional encryption techniques use 0s and 1s for data encryption, DNA computing uses the four bases of DNA, i.e. A, G, T and C, for data encryption. DNA computing has numerous applications in terms of encrypting the data, ranging from textual data to multimedia data like images, videos, audios, etc. The authors of Ref. [1] have proposed a DNA-based encryption technique to encode multimedia content. Here, the owner of the data generates a random DNA-based 1024-bit secret key. Then, multimedia is encrypted using this secret key and his own private key. Another layer of encryption is performed in this scheme using the public key of the cloud service provider. This scheme takes less time to generate key and key retrieval, as well as multimedia file data encryption and decryption. Lai et al. [2] have proposed an asymmetric encryption method using DNA computing. Their proposed technique uses the two pairs of keys for the encryption and signature, respectively. The keys and the ciphertext are generated in the form of DNA molecules. Namasudra et al. [3] have used DNA cryptography to encrypt big data. This scheme proposes the generation of a 1024-bit secret key using the user attributes, MAC address, decimal encoding rule, DNA bases and complementary rules. In this scheme, data are encrypted in the first stage, and the second stage involves encrypting the data with the data owner's private key. Finally, it is encrypted using the public key of the cloud service provider. Vijayakumar et al. [4] have proposed a DNA computing-based elliptic curve cryptography technique. This technique performs the encoding of raw text using DNA computing, and the encoded text is encrypted using elliptic curve cryptography. The scheme of Vijayakumar et al. [4] performs better as compared to traditional techniques in terms of robustness. Sukumaran and Mohammed [5] have proposed the client end encryption of data using DNA computing. This technique encrypts the data before uploading the same to the cloud. It is secured, but the data access time is much high as compared to traditional techniques. Sohal and Sharma [6] have proposed a symmetric key encryption scheme using DNA computing before uploading the encrypted data on the cloud server. This technique consumes high access time of data, which leads to delay in data accessing.

One of the key challenges for the data retrieval of encrypted data in cloud computing is the delay. In Ref. [7], a data accessing scheme is proposed to reduce the access time significantly. Here, the cloud service provider maintains a table based on the data or file size of the data for different data owners. When a user initiates an access request, the cloud service provider initiates a request based on the desired size in that table to find that specific data owner preventing the need to search through the complete database. Hu et al. [8] have proposed a robust and secure method to encrypt a color image using a Fibonacci chaotic sequence. This dynamic key-based encryption method scrambles the pixel of an image using the chaotic sequence. This scheme is highly robust and secure as well, however, it consumes a large amount of memory to store the same over the cloud environment. Tobin et al. [9] have proposed a process to personalize the One-Time Pad (OTP). The OTPs are generated for each data, and encrypted data is uploaded to the cloud for improving the data security. Singh and Saroj [10] have proposed a public auditing scheme for data in the cloud environment, which uses the Advanced Encryption Standard (AES-256) for encryption, SHA-512 for the integrity checking and RSA-15360 for the public key encryption. Their scheme supports efficient auditing. However, the storage issue over the cloud environment has not been addressed by the authors. Sighom et al. [11] have proposed a model to enhance data security by combining AES and SHA. The data accessing time of this scheme is impressive, but consumes large amount of memory for storing all the details. Moreover, this technique is vulnerable against man–in–the–middle attack.

Most of the current schemes for encrypting the data for the cloud environment try to improve the security over the previously existing schemes. Some of them have also tried to reduce the access time of data. However, performance is not much high. Thus, there are no existing algorithms that provide an encryption technique, which is equally fast and secure. To minimize the problems of the aforementioned existing schemes, a novel data encryption scheme is suggested in this chapter for the cloud environment by using DNA computing, SHA-512 and chaotic map. The proposed scheme provides the two-layer of security by using chaos theory and DNA computing-based encryption.

The control parameter of the chaos-based encryption is determined based on the value generated by SHA-512. This two-layer encryption process is applied to raw data to generate ciphertext in the form of DNA molecules. This ciphertext is stored on the cloud server that ensures improved data security and consumes less time. The main contributions of this chapter include:

1. To significantly improve the data security of the cloud environment, a two-layer encryption algorithm based on chaotic maps and DNA computing is introduced in this chapter.
2. SHA-512 is used in the proposed scheme to choose the control parameters for the chaos-based cryptosystems. This supports more randomness in the data encryption phase.
3. The proposed approach can resist many security attacks, namely masquerade attack, brute force attack, man-in-the-middle attack, collision attack and statistical attack, of the cloud computing environment.
4. Experimental analysis and comparisons of the cryptosystem's features are also presented in the chapter to show the effectiveness of the work over the well-known existing encryption algorithms for the cloud computing environment.

This chapter is organized as follows: The literature review has been presented in Section 2. Section 3 discusses the preliminary studies on chaotic maps and SHA-512. Section 4 deals with the system requirements, objectives, and entities involved in the proposed approach. The proposed technique is detailed in Section 5. Security analysis and results are represented in Sections 6 and 7, respectively. Finally, the concluding remarks and future research directions are presented in Section 8.

2. Literature review

The initial security schemes focused on developing more secure data systems rather than focusing on speed. In this section, many existing schemes are discussed in detail.

One of the popular applications of DNA computing has been presented in Ref. [12]. Here, Clelland et al. have proposed a scheme to hide the message in microdots. Microdot was a technology developed by the Germans during World War 2 to communicate secret information. A microdot is an extremely reduced photograph of a page written with the help of a typewriter and pasted on the full stop of a friendly letter. This scheme first encrypts the message using a large and complex human genomic DNA and this information is further hidden in a microdot. However, this scheme is slow.

Amin et al. [13] have used DNA computing for its large storing capacity and property of parallelism, which allows strengthening other cryptographic techniques. In this algorithm, the binary data is first encoded in a sequence of DNA bases. Then, an efficient binary searching algorithm locates multiple positions for each DNA base in the sequence in a selected genome. Once these positions are located, they are assembled in a file of pointers to the location of each DNA base of the genome.

In Ref. [14], a symmetric key cryptographic scheme has been proposed by using DNA computing. The encryption and decryption keys are generated in this scheme using DNA computing. This method hybridizes and identifies millions of DNA probes at the same time. Therefore, the encryption and decryption processes of the message are performed in a parallel way. Despite parallel functioning, the use of on-the-fly hybridization makes the algorithm slower.

In 2000, Leier et al. [15] have proposed another DNA encryption approach, where a two-stage architecture is used. The first stage deals with using DNA strands for steganography. This allows fast encryption and decryption. The second stage performs a graphical subtraction to strengthen the security of the overall system. However, the security of this system relies on computation power. Leier et al. have assumed that the attacker has the same "technological capabilities" as the sender and receiver of the message. But, due to the significant improvement of the hardware technologies, it allows users to use more computation power than before. So, the scheme fails in the present scenario, where there is a lack of computational power.

Instead of using actual DNA computing techniques, in Ref. [16], fundamental principles of molecular biology like transcription, splicing and translation are used to develop a novel scheme. In this scheme, once a DNA segment is spliced, the sequence is transcribed onto a single-strand mRNA (messenger ribonucleic acid). Finally, this mRNA sequence is translated into a sequence of amino acid. This scheme differs from the original splicing technique in terms of random noncontinuous start and end codes. The partial information in the cipher makes this technique weak.

Borda and Tornea [17] have discussed secret message writing method that includes OTP, DNA XOR rule and chromosome indexing methods, which are base for biomolecular computation. They have discussed the features and advantages of DNA cryptography. The authors have validated the advantages and limitations of DNA cryptography using laboratory experiments. Their work also exhibits that chromosome indexing is not a proper DNA cryptography algorithm, but utilizes high randomness of DNA.

Key exchange is another major issue with encrypting or decrypting the message as attackers can attack in the middle of the communication [18]. The authors of Ref. [19] have proposed a public key encryption system, including DNA computing as a one-way function to effectively distribute the keys among the stakeholders. The secret key information is stored in a random DNA sequence, and then, it is retrieved through Polymerase Chain Reaction (PCR) amplification followed by sequencing. This scheme is slow for multimedia data and it is not secured against the password guessing attack.

In Ref. [20], a secret sharing method has been proposed based on a probabilistic approach using the XOR function with DNA computing. The data in this method or algorithm is shared in three different batches and a cover image is used as a secret key. This algorithm is prone to a side-channel attack. Wang et al. [21] have used the same concept of DNA XOR reversible data hiding technique to encrypt sensitive multimedia data. Unlike other encryption techniques, this technique restores the exact information as the original information. One major drawback of this technique is that if an attacker retrieves the binary encoding rule and/or DNA XOR rule, s/he can easily retrieve the original data content.

Since the bases of the DNA can be used to generate millions of combinations to mask the data, it has been seen as an ideal approach for current encryption needs. But, selecting the ideal mask from those combinations is a tough task. Enayatifar et al. [22] have proposed chaos-based image encryption based on logistic mapping, DNA sequencing and genetic algorithm. In the first stage, many different DNA masks are generated through DNA sequencing and logistic maps. The next stage involves using a genetic algorithm to find the best DNA mask developed in the first stage. The authors of Ref. [23] have attempted to reduce the encryption and decryption time using parallel cryptography based on DNA computing, OTPs and DNA hybridization. The OTP is selected as a single-strand DNA string for encryption. Here, the decryption process is executed in a parallel computing environment to hybridize the DNA in parallel. This allows a reduced time for data decryption. However, this technique is not highly secured and weak against attacks like password guessing attacks and masquerade attacks.

3. Preliminary studies

This section covers the chaotic maps and the SHA-512 algorithm.

3.1 Chaotic maps

A chaotic map is a cryptosystem used to generate nonlinear and sophisticated random sequences to encrypt the original data. One of the initial works using chaotic maps for cryptography was developed in 1989 [24]. The work discusses how nonlinear iterative functions can be used to generate random sequences of numbers, i.e. chaotic sequence. The output from the chaotic maps depends highly on the control parameters and initial settings. These control parameters are treated as secret keys in cryptography, making chaotic maps a strong cryptosystem. The chaotic maps are classified by dimensions as mentioned below.

3.1.1 One-dimensional map (ODM)

ODMs are those chaotic maps that use a single parameter. This dimension is time, i.e. the physical system can be modeled as a function of discrete-time. These maps use an interval of [0, 1]. ODMs can be defined as Eq. (1) [25].

$$m_N(t + 1) = \frac{\left(\delta * C_N * \sqrt{m(t)}\right)^2}{1 + \left(\delta^2 - 1\right)\left(C_N\left(\sqrt{m(t)}\right)^2\right)} \tag{1}$$

where t represents time, N is any integer greater than 1, m_N represents $(N-1)$ model map, i.e. this function has $(N-1)$ critical points in the interval [0, 1], δ is the control parameter and C_N represents Chebyshev polynomials of type 1. Some examples of the one-dimensional chaotic map include:

1. **Logistic map:** It is the most basic type of chaotic map [26]. It is a nonlinear map with a nonlinearity factor of 2, i.e. quadratic. The logistic map can be expressed as Eq. (2).

$$m(t + 1) = \delta * m(t) * (1 - m(t)) \tag{2}$$

where t represents time, m represents the number of critical points in the interval [0, 1] and δ is the control parameter. After a certain threshold value of δ (between 3.5 and 4), the map starts to behave chaotically [27].

2. **Tent map:** A tent map also known as a triangle map is a single dimension iterative chaotic map [28]. The name tent map comes from the fact

that the shape of the graph for the function in the interval [0, 1] forms a tent. The function for the tent map is given in Eq. (3).

$$m(t+1) = \begin{cases} \delta.m(t), & 0 \leq m(t) \leq 0.5 \\ \delta.(1-m(t)), & 0.5 < x < 1 \end{cases} \tag{3}$$

where δ is the control parameter and $m(t)$ represents the chaotic map function, which calculates the critical points. Even though the function is linear, it tends to show chaotic behavior for some control parameter values.

3. **Cosine polynomial map:** Cosine Polynomial Map is an example of the one-dimensional chaotic map proposed by Talhaoui et al. [29] in 2020. The map can be represented using Eq. (4).

$$m(t+1) = \cos\left(\delta\left(m^3(t) + m(t)\right)\right) \tag{4}$$

where δ is the control parameter. The negative values of δ are ignored because of the use of even cosine function. This map shows chaotic behavior for most values of δ thus provides an infinitesimally large chaotic space for positive real numbers.

3.1.2 Two-dimensional map (TDM)

Unlike the ODMs, TDMs are represented as a function of two variables. The map can be expressed by Eq. (5) [30].

$$f(x, y) = \left(\frac{x}{\alpha}\right)^\gamma + \left(\frac{y}{\beta}\right)^\gamma - \sigma^2 \tag{5}$$

where α, β, γ and σ are control parameters. x, y are the dimensional parameters on which the map function is dependent. The equations for the two dimensions are shown in Eqs. (6) and (7) [31].

$$x(t+1) = \frac{\left(\delta * C_N * \sqrt{x(t)}\right)^2}{1 + \left(\delta^2 - 1\right)\left(C_N\left(\sqrt{x(t)}\right)^2\right)} \tag{6}$$

$$y(t+1) = \frac{\left(\delta * C_N * \sqrt{y(t)}\right)^2}{1 + \left(\delta^2 - 1\right)\left(C_N\left(\sqrt{y(t)}\right)^2\right)} \tag{7}$$

where t represents time, N is any integer greater than 1, $x(t+1)$ and $y(t+1)$ represent $(N-1)$ model map, i.e. the function has $(N-1)$ critical points in the interval [0, 1], δ is the control parameter and C_N represents Chebyshev polynomials of type 1.

1. **Hénon map:** Hénon map was developed by M. Hénon in 1976, is an example of two-dimensional maps [32]. It is a time-dependent nonlinear map. The equations for x and y variables in Eqs. (6) and (7) for the Hénon map can be expressed as given in Eq. (8).

$$\begin{cases} x(t+1) = 1 - \alpha x^2(t) + y(t) \\ \qquad y(t+1) = \beta x(t) \end{cases} \tag{8}$$

2. **Baker map:** Baker map is another two-dimensional chaotic map proposed by Jiri Fridrich in 1997–1998 [33,34]. It is a simple two-dimensional function, which outputs a map that is a unit square chaotic bijection on itself. The Baker map can be described with the help of Eq. (9).

$$f(x, y) = (x(t+1), y(t+1))$$
$$= \begin{cases} \dfrac{x(t)}{\rho}, \rho y(t) & 0 < x \le \rho \\ \dfrac{x(t) - \rho}{1 - \rho}, (1-\rho)y(t) + (1-\rho) & \rho < x \le 1 \end{cases} \tag{9}$$

where ρ is the control parameter that lies between 0 and 1. When the value of ρ is 0.5, it is known as the standard Baker map.

3. **Arnold's cat map:** Arnold's cat map, or simply the cat map, was first proposed for cryptographic use by Gabriel Peterson in 1997 [35]. It is based on the observation made by a Russian mathematician Vladimir I Arnold. He discovered a chaotic pattern on a cat. This transformation is applied to the image to randomize the pixels of the image in such a way that the original orientation of pixels, i.e. original image can be retrieved by applying a few iterations of the same transformation. In other words, the image can be encrypted and decrypted using the same transformation function is given in Eq. (10) [36].

$$\begin{cases} x(t+1) = [\alpha x(n) + \beta y(n)] \, mod \, N \\ y(t+1) = [\gamma x(n) + \varepsilon y(n)] \, mod \, N \end{cases} \tag{10}$$

where α, β, γ *and* ε are the control parameters and N is an integer value.

3.2 Chaotic maps as cryptosystem

These chaotic maps can be used to encrypt the data. The chaotic maps exhibit "avalanche" behavior that gives different results on a minor change in the initial condition or control parameters. This acts as a catalyst to use chaotic maps for encryption. Some steps need to be followed to use the chaotic maps to develop cryptosystems [34].

Step 1 **(Designing the chaotic map):** The chaotic map is a one-to-one function that can be selected from a large number of available chaotic maps. There are several conditions on choosing the map like the map should be simple, so that the encryption and decryption process can be quick. The selected map should allow parameterization, i.e. it must be able to create a short cipher key from a large number of possible cipher keys.

Step 2 **(Generalizing the map):** Generally, the maps are described geometrically because parameterizing them is very easy. The second step involves introducing a set of initial parameters to create a specific part of the cipher key. For example, a two-dimensional chaotic map can be identified by a sequence of numbers.

Step 3 **(Discretizing the map):** In the third step, the map is generalized according to the data format. For example, if the data is an image, the map needs to be converted to a finite lattice of points (number of pixels in this case). Now, the discretized map assigns a pixel to another random pixel in a bijective fashion. Similarly, discretizing the map for a normal text message forms a one-dimensional lattice (number of binary bits in this case). The map then assigns each binary bit to another binary bit in the same bijective fashion.

Step 4 **(Increasing the cipher strength):** Up to Step 3, a permutation cipher is already achieved, but this can be converted to a strong substitution cipher by increasing the dimension of the map. By increasing the cipher's complexity a little, a very strong substitution cipher can be achieved.

Step 5 **(Adding diffusion and confusion to the mix):** Though the substitution cipher achieved from Step 4 provides decent security, it lacks diffusion and confusion, two of the main properties of a strong cipher [37]. Confusion refers to the property that each bit in the ciphertext should be dependent on several parts of the key, thus increasing the ambiguity and makes it nearly impossible to extract the key from the ciphertext's bits. Diffusion refers to the property by which changing a single bit in the plaintext effects the ciphertext marginally and vice versa. Statistically, the ideal value for this change should be half the number of the total bits. Techniques like substitution boxes, permutation boxes and linear feedback registers are used to achieve these properties. Finally, a strong product cipher is produced using the chaotic map.

3.3 SHA-512

Secure Hash Algorithm or SHA is a hashing algorithm, which was developed by the National Institute of Standards and Technology in 1993 as SHA-1. Then, many SHA algorithms are released with different variants like SHA-256, SHA-384 and SHA-512. SHA-512 works on the input of 2128 bits and produces a 512-bit message digest. The input message is processed in blocks of 1024 bits. The working of SHA-512 is discussed in this section.

Step 1 **(Padding message with extra bits):** The message's length is made congruent to 896 modulo 1024 using the padding bits. The number of padding bits ranges between 1 and 1024.

Step 2 **(Append 128-bit unsigned integer):** A 128-bit block is appended to the string produced in Step 1. This block is an unsigned 128-bit integer and contains the length of the original message.

The produced output is a message that is an integer multiple of 1024. This message is then expressed as sequences of N 1024-bit blocks represented by $M_1, M_2, M_3, ..., M_N$. So, the length of the original message becomes $N*1024$ bits.

Step 3 **(Initiate buffer value for hashing):** A 512-bit buffer corresponding to each 1024-bit block is initialized to store the intermediate and final values achieved by the hash functions. These buffers are identified as eight 64-bit registers named a, b, c, d, e, f, g and h.

Step 4 **(Generate hash code for 1024-bit blocks):** In this step, the entire message formed in Step 2 is processed using operation "F" for hashing. The step is repeated for 80 rounds. Each round takes the 512-bit buffer value as input and updates the contents of the buffer.

The intermediate hashes are represented by $H_1, H_2, ..., H_{N-1}$ and the final hash code is represented by H_N.

Since each round takes the intermediate hash value from the previous round as input, the input of the first round has to be taken in advance known as Initial Vector (IV). Each round R in the algorithm makes use of a 64-bit value, W_R derived from the block being processed and a constant K_R. All the N blocks are processed in the same way and an output of 512 bits is received from the Nth iteration. Table 1 shows parameters of different SHAs and Fig. 1 depicts a workflow of the SHA-512 algorithm as explained above.

Table 1 Working parameters of different variants of SHA.

	SHA-512	SHA-384	SHA-256	SHA-1
Message digest size	512	384	256	160
Message size	$<2^{128}$	$<2^{128}$	$<2^{64}$	$<2^{64}$
Block size	1024	1024	512	512
Word size	64	64	32	32
No. of steps	80	80	80	80
Security	256	192	128	80

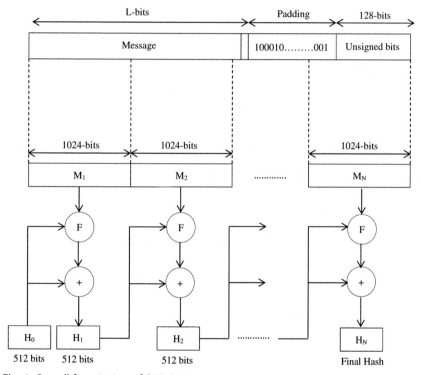

Fig. 1 Overall functioning of SHA-512.

4. Problem statement

In this section, system requirements, objectives and the system model of the proposed scheme have been discussed.

4.1 System model

The proposed scheme consists of three entities with their different responsibilities as discussed below:

1. **Cloud service provider:** This entity acts as an administrator for the cloud environment. It is responsible for providing the storage space and other cloud services to the users and for maintaining the overall cloud infrastructure, which includes having high computation power, abundant memory space and many more.
2. **Data owner:** This entity acts as a source of the data stored in the cloud server. Its responsibility is to provide access rights to the intended users or customers of the cloud environment.
3. **User:** The user is the entity that wants to utilize the cloud services from a cloud service provider. The users must go through the credential check phase, before accessing stored data from the cloud server.

The user's data is the most important aspect of any cloud environment. This data can be confidential for any company or individuals. For both aspects, it is necessary to protect the data from any unauthorized access. This data can be anything like company's sales of a certain period (time series data) [38], company's employees' details like designation, salary details, project details, etc.; employees' details like name, address, date of birth, sex, ethnicity, religious beliefs, etc. Cloud technology can also be used for any individual work like for storing banking details, images, videos, music and many more.

4.2 System requirements and objectives

The main objectives and system requirements of the proposed technique are discussed below:

1. **Security:** Security is the most important and most expected aspect of the cloud environment. The cloud computing environment utilizes internet-based processes almost for every task that open up a lot of possibilities for hackers and malicious users to steal sensitive and confidential information. In order to avoid security breaches, the cloud service

providers must maintain a cloud system that can ensure the security of all types of data. Though the Trusted Third Party (TTP) is indeed trusted, it is better to share further encrypted data to the cloud storage, so that the security of the cloud server is improved significantly. Thus, improving security is one of the major goals of the proposed work.

2. **Access control:** The data owners must ensure that the data are not shared with the person, who is not intended for it. They must check for credentials given by the user to access data, and then, only allow the authorized users to access data from the cloud server.

3. **Reduce system overhead:** In order to securely allow the user for accessing their data, the data owners are expected to be online throughout the process, thus it increases the overhead of the system. A suitable solution to reduce the system overhead is always a major requirement of any cloud server, and in this chapter, an attempt is made to solve this issue of a cloud environment.

5. Proposed work

The proposed cryptosystem is based on chaotic maps and SHA-512. Here, an efficient encoding technique based on DNA computing has been developed for fast and secure data retrieval from the cloud environment. The proposed scheme first uses a one-dimensional cosine polynomial map to generate a chaotic sequence in order to encrypt the data. To improve the security even further, SHA-512 has been used to generate the control parameters of the chaotic map. Finally, the data is encoded using a 1024-bit key generated using the bases of DNA, namely adenine, guanine, cytosine and thymine. Then, this encrypted data is stored in the cloud environment. To access the data, the user sends a data access request to the cloud service provider. The service provider gives the public key of the data owner to the user. The user sends the request to the data owner to get the secret key and encrypted certificate. Once the data owner verifies the credentials, data access from the cloud is granted to the user. The overall architecture of the proposed scheme is shown in Fig. 2.

5.1 System setup

In the system setup, the first step is to select the control parameter δ used by the map. For achieving the chaotic behavior of the map, the value of δ should be greater than 1.05. This value is generated using an input of the user's choice of the SHA-512 hashing algorithm. It produces a 512-bit

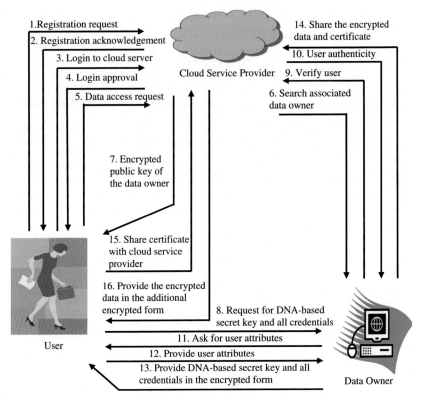

Fig. 2 Overall architecture of the proposed scheme.

digest, which is then used to generate δ. Once the value is set, the data is encrypted using the selected δ value that means data is encrypted using the chaotic method. Then, a big prime number (ρ) is chosen by the service provider to identify a multiplicative group Z_ρ^*. The cloud service provider selects a pair of the public and private key from Z_ρ^*. Then, public and private key pairs are also selected from Z_ρ^* by the service provider for both the data owner and the user. These key pairs are given to the entities at the time of their registration on the cloud platform. The users' public keys are publicly available, and the authorized users know the public key of the data owner. Only the authorized entities know the service provider's public key and an individual entity's private key is kept secret from the other entity.

5.2 User registration on cloud platform

The user requests the cloud service provider to get registered on the cloud platform. On receiving the request, the service provider asks for certain personal information from the user like their name, date of birth, residence, etc. to create a profile on the user's behalf. A Secure Socket Layer (SSL) is created to share the user's key pair, registration reply and a duplicate identity. Only the user can extract the real identity from the duplicate identity, and any attacker trying to fetch details from the duplicate identity must have the secret key. The same process takes place for the registration of the data owner on the cloud platform.

5.3 User login

Once the user is registered to the cloud service, the user is allowed to login into the server in order to access the data stored in the cloud storage. When the request to access the data is generated by the user, the service provider sends the public key of that specific data owner in encrypted form to the user. The user sends a request to the data owner for DNA computing-based secret key and the certificate. Once the data owner verifies the user's credentials, the data owner provides the user with the requested data's secret key and certificate in the encrypted form and the user can then decrypt it using his own private key and data owner's public key [39].

5.4 DNA-based key generation

Once the data owner verifies the authenticity of the user, it creates a secret key using the user's attributes based on decimal encoding and DNA computing. The attributes are first decimal encoded using the reference table shown in Table 2.

This encoding rule provides randomness to the overall algorithm as this decimal encoding rule is not general and could be changed as and when needed. The decimal obtained from here is converted into their corresponding 8-bit binary numbers and further these 8-bit binary numbers are converted to the corresponding ASCII value using Table 3.

This is done to assign a random integer value to the 8-bit binary number obtained. These ASCII values are again converted to an 8-bit binary number. If the length of the output produced is less than 1024-bits, extra bits (0s) are padded to make the length equal to 1024-bits. If the length of the output is more than 1024-bits, the extra bits from the right end are deleted to make the length equal to 1024-bits. The bits are deleted from the right end because most of the systems follow "little-endian" architecture where the least significant bits (LSB) are found on the right end of the number. A 1024-bit

Table 2 Decimal encoding rule.

Digit/ symbol/ alphabet	Decimal value	Digit/ symbol/ alphabet	Decimal value	Digit/ symbol/ alphabet	Decimal value	Digit/ symbol/ alphabet	Decimal value
1	01	`	11	A	25	a	51
2	02	~	12	B	26	b	52
3	03	!	13	C	27	c	53
4	04	@	14	D	28	d	54
.	.	.	.	.	.	.	.
.	.	.	.	.	.	.	.
.	.	.	.	.	.	.	.
8	08	(	22	X	48	x	74
9	09	)	23	Y	49	y	75
0	10	Blank space	24	Z	50	z	76

Table 3 ASCII catalog.

8-bit Binary number	ASCII
00000000	5
00000001	151
.	.
.	.
.	.
11111110	122
11111111	198

output is produced and is divided into 4 equal sections of 256 bits. Then, each group is mapped to the 4 DNA bases, i.e. A, G, T and C.

A randomly selected DNA reference key generated from the publicly available DNA bases is also included in the above sequence by the data owner. Finally, a complementary rule is applied to the DNA bases produced to add the confusion property to the resultant sequence. The complements of the bases are shown in Table 4.

Table 4 Complements of the DNA bases.

Bases	Complement
A	C
C	G
T	A
G	T

This whole process of key generation could be explained as an algorithm shown below:

Step 1: Once the user is verified, the user attributes are taken into account by the data owner.

Step 2: Decimal encoding of the string generated from Step 1 is done using Table 2.

Step 3: Each decimal encoded number is converted to its equivalent 8-bit binary representation.

Step 4: The 8-bit binary numbers are converted to the ASCII value with reference from Table 3.

Step 5: These ASCII values are again converted to their 8-bit binary representation.

Step 6: Padding or deletion to the output from Step 5 is performed making the length equivalent to 1024 bits.

Step 7: Break the resultant string into 4 equal parts of 256 bits each and assign a DNA base to each block.

Step 8: A randomly selected DNA reference key generated from the publicly available DNA bases is also included in the DNA sequence by the data owner.

Step 9: Each DNA base in the resultant sequence is complemented using Table 4.

A message is generated by the data owner to share the DNA-based secret key generated by Step 8 encrypted by the data owner's private key and user's public key, Tables 2–4, DNA reference key and the complementary rule. This message is decrypted by the user using its private key and the data owner's public key.

5.5 Double layer data encryption

Data encryption phase is executed by a data owner, which consists of two layers. Layer 1 uses the encryption of data using the one-dimensional cosine polynomial chaotic sequence. The value of δ, i.e. control parameter for the chaotic sequence is computed by SHA-512 algorithm. The benefit of using SHA-512 for deciding the value of δ is that SHA-512 hash is irreversible. The value of δ should also be hidden from the malicious users in order to keep the chaotic sequence safe. Choosing a random value from the real numbers can be easy for attackers to hack by using a brute force attack.

5.5.1 Layer 1 encryption: One-dimensional cosine polynomial chaotic based encryption

In this phase, at first, the user provides a random string as input; SHA-512 algorithm produces a 512-bit hexadecimal hash H of the input string using the encryption algorithm shown in Section 3.3. Once a hexadecimal string of 512 bits is produced, it can be rewritten as 128 nibbles. Out of those 128 nibbles, 2 different subsets of 96 nibbles (n_1) and 32 nibbles (n_2) are created. The control parameter δ for the one-dimensional cosine polynomial chaotic map is calculated using Eq. (11).

$$\delta = \frac{h2d(n_1)}{h2d(n_2)} + h2d(n_1)mod(h2d(n_2)) \tag{11}$$

where $h2d()$ is a function that converts a hexadecimal string to its equivalent decimal format. The hexadecimal digest produced by SHA-512 is converted into the decimal since the control parameter for the chaotic map should be a real decimal value. The steps in layer 1 encryption are as follows:

Step 1: Read the data D in the form of a matrix, round keys m_{ir}, δ where $i \in \{1,2,3,4\}$, $r \in \{1,2,...,R\}$ and R is the number of encryption rounds. It is worth mentioning here that δ is generated using SHA-512 while m_{1r}, m_{2r}, m_{3r}, m_{4r} and R are given by the user.

Step 2: Then, the round counter r is initialized to 1.

Step 3: Initialize X and Y as the number of rows and number of columns.

Step 4: Generate the encryption sequence S of length X using the cosine polynomial map given in Eq. (4) with m_{1r}, and δ are set as initial conditions.

Step 5: In the fifth step, the encryption sequence value S is normalized using Eq. (12).

$$S = \{S_i \mid S_i = (S_i * 10^7) \ mod \ X\} \tag{12}$$

where S_i is the ith member of encryption sequence.

Step 6: The round key m_{5r} is calculated using Eq. (13).

$$m_{5r} = m_{1r} + \mu\left(\frac{D}{\{D_1,\ D_{S_1}\}}\right) \tag{13}$$

where $\mu\left(\frac{D}{\{D_1,\,D_{S_1}\}}\right)$ is the average bit value for data D excluding the first bit (in case of image: first row) and the first encryption sequence. The m_{5r} is required for the initial conditions to generate a random sequence for the first row.

Step 7: In the seventh step, P and Q sequences of length Y are generated using the cosine polynomial map given by Eq. (4) with m_{3r}, m_{5r} and δ.

Step 8: P and Q are normalized using Eqs. (14) and (15).

$$P = \{P_i \mid P_i = (P_i * 10^7)\ mod\ 256\} \tag{14}$$

$$Q = \{Q_i \mid Q_i = (Q_i * 10^7)\ mod\ 256\} \tag{15}$$

Step 9: In the ninth, for each iteration, starting from the first bit to the last bit, encrypt D_i and D_{S_i} using Eqs. (16) and (17).

$$D_i = cs((D_i + f_i + \rho_i)\ mod\ 256, S_i) \tag{16}$$

$$D_{S_i} = (D_{S_i} + f_i + \rho_i)\ mod\ 256 \tag{17}$$

where $cs((D_i + f_i + \rho_i)\ mod\ 256,\ S_i)$ is a circular right shift operation on $((D_i + f_i + \rho_i)\ mod\ 256)$ for S_i times. f_i and ρ_i are calculated by Eqs. (18) and (19).

$$f_i = \begin{cases} P, & if\ i \neq 1 \\ Q, & if\ i = 1 \end{cases} \tag{18}$$

$$\rho_i = \begin{cases} D_X, & if\ i = 1 \\ D_{i-1}, & else \end{cases} \tag{19}$$

Step 10: The transpose operation is performed on the obtained matrix D for performing the permutation and substitution of the columns as well. Therefore, Steps 3–9 are repeated to encrypt the columns. m_{2r}, m_{4r} and δ are used as the initial conditions instead of m_{1r}, m_{3r} and δ for column operation.

Step 11: In the 11th step, another round of transpose operation is performed on the resultant ciphered data.

Step 12: Here, the round counter r is incremented and Steps 3–11 are repeated until $r > R$.

The output of layer 1 gives the cipher data encrypted using the one-dimensional cosine polynomial chaotic sequence.

5.5.2 Layer 2 encryption: DNA-based data encryption

Layer 1 encryption is followed by layer 2 encryption that uses DNA computing to encrypt the data. The encrypted data, i.e. the output of layer 1 is first broken into 1024-bit blocks and further each 1024-bit block is broken into 4 blocks of 256 bits. Each block is converted to corresponding ASCII values using the 8-bit binary sequence in the block. These ASCII values are further converted to the 8-bit binary representations. This is executed for each 256-bit block.

These 256-bit blocks are XORed (Exclusive OR) with a 256-bit block of the DNA key. After performing all the XOR operations on the 1024-bit block, every 2-bit binary number in the sequence is assigned a DNA base using Table 5. A total of 24 combinations are possible as shown in Table 5, the data owner can choose any of the combinations to use for encryption.

The same DNA reference key is used in the key generation phase and is added in the middle of the DNA base sequence generated from the above process. The same complementary rule is also applied to this sequence as well. The DNA sequence produced so far is encrypted using the data owner's private key and the cloud service provider's public key. Finally, this encrypted data is sent to the cloud server's database. Layer 2 encryption process consists of the following steps:

Table 5 DNA conversion of 2-bit binary numbers.

2-bit Binary number			
00	01	10	11
G	T	A	C
G	C	A	T
T	C	A	G
T	G	A	C
C	T	A	G
.	.	.	.
.	.	.	.
.	.	.	.
A	C	G	T
A	T	G	C

Step 1: Break the encrypted data into blocks of 1024 bits.

Step 2: In the second step, each 1024-bit block is broken into 4 blocks of 256 bits.

Step 3: Each block is processed in 8-bit binary numbers, and each 8-bit binary number is mapped to its corresponding ASCII value.

Step 4: Each ASCII value is again converted to an 8-bit binary number.

Step 5: Repeat Steps 3 and 4 for each 256-bit block.

Step 6: Every 256-bit block produced from Step 5 is XORed with a 256-bit block of the DNA key generated in the previous section.

Step 7: Group the 1024-bit output from Step 6 into 2 bits. Then, assign each 2-bit binary number with a DNA base using Table 5.

Step 8: Repeat Steps 2–7 for each 1024-bit block formed from the data to be encrypted.

The encrypted data using the two layers described above is stored over the cloud server after encrypting using the data owner's private key and service provider's public key. Thus, data security is improved significantly.

5.6 Data accessing from cloud server

There are many steps to access any data from the cloud server securely.

Step 1: The user generates a request to the cloud service provider encrypted by its own private key and service provider's public key.

Step 2: When the service provider receives the request in the encrypted format, it searches for the requested data owner's public key.

Step 3: In the third step, this public key is encrypted by the service provider's private key and the user's public key, and it is sent to the user.

Step 4: The user decrypts the message using the service provider's public key and his/her own private key.

Step 5: The user then generates a request to the data owner encrypted by using own private key and data owner's public key obtained from Step 4.

Step 6: The data owner ensures that the user is authorized by decrypting the request and by checking the user's credentials from the cloud service provider.

Step 7: Once assured, the service provider sends a user authentication confirmation reply to the data owner.

Step 8: In the eighth step, the data owner sends the DNA-based secret key and access certificate, R(number of rounds), m_{1r}, m_{2r}, m_{3r}, m_{4r} where $r \in \{1, 2, \dots, R\}$ to the user encrypted by using own private key and user's public key.

Step 9: Then, the requested data is encrypted by the data owner using the two layers encryption processes discussed in Section 5.5 and the encrypted data is stored along with the corresponding certificate of the data on the cloud server after encrypting by his/her own private key and service provider's public key.

Step 10: The user presents the certificate to the cloud service provider encrypted by using his/her private key and the service provider's public key.

Step 11: This certificate is decrypted at the service provider's end using its private key and user's public key.

Step 12: The decrypted certificate is verified with the certificate provided by the data owner for the requested data.

Step 13: Once the certificate is verified, the cloud service provider encrypts the requested data by its private key and user's public key, and then, it is sent to the user.

Step 14: The encrypted data is first decrypted using the private key of the user and the service provider's public key. Then, the user executes the decryption process discussed in the next section.

5.7 Decryption

The user performs the decryption phase in two layers. Layer 1 decryption is executed using DNA computing. Here, the ciphertext is decoded into bits using Table 5. The binary data is then broken into 1024-bit blocks, and further each 1024-bit block is broken into 4 blocks of 256 bits. Each block is converted to corresponding ASCII values using the 8-bit binary sequence in the block. These ASCII values are converted into corresponding characters for layer 1 decrypted data. Layer 1 decryption steps are as follows:

5.7.1 Layer 1 decryption

Step 1: Decode the DNA base for each 2-bit pair.

Step 2: Break the binary data into blocks of 1024-bits.

Step 3: In the third step, each 1024-bit block is broken into 4 blocks of 256 bits each.

Step 4: Every 256-bit block produced from Step 3 is XORed with the 256-bit block of DNA key.

Step 5: In the fifth step, each 8-bit binary number in the resultant 256-bit block is assigned to the ASCII value.

Step 6: At last, each ASCII value is converted back to the corresponding 8-bit binary number using Table 3.

5.7.2 Layer 2 decryption

Once the user decrypts the DNA encoded data, i.e. layer 1 decryption, the final layer of the decryption phase is executed, which involves decrypting the data using the chaotic cryptosystem, i.e. one-dimensional cosine polynomial map. There are the following steps to decrypt the chaotically encrypted data using a cosine polynomial map.

Step 1: Read the encrypted data D (i.e., output of layer 1 decryption) in the matrix form, round keys m_{ir} and δ, where $i \in \{1, 2, 3, 4\}$ and $r \in \{1, 2, \ldots\ldots, R\}$ and R is the number of encryption rounds to be performed. Variables m_{1r}, m_{2r}, m_{3r}, m_{4r} and R are based on user's choices.

Step 2: In the second step, initialize the round counter r to 1.

Step 3: Perform a transpose operation on the encrypted matrix D.

Step 4: Here, the values of X and Y are set as the number of rows and number of columns, respectively.

Step 5: Generate the encryption sequence S of length X using the cosine polynomial map given in Eq. (4) with m_{2r}, and δ is set as the initial conditions.

Step 6: In the sixth step, normalize the encryption sequence value S using Eq. (12).

Step 7: Here, sequence P of length Y is generated using the cosine polynomial map with m_{4r} and δ.

Step 8: Normalize the sequences P using Eq. (14).

Step 9: For each iteration, starting from the last bit going to the first bit decrypt D_i and D_{S_i} using Eqs. (20) and (21).

$$D_i = cs(D_i, (S_i - P - \rho_i)) \bmod 256 \qquad (20)$$
$$D_{S_i} = (D_{S_i} - P - \rho_i) \bmod 256 \qquad (21)$$

Step 10: In the 10th step, m_{5r} is calculated using Eq. (13).

Step 11: Now, sequence Q of length Y is generated using the cosine polynomial map with m_{5r} and δ.

Step 12: Normalize the sequence Q using Eq. (15).

Step 13: In this step, D_1 and D_{S_1} are decrypted using Eqs. (22) and (23).

$$D_1 = cs\left(D_1', (S_1 - Q - D_X)\right) \bmod 256 \qquad (22)$$
$$D_{S_1} = \left(D_{S_1}' - Q - D_X\right) \bmod 256 \qquad (23)$$

where D_1' and D_{S_1}' are the encrypted sequences.

Step 14: Perform the transpose operation on the matrix D obtained and repeat Steps 4–13 to decrypt the rows. Instead of using m_{2r}, m_{4r}, δ, user utilizes m_{1r}, m_{3r}, δ.

Step 15: Decrement the round counter r and repeat Steps 3–14 until $r < 1$. The output of layer 2 gives the original data.

6. Security analysis

The proposed algorithm uses a two-layer encryption approach for the cloud environment, which can resist many security attacks. A security analysis of this algorithm has been performed in this section showcasing its robustness against some common and powerful attacks.

1. **Brute force attack:** This is the most common approach to break the encryption, where the attacker tries all possible combinations to break the encryption. This attack uses trial-and-error to guess confidential login information, security keys, etc. Here, the attacker uses an exhaustive computation to try out all the different possible permutations and combinations of encryption keys [40,41]. Despite being one of the oldest and most primitive techniques, it is still very popular among attackers and malicious users. The ability to break the scheme majorly depends on the size of the key. In the proposed scheme, the secret key is randomly generated, encoded by DNA computing and further encoded by the ASCII value. Though the key in the proposed scheme is a fixed size 128-bit key, the attacker will need to run 256! permutations just to get the ASCII table. Even after that, the key is still encrypted using the DNA-based encryption scheme. Thus, this attack becomes ineffective in the proposed scheme.

2. **Collision attack:** A collision attack tries to reduce the number of search space of the keys by finding two different inputs, which results in the same hash value [42]. In the proposed scheme, the secret key is prepared from a combination of user attributes like user password, user ID, address, date of birth, etc. This results in a unique secret key for all the users. In order to get a key, the attacker has to gather all these details. Even if the hacker tries to combine inputs from two different users, he/she must need the complementary rule and all the tables used in the proposed approach to retrieve the data. In the proposed scheme, a single wrong user attribute results in a drastically different secret key, and thus, increases the hackers' efforts to get the keys and other secret details. This renders the collision attack useless against the proposed scheme.

3. **Masquerade attack:** Identity and document forgery have also become a common practice to get access to unauthorized data. This is how a masquerade attack works. The attacker tries to pretend to be someone else and tries to forge as much private information as possible [43]. The malicious users try to invoke an active communication with the authorized user in the form of an email, phone call or message. The main aim is to collect some crucial information like a one-time password. The nature of this attack is dynamic, thus it becomes really difficult to trace this attack. In the proposed scheme, even in the worst case, attackers and malicious users can only login to the cloud server pretending to be an authorized user. If s/he manages to pass through the user authentication stage, s/he can get access to the data that must be decrypted using the DNA-based secret key and other credentials. As all the secret information is shared with the authentic user, the decryption process by the attackers results in a random data. Moreover, an extra layer using the chaotic map improves the security of the proposed scheme significantly. Thus, the proposed scheme can resist this attack.

4. **Statistical attack:** This attack is generally attempted for the image data. The hacker tries to establish a relation between the encrypted and the original images using pixel intensity analysis [44]. A strong encryption algorithm for image data must ensure that the encrypted image's pixels are uniform and equally distributed across the image. The polynomial cosine chaotic map technique used in the proposed scheme allows encrypting the images in such a way that the image looks flat and uniformly distributed. Therefore, the hacker cannot establish any relationship of the encrypted image to the original image.

5. **Man-in-the-middle attack:** In this type of attack, the hacker intercepts the communication happening between two authorized parties. The intruder tries to alter or relay information that is passed from one authorized channel to another authorized channel. Usually, this is done in a network either by serving false information to an authorized user or directly intercepting the communication. In the proposed scheme, all information is communicated in the encrypted form in the cloud environment. So, even if the intruder gets the secret key from the communication channel, s/he must decrypt it using the secret information, such as ASCII encoding, complementary rule, etc. These secret details are only shared with authorized users. In addition, secret details are also required during the data decryption phase. Gathering all the secret information is nearly impossible, and thus, the proposed encryption algorithm also tackles this popular attack.

7. Results and discussion

In this section, simulation environment and results are presented in detail.

7.1 Experimental setup

In order to perform the experiments based on the proposed scheme, a simulation of the cloud environment has been made using CloudSim 4.0 [45]. The CloudSim simulator consists of entities like Virtual Machine (VM) and its Manager (VMM), Cloud Information Service (CIS), data center, cloudlets, brokers and hosts.

Cloudlets are small tasks, which are waiting in queue to be performed and are handled by the brokers. The data center sends a request to the CIS to get its resources registered in the cloud environment. Each data center has several hosts, and each host holds several VMs with their own set of specifications like RAM, bandwidth, etc. The broker, which handles the cloudlets asks for the details of resources available from CIS in order to run its pending cloudlets. Once the broker acknowledges the availability of resources, VMs are created for the broker and an acknowledgment of the VM creation is sent from the data center to the broker. After receiving the acknowledgment, the broker sends the cloudlets to the data center, and finally, the data center allocates the VMs for the cloudlets and sends another acknowledgment about the cloudlets being executed. CloudSim simulator has been installed on an Acer Nitro 7 machine with configurations listed below:

1. RAM: 16 GB
2. Storage: 1 TB SSD
3. Processor: 9th generation Intel Core i7 9th generation
4. Graphics: 6 GB NVIDIA GEFORCE 1660 Ti
5. Operating system: Windows 10 Professional

For simulating the proposed scheme, a heterogeneous cloud environment is set up by creating 10 data centers in the CloudSim toolkit. Each data center consists of 2000 physical nodes, 4 GB of storage capacity and 1 GB/s network bandwidth. Four different VMs each with 1 GB/s network bandwidth are used as listed below:

1. 2000 MIPS, 2 GB
2. 1500 MIPS, 1.5 GB
3. 1000 MIPS, 1 GB
4. 500 MIPS, 0.5 GB

7.2 Simulation results

Simulation is performed on different data to ensure the workability of the scheme. Table 6 shows the data used for encryption, the corresponding SHA-512 key used as the control parameter of the chaotic map and the achieved encoded text.

The developed algorithm has experimented with other existing encryption schemes. For evaluating purposes, three existing schemes, namely Reversible Data Hiding Scheme [21], Cross Tenant Access Control [46] and Chaos-based Image Encryption scheme [22] are considered. These schemes or studies are also based on DNA computing and focus on improving the overall encryption architecture of the cloud computing environment. The proposed scheme extends the overall basic architecture by combining chaos-based encryption with DNA-based encryption.

This proposed work is compared based on four different parameters, i.e. key generation time, key retrieval time, data encryption time and data decryption time. The results shown below are the average of 50 experiments.

Fig. 3 shows the average key generation time of the different schemes. In the proposed scheme, first the user has to initiate a request to the data owner to get the secret key. The data owner verifies the credentials of the user, and then, based on the user's attributes the data owner generates the DNA-based

Table 6 Data encrypted using the proposed scheme.

Input message	Key	DNA encoded string
Hello, How are u?	0A12@34cdf	CATCCATTTAGGATTCAATCAAAACGGGAA CGGGCCATTGGCGCGGTAGAGTATCAGGT CACGGACGT
This data is confidential.	D@123#s	TCCGAGGGCAGCAGAGCAATGGTCTAGAA ACGCTAGCACACCCGTACAGATTACAGCC GTTTATGCGTTCAGTAAAAGAAGAAACGG CCAAGAGTTGGACAGCC
Welcome sir to our university.	s@34$npoi	AAATCGAGATTAAGTGAGCTGTGCAATGCG AAAGAGACTTACTCCCTGCCTATCTCTAGT CGAGCGTACTTATAAGAGGTCTGCCACTTA TTTATCAATGTGTGAAGGGCCAACTTCGAA
The data is being protected.	dna@#C007	TCCTAGCCAGCTTCCTGTTCTCATATAGC CGGTGCTTGCGGCCCCGAACTAGAATCTG TATGATGCGGACAGCACGTACTCAAGACA GAGCGTAGTGCCAGATCTCGTACTATACC

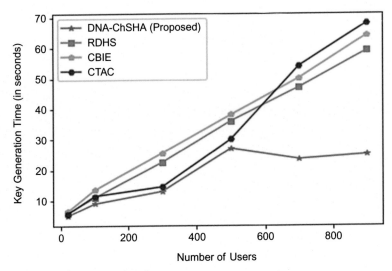

Fig. 3 Average key generation time.

secret key. This whole process is a little time taking, but once the key is generated and is shared with the user, the key retrieval time is not cumbersome. So, after a certain time, the proposed scheme results in lower-key retrieval times. While all the other existing schemes require users to generate the secret key every time the user has to access the data. Hence, these schemes show a linear increase for key generation.

The next simulation results show the key retrieval time. As explained earlier, once the key is generated, it is very easy and fast to retrieve the key. Other schemes are time-consuming because they generate and share the key every time the user wants to access the data from the cloud database. Fig. 4 shows the average key retrieval time for all the schemes in the cloud computing environment.

Encryption time is another parameter for assessing the proposed scheme. The proposed scheme has shown a significant improvement in the security of the data as shown in Fig. 5. Due to the use of a combination of chaos-based cryptography and DNA encoding, the proposed scheme consumes time. However, as the number of users increases, the proposed work consumes lower encryption times than the existing algorithms or schemes. CBIE is a time-consuming algorithm due to its iterative approach for finding the optimum results. RDHS technique uses the traditional histogram modification technique, which also increases the data encryption time.

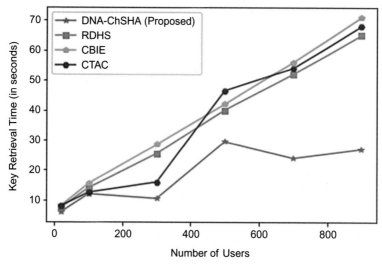

Fig. 4 Average key retrieval time.

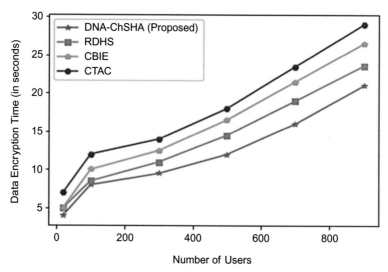

Fig. 5 Average encryption time.

Finally, the proposed scheme is evaluated based on the time consumption for data decryption. The overall decryption time of the proposed scheme is a little bit more as it involves decrypting another complex stage, i.e. chaos-based encryption. However, the proposed scheme performs well in terms of minimizing decryption time that includes the accessing time of data from the cloud and decoding time of encrypted data.

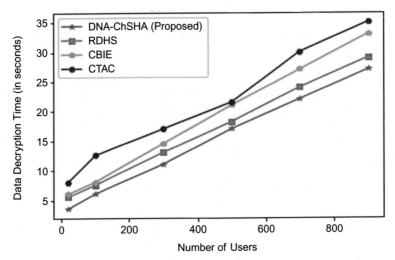

Fig. 6 Average decryption time.

In the existing schemes, there are many time expensive operations, including several XOR operations that make those algorithms very time-consuming, and hence, they result in high decryption times. Fig. 6 shows the average decryption time of all the considered schemes.

8. Conclusions and future works

With the rapid advancement of cloud computing, it faces data security issues. In this chapter, a novel encryption scheme for data is proposed, which uses a two-layer encryption scheme to protect the data from unauthorized access in the cloud computing environment. Here, at first, the data is encrypted using chaotic maps for which the control parameters are generated using SHA-512. Then, before sending or storing the data on the cloud, it is encrypted one more time using DNA computing. In this proposed work, a 1024-bit secret key uses user attributes and it is secured due to the randomness in assigning the DNA bases. Most of the existing encryption schemes for the cloud environment require the data owner to be online all the time, which increases the system overhead. However, in this scheme, once the data owner verifies the user's credentials and shares the DNA-based secret key with the user, it can go offline. Thus, it reduces the overall system overhead. The experimental results are shown to prove the effectiveness of the Chaos-based DNA encryption scheme for the cloud environment over the other well-known encryption schemes.

In this proposed scheme, the authenticity of the user is checked based on the shared password and certificate encrypted by using the public key cryptography technique. Public key cryptography is not as strong as other cryptographic schemes like elliptic curve cryptography, Advanced Encryption Standard (AES), etc. Therefore, in the future, a robust authentication scheme for the cloud computing environment can be developed for improving security.

References

[1] S. Namasudra, R. Chakraborty, A. Majumder, N.R. Moparthi, Securing multimedia by using DNA-based encryption in the cloud computing environment, ACM Trans. Multimed. Comput. Commun. Appl. 16 (3s) (2020) 1–19, https://doi.org/10.1145/3392665.

[2] X. Lai, M.X. Lu, L. Qin, J.S. Han, X.W. Fang, Asymmetric encryption and signature method with DNA technology, Sci. China Ser. F Inf. Sci. 53 (3) (2010) 506–514, https://doi.org/10.1007/s11432-010-0063-3.

[3] S. Namasudra, D. Devi, S. Kadry, R. Sundarasekar, A. Shanthini, Towards DNA based data security in the cloud computing environment, Comput. Commun. 151 (January) (2020) 539–547, https://doi.org/10.1016/j.comcom.2019.12.041.

[4] P. Vijayakumar, V. Vijayalakshmi, G. Zayaraz, DNA computing based elliptic curve cryptography, Int. J. Comput. Appl. 36 (4) (2011) 18–21.

[5] S.C. Sukumaran, M. Mohammed, DNA cryptography for secure data storage in cloud, Int. J. Netw. Secur. 20 (3) (2018) 447–454, https://doi.org/10.6633/IJNS.201805.20 (3).06.

[6] M. Sohal, S. Sharma, BDNA-A DNA inspired symmetric key cryptographic technique to secure cloud computing, J. King Saud Univ. Comput. Inform. Sci. (2018) 1417–1425, https://doi.org/10.1016/j.jksuci.2018.09.024.

[7] S. Namasudra, Fast and secure data accessing by using DNA computing for the cloud environment, IEEE Trans. Serv. Comput. (2020) 1–12, https://doi.org/10.1109/TSC.2020.3046471.

[8] X. Hu, L. Wei, W. Chen, Q. Chen, Y. Guo, Color image encryption algorithm based on dynamic chaos and matrix convolution, IEEE Access 8 (2020) 12452–12466, https://doi.org/10.1109/ACCESS.2020.2965740.

[9] P. Tobin, L. Tobin, M. Mc Keever, J. Blackledge, Chaos-based cryptography for cloud computing, in: 2016 27th Irish Signals and Systems Conference (ISSC), 2016, https://doi.org/10.1109/ISSC.2016.7528457.

[10] P. Singh, S.K. Saroj, A secure data dynamics and public auditing scheme for cloud storage, in: 2020 6th International Conference on Advanced Computing and Communication Systems, ICACCS 2020, 2020, pp. 695–700, https://doi.org/10.1109/ICACCS48705.2020.9074337.

[11] J.R.N. Sighom, P. Zhang, L. You, Security enhancement for data migration in the cloud, Future Internet 9 (3) (2017) 1–13, https://doi.org/10.3390/fi9030023.

[12] C.T. Clelland, V. Risca, C. Bancroft, Hiding messages in DNA microdots, Nature 399 (6736) (1999) 533–534, https://doi.org/10.1038/21092.

[13] S.T. Amin, M. Saeb, S. El-Gindi, A DNA-based implementation of YAEA encryption algorithm, Comput. Intell. (2006) 1–6.

[14] M.X. Lu, X.J. Lai, G.Z. Xiao, L. Qin, Symmetric-key cryptosystem with DNA technology, Sci. China Ser. F Inf. Sci. 50 (3) (2007) 324–333, https://doi.org/10.1007/s11432-007-0025-6.

[15] A. Leier, C. Richter, W. Banzhaf, H. Rauhe, Cryptography with DNA binary strands, Biosystems 57 (1) (2000) 13–22, https://doi.org/10.1016/S0303-2647(00)00083-6.

[16] K. Ning, A pseudo DNA cryptography method, Arxiv (2009) 1–21, https://doi.org/10.1016/j.compeleceng.2012.02.007.

[17] M. Borda, O. Tornea, DNA secret writing techniques, in: 2010 8th International Conference on Communications, COMM 2010, 2010, pp. 451–456, https://doi.org/10.1109/ICCOMM.2010.5509086.

[18] S. Namasudra, S. Sharma, G.C. Deka, P. Lorenz, DNA computing and table based data accessing in the cloud environment, J. Netw. Comput. Appl. 172 (April) (2020) 102835, https://doi.org/10.1016/j.jnca.2020.102835.

[19] K. Tanaka, A. Okamoto, I. Saito, Public-key system using DNA as a one-way function for key distribution, Biosystems 81 (1) (2005) 25–29, https://doi.org/10.1016/j.biosystems.2005.01.004.

[20] T. Tuncer, E. Avci, A reversible data hiding algorithm based on probabilistic DNA-XOR secret sharing scheme for color images, Displays 41 (2016) 1–8, https://doi.org/10.1016/j.displa.2015.10.005.

[21] B. Wang, Y. Xie, S. Zhou, C. Zhou, X. Zheng, Reversible data hiding based on DNA computing, Comput. Intell. Neurosci. 2017 (2013) 1–9, https://doi.org/10.1155/2017/7276084.

[22] R. Enayatifar, A.H. Abdullah, I.F. Isnin, Chaos-based image encryption using a hybrid genetic algorithm and a DNA sequence, Opt. Lasers Eng. 56 (2014) 83–93, https://doi.org/10.1016/j.optlaseng.2013.12.003.

[23] S. Pramanik, S.K. Setua, DNA cryptography, in: 2012 7th International Conference on Electrical and Computer Engineering, ICECE 2012, 2012, pp. 551–554, https://doi.org/10.1109/ICECE.2012.6471609.

[24] R. Matthews, On the derivation of a 'chaotic' encryption algorithm, Cryptologia 13 (1) (1989) 29–42, https://doi.org/10.1080/0161-118991863745.

[25] S. Behnia, A. Akhshani, H. Mahmodi, A. Akhavan, A novel algorithm for image encryption based on mixture of chaotic maps, Chaos Solitons Fractals 35 (2) (2008) 408–419, https://doi.org/10.1016/j.chaos.2006.05.011.

[26] R.M. May, Simple mathematical models with very complicated dynamics, in: B.R. Hunt, T.-Y. Li, J.A. Kennedy, H.E. Nusse (Eds.), The Theory of Chaotic Attractors, Springer New York, New York, NY, 2004, pp. 85–93.

[27] T.S. Ali, R. Ali, A novel medical image signcryption scheme using TLTS and Henon chaotic map, IEEE Access 8 (2020) 71974–71992, https://doi.org/10.1109/ACCESS.2020.2987615.

[28] M. Crampin, B. Heal, On the chaotic behaviour of the Henon map, Teach. Math. Its Appl. 13 (2) (1994) 83–89.

[29] M.Z. Talhaoui, X. Wang, M.A. Midoun, A new one-dimensional cosine polynomial chaotic map and its use in image encryption, Vis. Comput. 37 (3) (2021) 541–551, https://doi.org/10.1007/s00371-020-01822-8.

[30] H. Jiang, Y. Liu, Z. Wei, L. Zhang, A new class of two-dimensional chaotic maps with closed curve fixed points, Int. J. Bifurc. Chaos 29 (7) (2019) 1–10, https://doi.org/10.1142/S0218127419500949.

[31] A. Akhshani, S. Behnia, A. Akhavan, M.A. Jafarizadeh, H. Abu Hassan, Z. Hassan, Hash function based on hierarchy of 2D piecewise nonlinear chaotic maps, Chaos Solitons Fractals 42 (4) (2009) 2405–2412, https://doi.org/10.1016/j.chaos.2009.03.153.

[32] M. Hénon, A two-dimensional mapping with a strange attractor, in: B.R. Hunt, T.-Y. Li, J.A. Kennedy, H.E. Nusse (Eds.), The Theory of Chaotic Attractors, Springer New York, New York, NY, 2004, pp. 94–102.

[33] J. Fridrich, Two-dimensional chaotic maps, Int. J. Bifurc. Chaos 8 (6) (1998) 1259–1284.

[34] J. Fridrich, Image encryption based on chaotic maps, in: Proceedings of the IEEE International Conference on Systems, Man and Cybernetics, 2, 1997, pp. 1105–1110, https://doi.org/10.1109/icsmc.1997.638097.

[35] G. Peterson, Arnold's cat map, in: Math 45. Linear Algebra, 1997, pp. 1–7.

[36] X. Wu, Z.H. Guan, A novel digital watermark algorithm based on chaotic maps, Phys. Lett. A Gen. At. Solid State Phys. 365 (2007) 403–406, https://doi.org/10.1016/j.physleta.2007.01.034.

[37] C.E. Shannon, A Mathematical Theory of Cryptography, 1945, pp. 1–137. [Online]. Available: https://www.iacr.org/museum/shannon45.html.

[38] D. Agrawal, S. Minocha, A.K. Goel, Gradient boosting based classification of ion channels, in: Proceedings—IEEE 2021 International Conference on Computing, Communication, and Intelligent Systems, ICCCIS 2021, 2021, pp. 102–107, https://doi.org/10.1109/ICCCIS51004.2021.9397161.

[39] S. Namasudra, P. Roy, P. Vijayakumar, S. Audithan, B. Balusamy, Time efficient secure DNA based access control model for cloud computing environment, Future Gener. Comput. Syst. 73 (2017) 90–105, https://doi.org/10.1016/j.future.2017.01.017.

[40] W. Diffie, M.E. Hellman, Exhaustive cryptanalysis of the NBS data encryption standard, Computer 10 (6) (1977) 74–84, https://doi.org/10.1109/C-M.1977.217750.

[41] P. Pavithran, S. Mathew, S. Namasudra, P. Lorenz, A novel cryptosystem based on DNA cryptography and randomly generated mealy machine, Comput. Secur. 104 (2021) 102160, https://doi.org/10.1016/j.cose.2020.102160.

[42] K. Schramm, T. Wollinger, C. Paar, A new class of collision attacks and its application to DES, Lect. Notes Comput. Sci. 2887 (2003) 206–222, https://doi.org/10.1007/978-3-540-39887-5_16.

[43] M. Ben Salem, S.J. Stolfo, Decoy document deployment for effective masquerade attack detection, Lect. Notes Comput. Sci. 6739 (2011) 35–54, https://doi.org/10.1007/978-3-642-22424-9_3.

[44] P. Kocher, J. Jaffe, B. Jun, Introduction to differential power analysis and related attacks, Cryptogr. Res. (1998) 1–5.

[45] R.N. Calheiros, R. Ranjan, A. Beloglazov, C.A.F. De Rose, R. Buyya, CloudSim: a toolkit for modeling and simulation of cloud computing environments and evaluation of resource provisioning algorithms, Softw. Pract. Exper. 41 (2011) 23–50, https://doi.org/10.1002/spe.995.

[46] Q. Alam, S.U.R. Malik, A. Akhunzada, K.K.R. Choo, S. Tabbasum, M. Alam, A cross tenant access control (CTAC) model for cloud computing: formal specification and verification, IEEE Trans. Inf. Forensics Secur. 12 (6) (2017) 1259–1268, https://doi.org/10.1109/TIFS.2016.2646639.

About the authors

Divyansh Agrawal is currently working at PricewaterhouseCoopers US Advisory as an Associate-II in the Data & Analytics Technologies vertical. He completed his Bachelor of Technology in Computer Science and Engineering from Galgotias University, India in 2021. His research interests include Natural Language Processing, Computer Vision, ensemble methods, innovations in Deep Learning and encryption techniques in DNA computing.

Dr. Sachin Minocha is working as an Assistant professor in School of Computing Science and Engineering, Galgotias University. He has received his Ph.D. degree from Department of Computer Science and Engineering, Sant Longowal Institute of Engineering and Technology. Dr. Minocha has 5 patents, 10 publications in conference proceedings, book chapters, and refereed journals like CAEE, JARS, Expert Systems, CIBM. His research area is Machine Learning, Nature Inspired Techniques, Hyperspectral Images, and DNA computing.

CHAPTER NINE

Secure data communication using DNA computing adaptable to wireless sensor network

Sathish Gunasekaran[a] and Manish Kumar[b]

[a]Verena Haptic & VR Systems Pvt Ltd, Chennai, Tamil Nadu, India
[b]Department of Mathematics, Birla Institute of Technology and Science-Pilani, Hyderabad, Telangana, India

Contents

Advances in Computers, Volume 129
ISSN 0065-2458
https://doi.org/10.1016/bs.adcom.2022.08.008
317

Abstract

Due to the proliferation of electronic devices such as mobile phones, tablets, laptops, desktops, e-drives, hard drives, etc., transferring information in the form of images through the Internet has become an essential practice in this digital era. Red, green, blue (RGB) images are used to store or share information, such as medical imaging for disease diagnosis, e-learning, online shopping, defense services, personal images, and much more. Adversaries are trying to steal this critical information from images for their gain or personal interest. Hence, we need to prevent an adversary from misusing this crucial information using encryption algorithms. This chapter aims to provide a new, secure, and fast encryption algorithm for RGB images in deoxyribonucleic acid (DNA)-encoded domain involving random sequences generated by arithmetic progression (AP) and logistic map associated with generalized Vigenère-type table suitable for wireless sensor network (WSN). The randomness of generated sequences is performed using the National Institute of Standards and Technology (NIST) statistical suite test, and the results are provided in Appendix. The proposed encryption algorithm provides a huge keyspace and is robust against brute-force, dictionary, side-channel, correlation, differential, cropping, noise, chosen cipher, and plain image attacks. The various security analysis is performed on standard test images to demonstrate the proposed algorithm's efficiency and security. Further, a comparison with other competing existing algorithms is shown. The analysis results conclude that the proposed algorithm can opt for practical application.

Abbreviations

AP	arithmetic progression
ARMAC	arithmetic random matrix affine Cipher
BST	binary search tree
DNA	deoxyribonucleic acid
ECC	elliptic curve cryptography
ECDH	elliptic-curve Diffie–Hellman
FFT	fractional Fourier transform
MATLAB®	MATrix LABoratory
MSE	mean squared error
NIST	National Institute of Standards and Technology
PSNR	peak signal–to–noise ratio
RC4	Rivest Cipher 4
RGB	red, green, blue
SHA	Secure Hash Algorithm
SSL	secure sockets layer
UACI	unified average change intensity
WSN	wireless sensor network

1. Introduction

Image encryption plays an inevitable role in information security/cybersecurity, where more images are involved in websites, online marketing, graphics and video telecasting, etc. Digital images over communication channels are highly voluminous and need to be secured due to security breaches. Most image encryption algorithms use image scrambling and XOR operations in the transformed domain, which provides robust diffusion and confusion against applying scramble and XOR operations over original images only. Working directly with the original images would decrease the security level of the algorithms. In the past few decades, research done on various DNA-based image encryption schemes on different DNA operations shows the importance of information security in the DNA domain. DNA computing replaces the conventional hardware silicon chip-based computation with biomolecules [Adenine (A), Guanine (G), Cytosine (C), and Thymine (T)]. While computing any operation in the DNA domain, the original data will be transferred and executed in the DNA domain. There are specified rules defined in the DNA domain: encoding, complementary, addition, and subtraction rules to achieve specific goals like solving more complex computational problems, drug discovery, introducing randomness in cryptography, etc.

This chapter especially focuses on DNA-related work. Encryption of RGB images in the DNA domain can be seen in Refs. [1–26]. The algorithm proposed in Ref. [1] introduces bitwise scrambling, DNA addition, complementation, and bitwise XOR operations with key sequences generated in the DNA domain. It provides a notion of using various combinations of DNA operations. In Ref. [2], a multiple image encryption algorithm is proposed. The pixel values are permuted in the DNA domain using S-Box (generated from the commutative chain ring and mixed chaotic equation). The algorithm proposed in Ref. [3] uses permutations shuffling on image components of blocks decomposed ($m \times n$ blocks). DNA encoding, DNA decoding, DNA complement, and bitwise XOR are used alongside the keys generated with a Hash of the original image. A multiple-image encryption algorithm is proposed in Ref. [4], the algorithm composes multiple images into a one-dimensional array, and half of the pixels are used to permute the array along with the help of a DNA sequence. The algorithm's running time was the major concern when it was developed. The algorithm proposed in Ref. [5], converted the original image into the DNA domain, applied random masks generated by the Lorentz sequence, and finally applied FFT three times. The algorithm proposed in Ref. [6] encrypts images in the DNA domain by applying a new DNA permutation followed by a DNA diffusion process using a Hash of the plain image and a chaotic sequence. The original image in the DNA domain cyclic shifted and scrambled through various sequences, followed by DNA addition, XOR, and subtraction in Ref. [7]. A new combination of DNA addition and subtraction is proposed with image scrambling using the pseudo-random sequence in Ref. [8]. Laser and optical feedback technologies are used as a carrier signal to introduce randomness with DNA-encoded images, followed by DNA operations introducing robust and novel encryption algorithm for images proposed in Ref. [9]. Ref. [10] proposes row, and column-wise DNA additions on the original image to get the encrypted image. A reasonably long image encryption algorithm proposed in Ref. [11] involves steps like scrambling, confusion, and diffusion associated with the DNA domain. The algorithm proposed in Ref. [12] combines chaotic sequence, hamming distance, and DNA sequence operations. The suitable chaotic equation is chosen based on image properties, DNA XOR, conditional shifting, and diffusion operations introduced in the DNA domain. An algorithm proposed in Ref. [13] which involves simple DNA addition and complementation shows the effect of DNA domain and chaotic sequences. DNA chain is used for DNA addition.

Various other cryptographic formats and operations are integrated with the DNA domain to provide more robustness to the encryption algorithms. For instance, ECDH cryptography, ciphers, chaotic maps, and more advanced methods are integrated with the DNA domains are discussed in Ref. [14]. RC4 with DNA encoding is introduced in Ref. [15] to encrypt images effectively in the DNA domain. A combination of pixel scrambling and Feistel transformation is applied on DNA-encoded images sequentially, proceeded by Hill cipher, and followed by image diffusion in Ref. [16]. Image encrypted using DNA diffusion and permutation shuffling in the DNA domain is proposed in Ref. [17]. The algorithm proposed in Ref. [18] ultimately introduces a new technique for image encryption using binary search tree (BST) with a DNA domain where BST properties are used to attain security in the encryption algorithm. The concept of cellular automata is introduced in Ref. [19] for gray-scale images in the DNA domain. WSN has its challenges in design, deployment, physical sizes of devices, power consumption, data transfer mechanisms, energy constraints, data channel reliability, communication channel limitations like cost, timeout or delay, dynamics, and various types of attacks, memory, and computational limitations. Building data security is a highly challenging part of wireless sensor networks. The above limitations addressed in Refs. [27–30] can be resolved by introducing robust mechanisms to make the network more secure. Using encoding schemes like DNA encoding to introduce more confusion and diffusion in cryptography is suggested. The above references indicate that the DNA encoding scheme is more adaptable to address WSN limitations. Implementing the DNA encoding scheme is handled effectively in this chapter which is suitable for wireless sensor networks.

Many DNA-based encryption algorithms involve operations (DNA addition, complement, and introduce some specific processes in the DNA domain) such as XOR operation with coupled maps is proposed in Ref. [20]. A jumble of pixels and diffusion method is discussed in Ref. [21]. Various new diffusion mechanisms include longer steps, permutation, and discrete logarithm problem for security proof, DNA addition is introduced in Ref. [22], and DNA operations are executed along with keys generated in Refs. [23, 24]. A new image encryption algorithm is introduced in Ref. [25] using elliptic curve cryptography (ECC) in the DNA domain and involves the following operations: DNA addition, shifting of pixels, interlayered shifting of pixels, followed by applying ECC in the DNA domain. The algorithm gives an innovative notion to introduce conventional cryptography in the DNA domain to enhance security. Further, a few more essential research works are addressed based on the DNA domain in computing and cryptography. The 1024-bit DNA-based secret key is used to achieve secure and fast data accessing and control in cloud computing proposed in Refs. [31–34]. The blockchain application in healthcare and its interconnectivity with DNA computing is provided in Ref. [35]. The security, including confidentiality, integrity, authenticity, trust, privacy, and anonymity, is explained in Refs. [36–38]. Furthermore, the DNA cryptographic system using a mealy machine is proposed in Ref. [39]. In Ref. [40], the cryptanalysis is done to illustrate chosen plaintext attacks on DNA-based image encryption. It clearly depicts that the algorithm should be resistant to chosen-plaintext/ciphertext attacks. Ref. [41] talks about the various attacks that can be executed on image encryption algorithms. Most of the existing algorithms (discussed above) propose key generation from chaotic sequences or any other conventional methods, which does not describe all security tests discussed in this chapter.

In a few existing algorithms, the Hash of the original image is used in the key generation process of the few algorithms where it needs to be transferred before the decryption to the other end. Traditional key generation methods are used in some existing algorithms. Further, the number of encryption and decryption steps introduced in the DNA domain is less. Most of the algorithms cited in this paper lack in providing statistical security proof. Furthermore, the randomness of the keys generated in the existing algorithms is not proven. It is motivated by the drawbacks and weaknesses of the existing algorithms, as stated before, that suggestsproposing a new keyspace and robust encryption algorithm in an encoded domain (i.e., the DNA domain)rather than in the original domain.The proposed new key generation method in the DNA domain covers all possible security tests, including the National Institute of Standards and Technology (NIST) statistical test suit. Encryption over an encoded domain has an advantage that increases the encryption algorithm's efficiency compared to the same encryption algorithm without encoding. Applying wavelet transform or wavelet packets over the original image enhances the security level where the original image is in the wavelet domain andincreases the choice of introducing more randomness. The significant contribution of the proposed algorithm is as follows:

1. Most of the existing encryption algorithms stated in this chapter use the conventional key generation method, whereas the proposed algorithm generates its keys using a new key generation process.
2. To strengthen of security levels of the proposed algorithm, we have introduced new encryption steps in the DNA domain.
3. The proposed algorithm uses a robust DNA encoding (computing) scheme to add more security levels that offer confusion and cryptography diffusion. It maintains a good level of original image recovery when cropping is done in the encrypted image.
4. Implementing DNA encoding using MATLAB® is handled very effectively to reduce the run time of the proposed algorithm. It encodes and decodes RGB images very fast. The other existing algorithms do not provide any information about this.

The rest of the chapter is organized as follows: Section 2 involves background studies for the proposed algorithm. Section 3 is devoted to explaining the proposed scheme. In Section 3.2, the proposed algorithm is described. Sections 5 and 4 involves performance and security analysis through various statistical methods of the proposed algorithm. Section 5.3 contains a comparison between the proposed algorithm and different related algorithms presented previously. In Section 6 the conclusion drawn from the present work is mentioned. Finally, Appendix involves NIST statistical suite test results over Z_t generated in step 3 of the key generation procedure (Section 3.1).

2. Background studies

The image encryption algorithm proposed in Ref. [42] introduces arithmetic random matrix affine cipher (ARMAC), a new way of diffusion and shuffling images in the wavelet packet domain. The main aim of using ARMAC parameters is to provide more randomness in the encrypted image as proposed in Ref. [42], where the arithmetic progression (AP) series row and column are shifted by the parameter α and γ, and multiplied by parameters ξ and λ, respectively. For an image of size $m \times n$ the ARMAC works as follows:

$$X'(APSeriesRow(i), j) \equiv \xi X(APSeriesRow(i), j + \alpha(\mathrm{mod}\ n)),$$
$$X'(i, APSeriesColumn(j)) \equiv \lambda X(i + \gamma(\mathrm{mod}\ m), APSeriesColumn(j)) \quad (1)$$

The corresponding inversion steps are as follows:

$$X(APSeriesRow(i), j) \equiv \mu X'(APSeriesRow(i), j + n - \alpha(\mathrm{mod}\ n)),$$
$$X(i, APSeriesColumn(j)) \equiv \kappa X'(i + m - \gamma(\mathrm{mod}\ m), APSeriesColumn(j)), \quad (2)$$

where μ and κ are multiplayer parameters, $\xi\mu \equiv 1\ (\mathrm{mod}\ n)$, $\lambda\kappa \equiv 1, 0 < \alpha < n$, and $0 < \gamma < m$.

2.1 Improved arithmetic random matrix affine cipher

Kumar and Josyula have proposed the ARMAC in Ref. [42]. We have taken the multiplayer parameters as identity, and replaced α by α_i for every row i, and γ by γ_j for every column j, in Eqs. (1) and (2), and we have

$$X'(APSeriesRow(i), j) \equiv X(APSeriesRow(i), j + \alpha_i(\mathrm{mod}\ n)),$$
$$X'(i, APSeriesColumn(j)) \equiv X(i + \gamma_j(\mathrm{mod}\ m)APSeriesColumn(j)) \quad (3)$$

The corresponding inversion steps are as follows:

$$X(APSeriesRow(i), j) \equiv X'(APSeriesRow(i), j + n - \alpha_i(\mathrm{mod}\ n)),$$
$$X(i, APSeriesColumn(j)) \equiv X'(i + m - \gamma_j(\mathrm{mod}\ m), APSeriesColumn(j)), \quad (4)$$

where $0 < \alpha_i < n$ and $0 < \gamma_j < m$. Here in the improved version of the ARMAC parameters, AP series row and column are shifted by the parameters α_i/γ_j for each row i and for each column j; where i varies from 1 to m and j varies from 1 to n (here $m \times n \times 3$ is the image size to be encrypted) that enables the proposed algorithm to introduce more randomness as compared to the algorithm proposed in Ref. [42]. Affine cipher diffusion is introduced over $\mathbb{Z}_{2^{24}-1}$ to increase the security levels of the proposed algorithm. Both AP shuffling and affine cipher processes provide an elegant and efficient shuffling process as the operations are executed in the DNA domain.

3. Proposed scheme

The proposed scheme uses a DNA-encoded domain in which random sequences involving AP and a logistic map associated with a generalized Vigenère-type table are generated. The schematic diagram of the proposed algorithm is shown in Fig. 1. Notations and their meanings are defined for more convenience in Table 1, frequently used throughout the chapter.

3.1 Key generation

We explain how the key sequence is generated using the combination of AP and logistic sequence. NIST statistical test suite tests the randomness of the generated keyspace, and accomplished results are shown in Appendix.

The eight steps are performed to generate keys for the encryption algorithm. Let the RGB image to be encrypted is of size $m \times n \times 3$.

Step 1: Let us define AP as follows:

$$a_{\eta,\delta} = a_{\eta,1} + (\delta - 1)d_\eta \tag{5}$$

where $a_{\eta,1}$ is the initial term of an AP, and the common difference of successive members is d_η that are used as secret keys provided by users. Now, generate three AP by Eq. (5) for $\eta = 1, 2, 3$ and $\delta = 1,2,3,\ldots,N$, where $N = m \times n \times 3$.

Step 2: Users can select any random string (*KeyString*) of their own choice and compute SHA-256 of *KeyString* as *HashString* = SHA-256 (*KeyString*) which is 32 bytes of an integer array. For instance, *HashString* = $\{hs_1, hs_2, \ldots, hs_{32}\}$. Users can choose their own choice of integers $a_{1,1}^{user}, a_{2,1}^{user}, a_{3,1}^{user}, d_1^{user}, d_2^{user}$, and d_3^{user} to compute the following initial values and differences for the three AP as defined in Eq. (6):

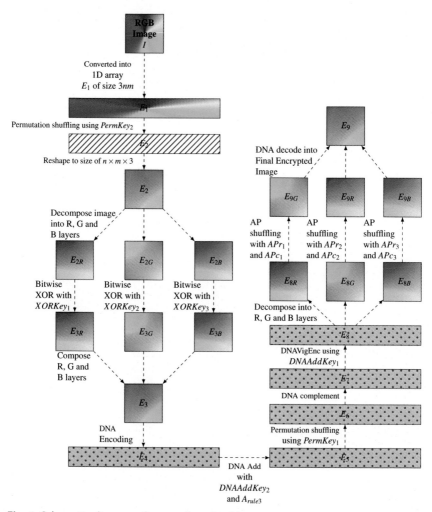

Fig. 1 Schematic diagram of encryption algorithm.

$$\left.\begin{array}{l} a_{1,1} = a_{1,1}^{user} + \sum_{k=1}^{15} hs_k \\[2mm] a_{2,1} = a_{2,1}^{user} + \sum_{k=11}^{25} hs_k \\[2mm] a_{3,1} = a_{3,1}^{user} + \sum_{k=12}^{32} hs_k \\[2mm] d_1 = d_1^{user} + \sum_{k=1}^{4} hs_k \\[2mm] d_2 = d_2^{user} + \sum_{k=3}^{6} hs_k \\[2mm] d_3 = d_3^{user} + \sum_{k=7}^{10} hs_k \end{array}\right\} \tag{6}$$

Table 1 Notations and their description.

Notations	Description
$a_{\eta,\,\delta}$	The ηth arithmetic progression (AP)
d_η	Common difference of successive terms for η^{th} AP
$KeyString$	String selected by users to apply Secure Hash Algorithm-256 (SHA-256)
$HashString$	Output values of the SHA-256 applied on $KeyString$
$a_{\eta,1}^{user}$	Initial values are chosen by users for the ηth AP
$d_{\eta,1}^{user}$	Common difference of successive terms chose by users for the ηth AP
$c_{\eta,\,\delta}$	The ηth logistic map
r_η	Parameter for the ηth logistic map
u_i	Positive integers are chosen by users
$DNAencode(X, DNARule)$	DNA encoding process on the sequence X using the $DNARule$ rule
$DNAadd(X, Y, DNARule)$	DNA addition mechanism of X and Y
$DNAcomplment(X)$	DNA complement of a sequence X
$rearrange(Z, C)$	Represents the rearrangement of the sequence Z based on the sequence C. At first C is sorted and the original positions of all the elements of sequence C before sorting are chosen. That is the new rearrangement index for sequence Z.
A_{rule}^{user}	Positive integers chosen by users to generate rules for the proposed algorithm
$XorKey$	Key used for bitwise XOR with the image
$DNAAddKey_l$	Keys used for DNA addition for $l = 1, 2$
$AP_{r1}, AP_{r2}, AP_{r3}$	Used for AP-based row shuffling
$AC_{r1}, AC_{r2}, AC_{r3}$	Positive integers chosen by users to generate rules for the proposed algorithm
$PermKey_l$	Keys used for permutation-based shuffling for $l = 1, 2$

The generated three AP series are called as $AP_1 = \{a_{1,\delta} : \delta = 1,2,3,\ldots,N\}$, $AP_2 = \{a_{2,\delta} : \delta = 1,2,3,\ldots,N\}$, and $AP_3 = \{a_{3,\delta} : \delta = 1,2,3,\ldots,N\}$. Note that the above sum is taken randomly; it is the user's choice to take any number of sums over Hash byte values.

Step 3: Take (mod 8) as there are eight rules for applying DNA encoding over the generated sequences in step 2 is given below:

$$\left.\begin{aligned}
K_{rule1} &= a_{1,1} \times u_1 \ (\mathrm{mod}\,8) + 1 \\
K_{rule2} &= a_{2,1} \times u_2 \ (\mathrm{mod}\,8) + 1 \\
K_{rule3} &= a_{3,1} \times u_3 \ (\mathrm{mod}\,8) + 1 \\
K_{rule4} &= d_1 \times u_4 \ (\mathrm{mod}\,8) + 1 \\
K_{rule5} &= d_2 \times u_5 \ (\mathrm{mod}\,8) + 1 \\
K_{rule6} &= d_3 \times u_6 \ (\mathrm{mod}\,8) + 1 \\
K_{rule7} &= max(a_{1,1}, a_{2,1}, a_{3,1}) \times u_7 \ (\mathrm{mod}\,8) + 1 \\
K_{rule8} &= min(a_{1,1}, a_{2,1}, a_{3,1}) \times u_8 \ (\mathrm{mod}\,8) + 1
\end{aligned}\right\} \tag{7}$$

Here, users can select any positive integer values u_1, u_2, u_3, u_4, u_5, u_6, u_7, and u_8 of their own choice.

Step 4: The rules stated in step 3 by Eq. (7) are modified to introduce more randomness using the following process provided in Eq. (8):

$$\left.\begin{aligned}
K_{rule1}^* &= \left(K_{rule1} + K_{rule2} + K_{rule3} + K_{rule4}\right) (\mathrm{mod}\,8) + 1 \\
K_{rule2}^* &= \left(K_{rule2} + K_{rule3} + K_{rule4} + K_{rule5}\right) (\mathrm{mod}\,8) + 1 \\
K_{rule3}^* &= \left(K_{rule3} + K_{rule4} + K_{rule5} + K_{rule6}\right) (\mathrm{mod}\,8) + 1 \\
K_{rule4}^* &= \left(K_{rule4} + K_{rule5} + K_{rule6} + K_{rule7}\right) (\mathrm{mod}\,8) + 1 \\
K_{rule5}^* &= \left(K_{rule5} + K_{rule6} + K_{rule7} + K_{rule8}\right) (\mathrm{mod}\,8) + 1 \\
K_{rule6}^* &= \left(K_{rule6} + K_{rule7} + K_{rule8} + K_{rule1}\right) (\mathrm{mod}\,8) + 1 \\
K_{rule7}^* &= \left(K_{rule7} + K_{rule8} + K_{rule1} + K_{rule2}\right) (\mathrm{mod}\,8) + 1 \\
K_{rule8}^* &= \left(K_{rule8} + K_{rule1} + K_{rule2} + K_{rule3}\right) (\mathrm{mod}\,8) + 1
\end{aligned}\right\} \tag{8}$$

Step 5: Compute $YP_1 = \{y_{1,\delta} : \delta = 1,2,3,...,N\}$, $YP_2 = \{y_{2,\delta}; \delta = 1,2,3,...,N\}$, and $YP_3 = \{y_{3,\delta} : \delta = 1,2,3,...,N\}$ where $y_{\eta,\delta} = a_{\eta,\delta}$ (mod 256) and $YP_4 = (YP_1 + YP_2 + YP_3)$ (mod 256).

Step 6: Define a logistic map as follows:

$$c_{\eta,\delta+1} = r_\eta\, c_{\eta,\delta}(1 - c_{\eta,\delta}), \tag{9}$$

where $c_{\eta,\delta} \in (0, 1)$, $r_\eta \in [3.54, 4]$ for each $\eta = 1, 2, 3$ and $\delta = 1,2,3,...,N$. We generate three logistic maps using Eq. (9), say $CM_1 = \{c_{1,\delta} : \delta = 1,2,3,...,N\}, CM_2 = \{c_{2,\delta} : \delta = 1,2,3,...,N\}$, and $CM_3 = \{c_{3,\delta} : \delta = 1,2,3,...,N\}$.

Step 7: Compute the following sequences given in Eq. (10):

$$
\left.
\begin{aligned}
Z_1 &= DNAencode(YP_1, K^*_{rule1}) \text{ and } Z'_1 = DNAcomplement(Z_1) \\
Z_2 &= DNAencode(YP_2, K^*_{rule2}) \text{ and } Z'_2 = DNAcomplement(Z_2) \\
Z_3 &= DNAencode(YP_3, K^*_{rule3}) \text{ and } Z'_3 = DNAcomplement(Z_3) \\
Z_4 &= DNAencode(YP_4, K^*_{rule4}) \text{ and } Z'_4 = DNAcomplement(Z_4) \\
Z_5^a &= DNAadd(YP_1, YP_2, K^*_{rule5}),\ Z'_5 = DNAcomplement(Z_5^a) \text{ and } Z_5 = rearrange(Z'_5, CM_1) \\
Z_6^a &= DNAadd(Z_5, YP_3, K^*_{rule6}),\ Z'_6 = DNAcomplement(Z_6^a) \text{ and } Z_6 = rearrange(Z'_6, CM_2) \\
Z_7^a &= DNAadd(Z_6, YP_4, K^*_{rule7}),\ Z'_7 = DNAcomplement(Z_7^a) \text{ and } Z_7 = rearrange(Z'_7, CM_3) \\
Z_8 &= DNAdecode(Z_7, K^*_{rule8})
\end{aligned}
\right\}
\tag{10}
$$

where $DNAencode(X, DNARule)$ is DNA encoding process on the sequence X using the rule $DNARule$,

$DNAadd(X, Y, DNARule)$ is DNA addition mechanism of X and Y, $DNAcomplment(X)$ represents DNA complement of a sequence X, and $rearrange(Z, C)$ represents the rearrangement of the sequence Z based on the sequence C provided in Table 1. DNA encoding and decoding schemes are implemented in MATLAB® so that the running time is significantly less (less than 0.2 s). The DNA encoding scheme adds more randomness, which is proved by the NIST statistical test suite in Ref. [43].

Step 8: Finally, the following keys are generated in Eq. (11) for the proposed encryption and decryption algorithm:

$$Y_5 = YP_1 + YP_2 + YP_3$$

$$XORKey = \left(y_{5,1}, y_{5,2}, \cdots, y_{5,m\times n\times3}\right) + \left(y_{5,(m\times n\times3)+1}, y_{5,(m\times n\times3)+2}, \cdots, y_{5,(m\times n\times3)+(m\times n\times3)}\right)$$

$$DNAAddKey_1 = \{z_{7,1}, z_{7,2}, \cdots, z_{7,m\times n\times12}\}$$

$$DNAAddKey_2 = \{z_{7,m\times n\times12+1}, z_{7,m\times n\times12+2}, \cdots, z_{7,(m\times n\times12)+(m\times n\times12)}\}$$

$$APr_1 = \{a_{1,1} \pmod{2m}, a_{1,2} \pmod{2m}, \cdots, a_{1,2m} \pmod{2m}\}$$

$$APr_2 = \{a_{1,2m+1} \pmod{2m}, \cdots, a_{1,4m} \pmod{2m}\}$$

$$APr_3 = \{a_{1,4m+1} \pmod{2m}, \cdots, a_{1,6m} \pmod{2m}\}$$

$$APc_1 = \{z_{8,1} \pmod{2n}, z_{8,2} \pmod{2n}, \cdots, z_{8,2n} \pmod{2n}\}$$

$$APc_2 = \{z_{8,2n+1} \pmod{2n}, z_{8,2n+2} \pmod{2n}, \cdots, z_{8,4n} \pmod{2n}\}$$

$$APc_3 = \{z_{8,4n+1} \pmod{2n}, z_{8,4n+2} \pmod{2n}, \cdots, z_{8,6n} \pmod{2n}\}$$

$$PermKey_1^* = \{(c_{3,1} + c_{1,1}) \pmod 1, (c_{3,2} + c_{1,2}) \pmod 1, \cdots, (c_{3,m\times n\times12} + c_{1,m\times n\times12}) \pmod 1\}$$

$$PermKey_1 = permutation(PermKey_1^*)$$

$$PermKey_2^* = \{(c_{3,1} + c_{2,1}) \pmod 1, (c_{3,2} + c_{2,2}) \pmod 1, \cdots, (c_{3,m\times n\times3} + c_{2,m\times n\times3}) \pmod 1\}$$

$$PermKey_2 = permutation(PermKey_2^*)$$

(11)

where *permutation* is defined in step 7 and $+$ represents coordinate-wise addition, $XORKey$ is reshaped into the matrix of size $512 \times 512 \times 3$, and finally, $XORKey$ is decomposed into three matrices $XORKey_1$, $XORKey_2$, and $XORKey_3$, respectively.

Thus, the generated keys ($XORKey$, $DNAAddKey_1$, $DNAAddKey_2$, APr_1, APr_2, APr_3, APc_1, APc_2, APc_3, $PermKey_1$, and $PermKey_2$) in this step are used for the proposed encryption/decryption algorithm.

3.2 Encryption

The encryption mechanism contains the nine steps and the size of the image to be encrypted is $m \times n \times 3$. Let the three rules that are used for encryption be $A_{rule3} = A_{rule3}^{user}$, $A_{rule2} = (A_{rule2}^{user} * A_{rule3}^{user})$ (mod 8)+1, and $A_{rule1} = (A_{rule1}^{user} * A_{rule2})$ (mod 8)+1. Here once again A_{rule3}^{user}, A_{rule2}^{user}, and A_{rule1}^{user} are users choice and these can be taken any positive integer.

Step 1: At first, the original RGB image (say, I) with R, G, and B layers reshaped into a one-dimensional array of size $3mn$ (say, E_1). Now, E_1 is shuffled based on the permutation key $PermKey_2$ provided in Eq. (12) as follows:

$$E_2(i) = E_1(PermKey_2(i)) \tag{12}$$

to get E_2 and now E_2 is again reshaped into the same size $m \times n \times 3$ as the original image.

Step 2: Decompose E_2 into three layers E_{2R}, E_{2G}, and E_{2B}, each of size $n \times m$, and do the following:

$$\left.\begin{array}{l} E_{3R} = (E_{2R} \oplus XORKey_1) \ (\text{mod } 256) \\ E_{3G} = (E_{2G} \oplus XORKey_2 \oplus E_{3R}) \ (\text{mod } 256) \\ E_{3B} = (E_{2B} \oplus XORKey_3 \oplus E_{3G}) \ (\text{mod } 256) \end{array}\right\} \tag{13}$$

Compose E_{3R}, E_{3G}, and E_{3B} given in Eq. (13) into E_3.

Step 3: Apply DNA encoding on E_3 to get $E_4 = DNAencode(E_3, A_{rule1})$.

Step 4: Do the DNA addition on E_4 to get $E_5 = DNAadd(E_4, DNAAddKey_2, A_{rule3})$.

Step 5: Now E_5 is shuffled based on the permutation key $PermKey_1$ given by Eq. (14) yielding E_6.

$$E_6(i) = E_5(PermKey_1(i)) \tag{14}$$

Step 6: Compute $E_7 = DNAcomplement(E_6)$.

Step 7: The generalized Vigenère-type encryption is proposed in Ref. [44]. In this chapter, the generalized Vigenère-type table entries presented in Ref. [44] are now replaced with DNA molecules A, T, C, and G, and all the possible permutations and combinations can be added to the generalized Vigenère-type table (based on DNA rules). Then the generalized Vigenère-type encryption is applied on E_7 to get E_8.

$$E_8(i) = DNAVigEnc(E_7, DNAAddKey_1) \tag{15}$$

Step 8: Improved AP shuffling: Decompose E_8 provided in Eq. (15) into three components E_{8R}, E_{8G}, and E_{8B}, each of size $2m \times 2n$. Now apply improved AP shuffling on E_{8R}, E_{8G}, and E_{8B} to get E_{9R}, E_{9G}, and E_{9B}, respectively.

$$\left.\begin{array}{l} E_{9R}(i_1, j) = E_{8R}(i_1, \ j_1) \\ E_{9G}(i_2, j) = E_{8G}(i_2, \ j_2) \\ E_{9B}(i_3, j) = E_{8B}(i_3, \ j_3) \end{array}\right\} \tag{16}$$

where $i_1 = APr_1(i)$ (mod $2m$), $i_2 = APr_2(i)$ (mod $2m$), $i_3 = APr_3(i)$ (mod $2m$), $j_1 = j + APc_1(j)$ (mod $2n$), $j_2 = j + APc_2(j)$ (mod $2n$), $j_3 = j + APc_3(j)$ (mod $2n$), and j_1, j_2, j_3 are in the place of α_i given by Eq. (3).

Step 9: Compose E_{9R}, E_{9G}, and E_{9B} given in (16) to get the image E_9. Finally, apply the DNA decoding process to get the encrypted image $E_{10} = DNAdecode(E_9, A_{rule2})$.

Finally, a schematic diagram of the proposed encryption algorithm is shown in Fig. 1.

3.3 Decryption

The decryption process is executed by applying encryption steps with minor changes in the reverse order from Steps 9 to 1, and the decryption process is explained here as follows:

Step 1: Apply DNA decoding process on the encrypted image $D_1 = DNAdecode(E_{10}, A_{rule2})$. Decompose D_1 to get the images D_{1R}, D_{1G}, and D_{1B} each of size $2m \times 2n$.

Step 2: Now apply improved AP reshuffling on D_{1R}, D_{1G}, and D_{1B} to get D_{2R}, D_{2G}, and D_{2B}, respectively, as follows:

$$\left.\begin{array}{l} D_{2R}(i_1, \ j_1) = D_{1R}(i_1, \ j) \\ D_{2G}(i_2, \ j_2) = D_{1G}(i_2, \ j) \\ D_{2B}(i_3, \ j_3) = D_{1B}(i_3, \ j) \end{array}\right\} \tag{17}$$

where $i_1 = APr_1(i)$ (mod $2m$), $i_2 = APr_2(i)$ (mod $2m$), $i_3 = APr_3(i)$ (mod $2m$), $j_1 = j + 2n - APc_1(j)$ (mod $2n$), $j_2 = j + 2n - APc_2(j)$ (mod $2n$), $j_3 = j + 2n - APc_3(j)$ (mod $2n$), and j_1, j_2, and j_3 are in the place of α_i provided in Eq. (4). Compose D_{2R}, D_{2G}, and D_{2B} given in Eq. (17) into D_2.

Step 3: Now, generalized Vigenère decryption is applied on D_2 to get D_3 given by Eq. (18) as follows:

$$D_3(i) = DNAVigDec(D_2, DNAAddKey_1) \tag{18}$$

Step 4: Compute $D_4 = DNAcomplement(D_3)$.

Step 5: Then D_4 is reshuffled based on permutation key $PermKey_1$, which provides D_5 as below:

$$D_5(PermKey_1(i)) = D_4(i) \tag{19}$$

Step 6: DNA subtraction is executed on D_5 given in Eq. (19) to get $D_6 = DNAadd(D_5, DNAAddKey_2, A_{rule3})$.

Step 7: Apply DNA decoding on D_6 to get $D_7 = DNAdecode(D_6, A_{rule1})$.

Step 8: Decompose D_7 into three layers D_{7R}, D_{7G}, and D_{7B}. Then do the following:

$$\left. \begin{array}{l} D_{8R} = (D_{7R} \oplus XORKey_1) \ (\mathrm{mod}\ 256) \\ D_{8G} = (D_{7G} \oplus XORKey_2 \oplus D_{7R}) \ (\mathrm{mod}\ 256) \\ D_{8B} = (D_{7B} \oplus XORKey_3 \oplus D_{7G}) \ (\mathrm{mod}\ 256) \end{array} \right\} \tag{20}$$

Compose D_{8R}, D_{8G}, and D_{8B} provided in Eq. (20) into D_8.

Step 9: The image (D_8) is reshaped into a one-dimensional array of size $3mn$ (D_9). Then D_9 is reshuffled based on permutation key $PermKey_2$ to get the image D_{10} as follows:

$$D_{10}(PermKey_2(i)) = D_9(i) \tag{21}$$

Then D_{10} given in Eq. (21) is reshaped into the decrypted image of size $m \times n \times 3$.

4. Security analysis

An encryption algorithm should resist vulnerabilities such as brute-force, chosen cipher, and plain image and statistical attacks while maintaining the trade-off between running time and security. The security of the algorithm is measured through the following tests.

- The histogram analysis is used to infer the pattern in the pixel intensity levels of the encrypted image. The constant level that occurs in the histogram provides no statistical information about an encrypted image.
- Mean squared error (MSE) depicts the squared difference between two images. The zero MSE value indicates no difference between original

and decrypted images. The positive MSE between original and encrypted images indicates a significant difference in both images.

- Peak signal-to-noise ratio (PSNR) depicts the impact of noise over the signal. The infinite PSNR value indicates no noise impact between original and decrypted images. Finite PSNR values indicate a tremendous impact of noise between original and encrypted images.
- Correlation analysis helps identify the distribution of highly correlated pixels (original image) in the encrypted domain. The results show no information is revealed in correlation analysis.
- The keyspace analysis is done to show the large keyspace and its resistance to brute-force and other key space attacks.
- Information entropy shows the high randomness that occurs in encrypted images.
- NPCR and UACI analyses show the change rate between original and encrypted images. The statistical test ensures NPCR and UACI values satisfy the threshold levels.
- Cropping and noise channel attacks analysis shows the resistance of the algorithm against these attacks.

The values in the tables are generated through programs explicitly written for security analysis. The programs use original, encrypted, and decrypted images to create corresponding results for MSE, PSNR, correlation, information entropy, NPCR, and UACI values presented in tables. MATLAB® provides some inbuilt functions to write these programs.

4.1 Histogram analysis

Histogram analysis to visualize and graphically analyze the pixel frequencies in the image for each pixel or intensity value between 1 and 255. The histogram is very useful for the adversary to analyze the data distribution graphical view (cipher space). An adversary tries to infer the distribution of pixel frequencies of encrypted images to apply any attacks. But, the proposed algorithm produced an encrypted image histogram of constant level for any original image. Hence getting any information from a histogram of the encrypted image is very difficult for an adversary. Histograms of Baboon, Airplane, Fruits, and Pepper images are shown in Figs. 2–5. Histograms shown in these figures depict statistically indistinguishable different encrypted images.

4.2 The mean squared error and peak signal-to-noise ratio

Let $p(i, j)$ be the original/plain image of size $m \times n$ and $c(i, j)$ be the encrypted/cipher image. Then MSE is defined by Eq. (22) as follows:

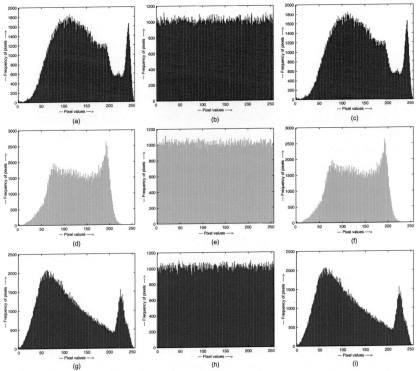

Fig. 2 Histogram for Baboon image: (A) original/plain red layer, (B) encrypted/ciphered red layer, (C) decrypted/deciphered red layer, (D) original/plain green layer, (E) encrypted/ciphered green layer, (F) decrypted/deciphered green layer, (G) original/plain blue layer, (H) encrypted/ciphered blue layer, and (I) decrypted/deciphered blue layer.

$$MSE = \frac{1}{m \times n} \sum_{i=1}^{m} \sum_{j=1}^{n} \left[p(i,j) - c(i,j) \right]^2 \qquad (22)$$

The MSE values between plain and deciphered images are given in Table 2. The zero values presented in Table 2 show no data loss during the encryption/decryption process. Table 3 shows MSE between plain and cipher images, and large values in this table reveal that the encrypted image is perfectly different from the original image.

As the name suggests, PSNR describes the ratio between peak signal value (max power) and peak distorting noise value. PSNR for the images between the plain and cipher/deciphered images are defined by Eq. (23) as follows:

$$PSNR = 20 \times \left[log \frac{Max_P^2}{\sqrt{MSE}} \right], \qquad (23)$$

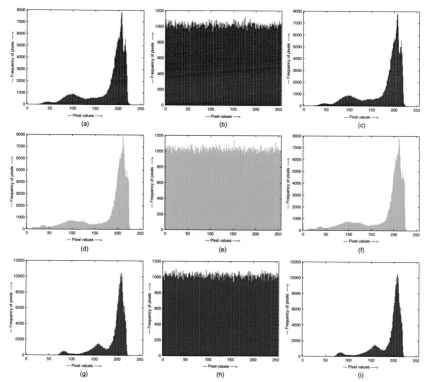

Fig. 3 Histogram for Airplane image: (A) original/plain red layer, (B) encrypted/ciphered red layer, (C) decrypted/deciphered red layer, (D) original/plain green layer, (E) encrypted/ciphered green layer, (F) decrypted/deciphered green layer, (G) original/plain blue layer, (H) encrypted/ciphered blue layer, and (I) decrypted/deciphered blue layer.

where Max_P represents the maximum pixel value of the image, and a significant small PSNR value between plain and ciphered images ensures robust security. The proposed algorithm provides 0 and ∞ values for MSE and PSNR for all plain and deciphered images. The MSE and PSNR between plain and ciphered images are given in Table 3.

4.3 Correlation analysis

Highly correlated pixel values across horizontal, vertical, and diagonal directions from original/plain are chosen for correlation analysis. Results are analyzed for corresponding pixels of the encrypted image in Tables 4–6. More than 1000 pairs of high correlated pixels are chosen from the original image

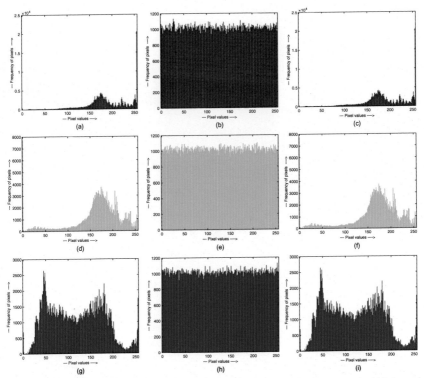

Fig. 4 Histogram for Peppers image: (A) original/plain red layer, (B) encrypted/ciphered red layer, (C) decrypted/deciphered red layer, (D) original/plain green layer, (E) Encrypted/ciphered green layer, (F) decrypted/deciphered green layer, (G) original/plain blue layer, (H) encrypted/ciphered blue layer, and (I) decrypted/deciphered blue layer.

to calculate correlation coefficients provided by Eq. (24). Here, r_{xy} represents a correlation value for selected pairs of pixels:

$$\left.\begin{aligned} r_{xy} &= \frac{Covariance(x, y)}{\sqrt{\sigma_x}\sqrt{\sigma_y}} \\ \sigma_x &= \frac{1}{K}\sum_{i=1}^{K}(x_i - E(x))^2 \end{aligned}\right\} \tag{24}$$

where x and y are an array of chosen adjacent pixels in every channel, and K is the cardinality of the sample array taken. The correlation values of encrypted images are shallow, clearly showing no information can be revealed to get high correlated pixel values chosen for the original image. The correlation plots are shown in Figs. 6 and 7.

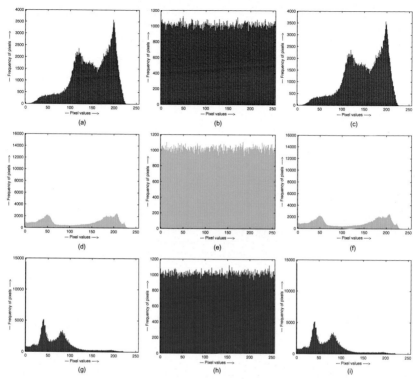

Fig. 5 Histogram for Fruits image: (A) original/plain red layer, (B) encrypted/ciphered red layer, (C) decrypted/deciphered red layer, (D) original/plain green layer, (E) encrypted/ciphered green layer, (F) decrypted/deciphered green layer, (G) original/plain blue layer, (H) encrypted/ciphered blue layer, and (I) decrypted/deciphered blue layer.

Table 2 MSE and PSNR between original/plain and decrypted images of (a) Baboon, (b) Airplane, (c) Fruits, and (d) Peppers.

Image	Baboon		Airplane		Fruits		Peppers	
Layers	MSE	PSNR	MSE	PSNR	MSE	PSNR	MSE	PSNR
R	0	∞	0	∞	0	∞	0	∞
G	0	∞	0	∞	0	∞	0	∞
B	0	∞	0	∞	0	∞	0	∞

Table 3 MSE and PSNR between original/plain and encrypted images of (a) Baboon, (b) Airplane, (c) Fruits, and (d) Peppers.

Image	Baboon		Airplane		Fruits		Peppers	
Layers	MSE	PSNR	MSE	PSNR	MSE	PSNR	MSE	PSNR
R	8.6301e+03	8.7706	9.9437e+03	8.1553	1.1123e+04	7.6686	7.9793e+03	9.1112
G	7.7432e+03	9.2416	1.0676e+04	7.8469	9.9683e+03	8.1446	1.1247e+04	7.6206
B	9.4732e+03	8.3658	1.0476e+04	7.9287	9.1211e+03	8.5304	1.1142e+04	7.6610

Table 4 Correlation analysis in the horizontal direction of (a) Baboon, (b) Airplane, (c) Fruits, and (d) Peppers images.

Image	Baboon		Airplane		Fruits		Peppers	
Layers	Original	Encrypted	Original	Encrypted	Original	Encrypted	Original	Encrypted
R	0.9046	3.8462e-04	0.9611	3.6713e-06	0.9568	1.3930e-06	0.9532	1.1569e-04
G	0.8444	3.7068e-04	0.9610	5.7264e-06	0.9575	2.1393e-05	0.9675	1.8990e-05
B	0.8736	7.2212e-07	0.9395	3.3574e-04	0.9298	7.4678e-05	0.9395	4.8433e-06

Table 5 Correlation analysis in the vertical direction of (a) Baboon, (b) Airplane, (c) Fruits, and (d) Peppers images.

Image	Baboon		Airplane		Fruits		Peppers	
Layers	Original	Encrypted	Original	Encrypted	Original	Encrypted	Original	Encrypted
R	0.9302	5.7550e-04	0.9523	1.5256e-04	0.9798	3.1296e-05	0.9480	4.4638e-07
G	0.8499	2.5326e-04	0.9703	8.2293e-04	0.9784	9.1080e-05	0.9691	7.1465e-05
B	0.8807	1.3071e-04	0.9351	1.7564e-05	0.9207	2.7062e-05	0.9450	1.3490e-04

Table 6 Correlation analysis in the diagonal direction of (a) Baboon, (b) Airplane, (c) Fruits, and (d) Peppers images.

Image	Baboon		Airplane		Fruits		Peppers	
Layers	Original	Encrypted	Original	Encrypted	Original	Encrypted	Original	Encrypted
R	0.9079	4.5575e-05	0.9342	4.0218e-06	0.9531	9.6309e-05	0.9262	2.1707e-05
G	0.8045	1.6667e-06	0.9418	4.5462e-05	0.9643	1.0296e-06	0.9604	6.0072e-05
B	0.8816	2.5288e-06	0.9007	2.0007e-05	0.9703	2.8311e-05	0.9165	2.9211e-04

4.4 Keyspace

In this section, the key space size is discussed. The following secret keys ($HashString$, $a_{1,1}^0$, $a_{1,2}^0$, $a_{1,3}^0$, d_1^0, d_2^0, d_3^0, r_1, r_2, r_3 and u_1, u_2, u_3, u_4, u_5, u_6, u_7, u_8) are used in the proposed algorithm.

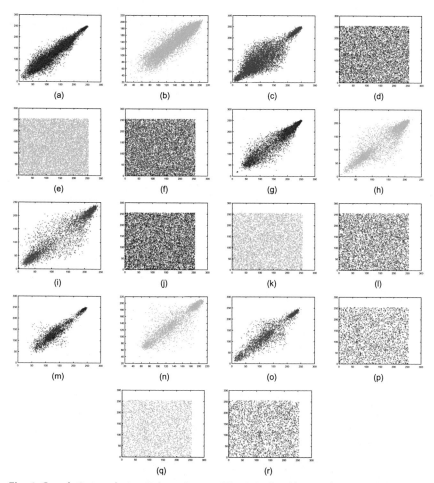

Fig. 6 Correlation analysis—Baboon image: (A) original red layer in horizontal direction, (B) original green layer in horizontal direction, (C) original blue layer in horizontal direction, (D) encrypted red layer in horizontal direction, (E) encrypted green layer in horizontal direction, (F) encrypted blue layer in horizontal direction, (G) original red layer in vertical direction, (H) original green layer in vertical direction, (I) original blue layer in vertical direction, (J) encrypted red layer in vertical direction, (K) encrypted green layer in vertical direction, (L) encrypted blue layer in vertical direction, (M) original red layer in diagonal direction, (N) original green layer in diagonal direction, (O) original blue layer in diagonal direction, (P) encrypted red layer in diagonal direction, (Q) encrypted green layer in diagonal direction, and (R) encrypted blue layer in diagonal direction.

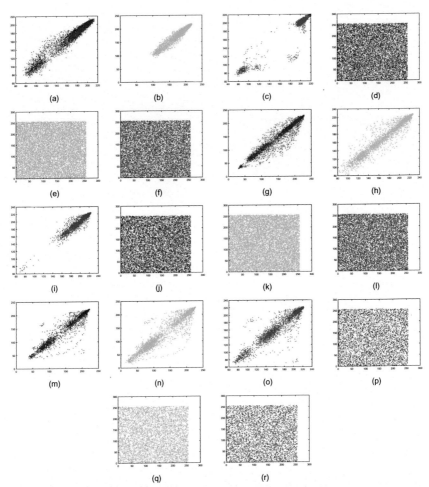

Fig. 7 Correlation analysis—Airplane image: (A) original red layer in horizontal direction, (B) original green layer in horizontal direction, (C) original blue layer in horizontal direction, (D) encrypted red layer in horizontal direction, (E) encrypted green layer in horizontal direction, (F) encrypted blue layer in horizontal direction, (G) original red layer in vertical direction, (H) original green layer in vertical direction, (I) original blue layer in vertical direction, (J) encrypted red layer in vertical direction, (K) encrypted green layer in vertical direction, (L) encrypted blue layer in vertical direction, (M) original red layer in diagonal direction, (N) original green layer in diagonal direction, (O) original blue layer in diagonal direction, (P) encrypted red layer in diagonal direction, (Q) encrypted green layer in diagonal direction, and (R) encrypted blue layer in diagonal direction.

1. The *KeyString* is taken with length up to 32 bytes (256 bits) for the proposed algorithm. So the number of possible strings are $2^{256} \approx 10^{77}$.
2. The keys $a_{1,1}^0$, $a_{1,2}^0$, $a_{1,3}^0$, d_1^0, d_2^0, d_3^0, r_1, r_2, and r_3 are sensitive up to 15 decimal places. So the number of possible values is 10^{135}.
3. Keys u_1, u_2, u_3, u_4, u_5, u_6, u_7, and u_8 all are positive integer arrays based on users' choice that are used for key generation. For the proposed algorithm, $u_i \in [1, 10^6]$. Since there are eight values, then the number of such keys will be 10^{48}.
4. Each key A_i^{user} for $i = 1, 2, 3$ is a positive integer array that is used for encryption/decryption algorithm. For the proposed algorithm, chosen range for each key A_i^{user} for $i = 1, 2, 3$ is $[1, 10^6]$. Since there are three values, the total number of such possible arrays is 10^{18}.

Hence, the total size of the keyspace is $\approx 10^{278}$.

4.5 Key change analysis

Assume the attacker is trying to infer useful information from the image by guessing keys, or they are generating keys themself and trying to decrypt an encrypted image. Suppose the intruder can access all the exact keys except only one, then the attacker cannot recover the correct picture, and the results can be viewed in Fig. 8.

4.5.1 Information entropy analysis

The information entropy is a useful tool to measure the uncertainty/variability and randomness of pixel values in the encrypted image. If the values are uniformly scattered, the information entropy will be high enough, showing the randomness. The information entropy is given by Eq. (25) as follows:

$$H(s) = \sum_{i=0}^{m \times n} prob(c_i) \log_2 \left(\frac{1}{prob(c_i)} \right) \qquad (25)$$

where $prob(c_i)$ represents the probability of the presence of the pixel value c_i. As the images contain 8-bit pixel values, the information entropy is close to $H(s) = 8$ bits. The information entropy values for the proposed encryption algorithm (given in Table 7) demonstrate the robustness against the entropy attack.

4.6 Differential attacks

Differential attacks show how differences in a marginal change of pixels in the input image can affect the resultant difference in the output image.

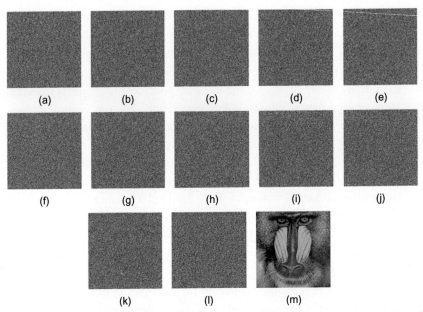

Fig. 8 Keyspace security analysis: incorrect image is decrypted with minute change of (A) last letter in *KeyString*, (B) 0.000000000000001 in $a_{1,1}^0$, (C) 0.00000000000001 in $a_{1,2}^0$, (D) 0.00000000000001 in $a_{1,3}^0$, (E) 0.00000000000001 in $d_{1,1}^0$, (F) 0.00000000000001 in $d_{1,2}^0$, (G) 0.00000000000001 in $d_{1,3}^0$, (H) 0.00000000000001 in r_1, (I) 0.00000000000001 in r_2, (J) 0.00000000000001 in r_3, (K) 1 in K_{rule1}, (L) 1 in A_{rule1}, and (M) correctly decrypted image.

Table 7 The result of information entropy.

Image components	Red	Green	Blue
Baboon	7.9994	7.9992	7.9993
Airplane	7.9992	7.9993	7.9994
Fruits	7.9994	7.9993	7.9993
Peppers	7.9993	7.9993	7.9994

Attackers often try to find any significant relationship between plain and cipher images by changing the value of any one pixel of the original image at a time and encrypting it using the same set of keys. The two encrypted images may reveal information about the original image if they have similar pixel values. The attacker can guess the location of the changed pixel in the encrypted image. So, it is necessary that a small change, even in one pixel, causes the cipher image to switch completely to counter the differential attacks. The NPCR and UACI analyses suggest that the proposed algorithm is resistant to differential attacks.

4.6.1 NPCR analysis

Number of pixels change rate (NPCR) represents pixel change rate between two encrypted images (C^1 and C^2) with few pixel values changed in respective original images. NPCR ($N(C^1, C^2)$) is calculated by Eq. (26) as follows:

$$N(C^1, C^2) = \frac{\sum_{i=1}^{n}\sum_{j=1}^{m} f(i,j)}{n \times n} \times 100\% \qquad (26)$$

where

$$f(i,j) = \begin{cases} 1, & \text{if } C^1(i,j) = C^2(i,j) \\ 0, & \text{otherwise.} \end{cases}$$

4.6.2 UACI analysis

Unified average change intensity (UACI) represents the average pixel value differences between two encrypted images. UACI ($U(C^1, C^2)$) is calculated by Eq. (27) as follows:

$$U(C^1, C^2) = \frac{\sum_{i=1}^{n}\sum_{j=1}^{m} \dfrac{C^1(i,j) - C^2(i,j)}{255}}{n \times m} \times 100\% \qquad (27)$$

NPCR and UACI values are shown in Table 8 for each layer (R, G, and B) between two encrypted images, and experimental results show the variance occurs in theoretical threshold values given in Ref. [45]. Hence, the proposed algorithm is immune to differential attacks.

4.6.3 Statistical test for number of pixels change rate (NPCR)

From Ref. [45], statistical threshold values are verified through NPCR for two encrypted images C^1 and C^2, and then two hypotheses (H_0 and H_1) along with significance levels α for $N(C^1, C^2)$ are defined by Eq. (28) as follows:

$$\left.\begin{array}{l} H_0 : N(C^1, C^2) = \mu_N \\ H_1 : N(C^1, C^2) < \mu_N \end{array}\right\} \qquad (28)$$

Reject H_0, if $N(C^1, C^2) < N_\alpha^*$; otherwise accept H_0, here

Table 8 Results of NPCR (%) and UACI (%) of the proposed algorithm.

Encrypted image	NPCR (in %)			UACI (in %)		
	Red	Green	Blue	Red	Green	Blue
Baboon	99.6140	99.6056	99.6056	33.4404	33.4940	33.4638
Airplane	99.6177	99.5998	99.6323	33.4931	33.4657	33.4529
Peppers	99.6001	99.6029	99.6002	33.5032	33.4198	33.5022
Fruits	99.6159	99.6128	99.6204	33.4409	33.4966	33.4227
Average of each component	99.6119	99.6053	99.6147	33.4694	33.4690	33.4604
Average for all images	99.6106			33.4663		

Table 9 Statistical test—number of pixels change rate.

			$F = 255$		
Encrypted image	μ_N	σ_N	$N^*_{0.05}$	$N^*_{0.01}$	$N^*_{0.001}$
Numerical values	99.6094	0.0066	99.5994	99.5952	99.5906
Baboon (99.6084)			PASS (✓)	PASS (✓)	PASS (✓)
Airplane (99.6166)			PASS (✓)	PASS (✓)	PASS (✓)
Fruits (99.6011)			PASS (✓)	PASS (✓)	PASS (✓)
Peppers (99.6164)			PASS (✓)	PASS (✓)	PASS (✓)

$$N^*_\alpha = \mu_N - \phi^{-1}(\alpha)\sigma_N = \frac{\left(Max - \phi^{-1}(\alpha)\sqrt{\frac{Max}{MN}}\right)}{Max + 1}$$

$$\mu_N = \frac{Max}{Max + 1} \quad \text{and} \quad \sigma_N^2 = \frac{Max}{(Max + 1)^2 MN}$$

where Max represents largest pixel value present in the original image.

One can observe from Table 9 that $N(C^1, C^2)$ values for Baboon, Airplane, and Peppers are greater than N^*_α values where $\alpha = 0.05, 0.01$, and 0.001. Hence, the null hypothesis (H_0) is accepted. Thus, the proposed algorithm passed the NPCR statistical test successfully.

4.6.4 Statistical test for UACI

Similarly, from Ref. [45], statistical threshold values are verified through UACI for two encrypted images E^1 and E^2, and then two hypotheses

$(H_0$ and H_1) along with significance levels α for $U(E^1, E^2)$ are defined by Eq. (29) as follows:

$$\left.\begin{array}{l} H_0 : U(E^1, E^2) = \mu_U \\ H_1 : N(E^1, E^2) < \mu_U \end{array}\right\} \tag{29}$$

Reject H_0, if $U(C^1, C^2) \notin (U_\alpha^{*+}, U_\alpha^{*-})$; otherwise accept H_0, here

$$\left.\begin{array}{l} U_\alpha^{*+} = \mu_U + \phi^{-1}(\alpha/2)\sigma_U \\ U_\alpha^{*-} = \mu_U - \phi^{-1}(\alpha/2)\sigma_U \end{array}\right\}$$

$$\mu_U = \frac{Max + 2}{3Max + 3} \text{ and } \sigma_U^2 = \frac{(Max + 2)(Max^2 + 2Max + 3)}{18(Max + 1)^2 M \times N \times Max}.$$

One can observe from Table 10 that $U(E^1, E^2)$ values for Baboon, Airplane, and Peppers fall within the interval values $(U_\alpha^{*+}, U_\alpha^{*-})$ where $\alpha = 0.05, 0.01$, and 0.001. Hence, the null hypothesis (H_0) is accepted. Thus, the proposed algorithm passed the UACI statistical test successfully.

4.7 Cropped attack analysis

Sometimes attackers are interested in making data loss so that the receiver would not receive information sent correctly. They do this by using a cropping attack on the encrypted image. The encrypted/cipher image is cropped (plugged zero pixel values in cropped location) with 50% data in each layer from the left and different places and then decrypted using the decryption algorithm and using the same keys as used during encryption. The results in Fig. 9 show that the original image is recovered with marginal information loss. With this proposed algorithm, the receiver could read the

Table 10 Statistical test—UACI.

Encrypted image	μ_U	σ_U	$U_{0.05}^{*+}/U_{0.05}^{*-}$	$U_{0.01}^{*+}/U_{0.01}^{*-}$	$U_{0.001}^{*+}/U_{0.001}^{*-}$
			F = 255		
Numerical values	33.4635	0.0231	33.4183	33.4040	33.3875
			33.5088	33.5231	33.5396
Baboon (33.4661)			PASS (✓)	PASS (✓)	PASS (✓)
Airplane (33.4706)			PASS (✓)	PASS (✓)	PASS (✓)
Fruits (33.4751)			PASS (✓)	PASS (✓)	PASS (✓)
Peppers (33.4663)			PASS (✓)	PASS (✓)	PASS (✓)

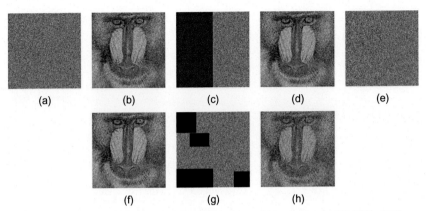

(a)　　　(b)　　　(c)　　　(d)　　　(e)

(f)　　　(g)　　　(h)

Fig. 9 Cropping attack resistance: Encrypted image with cropping in (A) 50% of R layer, (B) decrypted image, (C) 50% of G layer, (D) decrypted image, (E) 50% of B layer, (F) decrypted image, (G) different locations of R, G, and B layers, and (H) decrypted image.

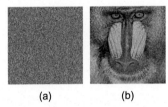

(a)　　　(b)

Fig. 10 Noise attack resistance via communication channel: (A) Salt and Pepper noise imposed encrypted image and (B) decrypted image.

information even though there is 50% data loss. Hence, the proposed encryption algorithm helps to prevent cropping attacks.

4.8 Communication channel noise analysis

The communication channel is subject to various types of noises. In a wireless medium, there are reasons for a loss in signals (a) multi-path propagation, (b) interference, (c) attenuation, and (d) near-far problem. Hence, the encrypted image is imposed into salt and pepper noise, and the decrypted images are depicted in Fig. 10.

5. Performance analysis

5.1 Experimental setup and computational complexity analysis

The encryption algorithm proposed in this chapter executes permutation shuffling, bitwise XOR, DNA encoding and decoding, addition, and complement operations. The algorithm is implemented to take a constant amount of time to execute each of its operations. RGB image in the DNA domain is of size $O(m \times n \times 12)$. Hence the proposed algorithm has a computational complexity of $O(m \times n \times 12)$. The encryption and decryption implementations for the image of size $512 \times 512 \times 3$ run on average 1.20 and 1.25 s. The proposed algorithm is tested with a configuration of 8 GB RAM, Intel® Core™ i5 processor with 8300H CPU at 2.30 GHz, Windows 10, and implemented in MATLAB® R2020a. It depicts that running time is very less for the proposed algorithm.

5.2 Results and discussions

This section discusses the procedure for encrypting/decrypting RGB images with keys used by the proposed algorithm and the results obtained during the encryption and decryption process.

5.2.1 Key generation

From Step 2 of Section 3.1, $KeyString = 'bsqbdipudvdsnj'$ is chosen and $HashString = SHA\text{-}256(KeyString)$ is computed where $HashString$ has 256 bits of length. Then $a_{1,1}^{user} = 346{,}821, a_{1,2}^{user} = 36{,}746{,}723, a_{1,3}^{user} = 342{,}367, d_1^{user} = 349, d_2^{user} = 7{,}671{,}147$, and $d_3^{user} = 4{,}193{,}793$ are chosen, and $a_{1,1}, a_{1,2}, a_{1,3}$, d_1, d_2, and d_3 are calculated using Eq. (6), followed by Eq. (5) that computes AP_1, AP_2, and AP_3. From Step 3 of Section 3.1, we choose $u_1 = 236, u_2 = 2323, u_3 = 1136, u_4 = 2316, u_5 = 2216, u_6 = 926, u_7 = 2636$, and $u_8 = 116$, and then $K_{rule2}, K_{rule1}, K_{rule3}, K_{rule4}, K_{rule5}, K_{rule6}, K_{rule7}$, and K_{rule8} are calculated by Eq. (7) followed by $K_{rule1}^{*}, K_{rule2}^{*}, K_{rule3}^{*}, K_{rule4}^{*}, K_{rule5}^{*}, K_{rule6}^{*}, K_{rule7}^{*}$, and K_{rule8}^{*} that are calculated by using Eq. (8). Step 5 computes YP_1, YP_2, and YP_3. In Step 6, we take $r_1 = 3.9232332325312977, r_2 = 3.9387984542351239, r_3 = 3.934221263229951, c_1 = 0.5127, c_2 = 0.8736$, and $c_3 = 0.65375$ followed by Eq. (9) that computes CM_1, CM_2, and CM_3. Eq. (10) generates the sequences $z_1, z_2, z_3, z_4, z_5, z_6, z_7$, and z_8. Finally, Eq. (11) computes keys $XORKey$, $DNAAddKey_1, DNAAddKey_2, APr_1, APr_2, APr_3, APc_1, APc_2, APc_3$, $PermKey_1$, and $PermKey_2$ used for encryption and decryption. Now, encryption keys are generated and then we do encryption explained in the next section.

5.2.2 Encryption and decryption process

RGB images of size $512 \times 512 \times 3$ as shown in Fig. 11 are chosen for encryption and decryption. The rules $A_{rule3}^{user} = 1, A_{rule2}^{user} = 6236$, and $A_{rule1}^{user} = 623{,}621$ are chosen to calculate A_{rule3}, A_{rule2}, and A_{rule1}.

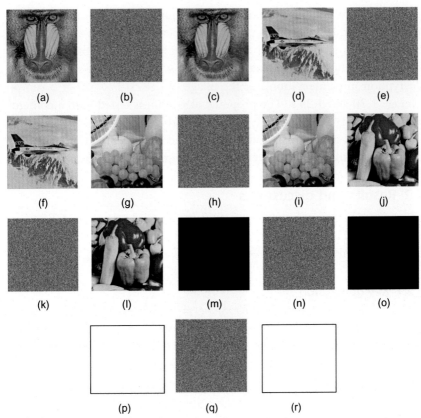

Fig. 11 Results of encrypted and decrypted images: (A) Baboon—original/plain image, (B) Baboon—encrypted/ciphered image, (C) Baboon—decrypted/deciphered image, (D) Airplane—original/plain image, (E) Airplane—encrypted/ciphered image, (F) Airplane—decrypted/deciphered image, (G) Fruits—original/plain image, (H) Fruits—encrypted/ciphered image, (I) Fruits—decrypted/deciphered image, (J) Peppers—original/plain image, (K) Peppers—encrypted/ciphered image, (L) Peppers—decrypted/deciphered image, (M) Black—original/plain image, (N) Black—encrypted/ciphered image, (O) Black—decrypted/deciphered image, (P) White—original/plain image, (Q) white—encrypted/ciphered image, and (R) white—decrypted/deciphered image.

1. The original RGB image of size $512 \times 512 \times 3$ converted into a one-dimensional array, say E_1 of size $1 \times 786, 432$, and $PermKey_2$ is used to reorders E_1, which yields E_2.

2. E_2 is converted into three layers each of size 512×512, then Eq. (13) applied by interlayer XOR using the keys $XORKey_1$, $XORKey_2$, and $XORKey_3$, and thereafter modulo is used in order to maintain 8 bits

per pixel value. This step yielded E_{3R}, E_{3G}, and E_{3B} each of size 512 × 512 then composed into a single image of size 512 × 512 × 3.

3. *DNAencode* operation is applied on E_3 using A_{rule1} to get E_4 of size 1 × 3, 145, 728.

4. DNA addition is executed on E_4 by *DNAADDKey$_2$*, which yields E_5.

5. In similar to the first step, permutation shuffling is applied over E_5 using the key *PermKey$_1$*, which is of size 1 × 3145728 because both the image and key are in DNA-encoded domain providing E_6.

6. DNA complement on E_6 provides E_7.

7. This step executes *DNAV igEnc* on E_7 by using the key *DNAA ddKey$_1$* giving E_8.

8. E_8 is converted layers each of size 1024 × 1024, and improved AP shuffling is used to get E_9. Finally, DNA decoding is applied to get E_{10}.

The decryption process is similar to the encryption process by taking the steps with minor changes in the reverse order. The encryption and decryption results are provided below:

- Fig. 11A, D, G, J, M, and P represent original images of Baboon, Airplane, Peppers, Fruits, Black, and White, respectively.
- Fig. 11B, E, H, K, N, and Q represent encrypted images of Baboon, Airplane, Peppers, Fruits, Black, and White, respectively. It can be observed that there is no information gathered from the encrypted images.
- Considering all the encrypted images in one place, one can find that it is indistinguishable.
- Fig. 11C, F, I, L, O, and R represent decrypted images of Baboon, Airplane, Peppers, Fruits, Black, and White, respectively. It can be observed—the original image is recovered without any loss of data.

5.3 Comparison

The proposed algorithm is compared against the best values of other existing algorithms for (a) keyspace—The proposed algorithm has a large keyspace (10^{278}) than other algorithms in Table 11, (b) NPCR and UACI values for the proposed algorithm are more close to average NPCR and UACI values (given in Tables 9 and 10, respectively) compared to other algorithms in Table 12, and (c) horizontal, vertical, and diagonal correlation values for the proposed algorithm is significantly less (near 10^{-5}) than other algorithms in Table 13.

Table 11 Comparison of keyspace of the proposed algorithm against existing algorithms.

Algorithms	Keyspace
Ref. [1]	10^{120}
Ref. [2]	10^{74}
Ref. [3]	$\approx 10^{126}$
Ref. [6]	10^{141}
Ref. [7]	10^{154}
Ref. [8]	10^{56}
Ref. [9]	10^{38}
Ref. [10]	$\approx 10^{113}$
Ref. [11]	$\approx 10^{158}$
Ref. [16]	10^{100}
Proposed algorithm	10^{278}

Table 12 Comparison of NPCR and UACI values of the proposed method against other algorithms.

	Average NPCR	Average UACI
Ref. [1]	59.7406	25.0487
Ref. [2]	99.6367 (Avg)	33.3933 (Avg)
Ref. [3]	99.6173	33.4756
Ref. [5]	99.5789	33.4549
Ref. [6]	99.6029	33.4329
Ref. [7]	99.5783 (Avg)	33.4283 (Avg)
Ref. [8]	99.6222 (Avg)	33.4233 (avg)
Ref. [9]	99.43 (Avg)	33.5 (Avg)
Ref. [10]	99.6111	33.46 (Avg)
Ref. [11]	99.6101	33.4693
Ref. [16]	99.6185	28.7344
Proposed algorithm	99.6106 (Avg)	33.4663 (Avg)

Table 13 Comparison of correlation against existing other algorithms.

Correlation	Horizontal	Vertical	Diagonal
Ref. [1]	0.0009	0.00063	0.00079
Ref. [2]	0.0002	0.0002	−0.0005
Ref. [3]	0.0013	−0.0049	0.0057
Ref. [5]	0.0693	0.0610	0.0242
Ref. [6]	−0.0139	0.0177	6.7947e-04
Ref. [7]	0.0013	0.0005	0.0006
Ref. [8]	0.0011	0.0009	0.0012
Ref. [9]	−0.0016	0.0012	0.0105
Ref. [10]	−0.0002	−0.0015	−0.0008
Ref. [11]	0.0023	0.0025	0.0024
Ref. [16]	0.0039	−0.0314	0.0158
Proposed algorithm	9.2808e-05	1.6837e-05	4.15e-05

6. Conclusion

This chapter proposes a new, secure, fast encryption algorithm for RGB images (data) using DNA computing adaptable to wireless sensornetworks. A new key generation algorithm is proposed using AP and logistic series in the DNA domain. The proposed algorithm involves permutation shuffling, generalized Vigenère-type encryption, AP-series-based shuffling, and different DNA encoding operations. Results in Appendix demonstrate that the generated sequence is random. The proposed encryption algorithm provides a huge keyspace and is robust against all types of common attacks (such as the differential, cropping, and communication channel to noise attacks). The various statistical tests are performed on standard test images to demonstrate the proposed algorithm's efficiency and security. Data presented in Tables 11–13 state that the proposed algorithm has better performance and security than others. Future work is to enhance the generalized Vigenère-type encryption by taking more than one DNA molecule as a blockfor SSL key exchange and the digital signature of wireless sensor networks.

Acknowledgments

The authors thank the anonymous reviewers and the handling editor for helpful and constructive comments that significantly contributed to improving this chapter. Further, the authors are also grateful to the Science & Engineering Research Board, Government of India, for providing financial support through project file no. YSS/2015/000930.

Appendix

In this section, the results obtained by the NIST statistical test suite on generated random sequences. All successful results are provided side by side in Figs. A.1–A.5. The randomness of the key sequence (Z_t) is generated and converted from decimal into a binary sequence. This binary sequence is chosen with a block size of each 400,000 bits. The 10 and 15 blocks are taken to test randomness using NIST statistical test suite. The results are provided in Appendix.

```
RESULTS FOR THE UNIFORMITY OF P-VALUES AND THE PROPORTION OF PASSING SEQUENCES
-----------------------------------------------------------------------------
   generator is <key binary data.txt>
-----------------------------------------------------------------------------
 C1  C2  C3  C4  C5  C6  C7  C8  C9 C10  P-VALUE   PROPORTION  STATISTICAL TEST
-----------------------------------------------------------------------------
  2   1   0   1   1   1   1   0   1   0  0.911413    10/10     Frequency
  1   0   0   1   5   1   1   0   0   0  0.017912    10/10     BlockFrequency
  3   1   1   1   2   0   0   2   1   1  0.350485     9/10     CumulativeSums
  2   1   1   0   1   1   1   1   1   1  0.739918    10/10     CumulativeSums
  1   1   2   1   0   2   1   0   0   1  0.739918     9/10     Runs
  2   1   2   1   1   2   0   0   2   1  0.534146    10/10     LongestRun
  2   1   1   1   2   1   0   1   0   1  0.739918    10/10     Rank
  1   1   1   0   0   2   2   0   1   2  0.739918    10/10     FFT
  1   0   1   0   2   1   1   1   1   2  0.534146    10/10     NonOverlappingTemplate
  1   0   2   0   2   2   0   1   1   1  0.534146    10/10     NonOverlappingTemplate
  0   1   1   1   3   1   1   0   1   1  0.911413    10/10     NonOverlappingTemplate
  1   1   0   1   3   0   1   1   2   0  0.350485    10/10     NonOverlappingTemplate
  1   0   3   1   2   0   2   0   1   0  0.350485    10/10     NonOverlappingTemplate
  1   1   0   0   1   2   2   0   3   0  0.534146     9/10     NonOverlappingTemplate
  4   0   3   1   1   0   0   0   0   1  0.350485    10/10     NonOverlappingTemplate
  0   0   0   2   1   1   2   2   1   1  0.350485    10/10     NonOverlappingTemplate
  1   1   1   2   1   1   0   0   1   2  0.122325     8/10     NonOverlappingTemplate
  2   0   1   1   1   2   1   2   0   0  0.739918    10/10     NonOverlappingTemplate
  0   1   1   1   1   2   2   0   0   2  0.911413    10/10     NonOverlappingTemplate
  2   2   1   2   2   0   1   0   0   0  0.534146    10/10     NonOverlappingTemplate
  1   1   1   1   0   2   1   3   0   0  0.122325    10/10     NonOverlappingTemplate
  1   1   1   2   1   2   0   0   1   1  0.911413    10/10     NonOverlappingTemplate
  4   0   3   1   1   1   0   0   0   0  0.066882    10/10     NonOverlappingTemplate
  0   0   1   2   0   2   1   2   2   0  0.739918    10/10     NonOverlappingTemplate
  1   1   1   1   2   1   0   1   1   1  0.350485    10/10     NonOverlappingTemplate
  1   1   1   2   0   2   1   0   1   1  0.911413    10/10     NonOverlappingTemplate
  1   0   1   1   2   2   0   1   1   1  0.911413    10/10     NonOverlappingTemplate
  2   2   1   2   0   0   1   1   1   0  0.534146    10/10     NonOverlappingTemplate
  1   1   0   0   1   1   2   1   1   2  0.739918    10/10     NonOverlappingTemplate
  2   2   1   0   0   2   1   1   1   0  0.213309    10/10     NonOverlappingTemplate
  1   0   1   0   1   1   1   2   2   1  0.534146    10/10     NonOverlappingTemplate
  1   2   0   1   1   0   0   1   2   2  0.991468    10/10     NonOverlappingTemplate
  1   1   2   1   1   0   1   2   0   1  0.534146     9/10     NonOverlappingTemplate
```

(a)

```
 C1  C2  C3  C4  C5  C6  C7  C8  C9 C10  P-VALUE   PROPORTION  STATISTICAL TEST
  0   1   2   1   2   0   2   2   1   0  0.534146    10/10     NonOverlappingTemplate
  1   2   1   1   3   1   1   0   1   2  0.350485    10/10     NonOverlappingTemplate
  1   1   1   2   1   1   1   0   0   0  0.350485    10/10     NonOverlappingTemplate
  0   1   2   0   2   0   1   0   4   0  0.739918    10/10     NonOverlappingTemplate
  1   1   1   1   0   1   2   1   0   1  0.213309    10/10     NonOverlappingTemplate
  1   1   1   1   1   0   0   1   1   2  0.911413    10/10     NonOverlappingTemplate
  0   0   1   2   3   2   1   2   0   3  0.213309    10/10     NonOverlappingTemplate
  0   3   1   1   2   0   2   1   0   1  0.350485    10/10     NonOverlappingTemplate
  3   0   0   1   0   1   0   2   2   1  0.739918    10/10     NonOverlappingTemplate
  0   1   2   1   0   1   3   1   2   3  0.017912    10/10     NonOverlappingTemplate
  1   2   0   1   0   2   1   1   1   1  0.534146    10/10     NonOverlappingTemplate
  1   1   2   2   1   1   1   0   0   1  0.122325    10/10     NonOverlappingTemplate
  3   0   0   2   0   1   2   2   0   0  0.213309     9/10     NonOverlappingTemplate
  2   1   1   1   1   2   0   1   1   0  0.350485    10/10     NonOverlappingTemplate
  2   0   1   2   0   0   2   2   1   0  0.122325    10/10     NonOverlappingTemplate
  0   2   0   1   1   0   1   2   3   0  0.534146    10/10     NonOverlappingTemplate
  1   0   2   2   1   2   1   0   1   0  0.991468    10/10     NonOverlappingTemplate
  1   1   0   1   1   2   1   2   0   1  0.911413     8/10     NonOverlappingTemplate
```

(b)

Fig. A.1 NIST statistical suite test result: (A) and (B) 10 blocks of binary data with block size of 400,000 bits. (A) Label 1; (B) label 2.

Fig. A.2 NIST statistical suite test result: (A) and (B) 10 blocks of binary data with block size of 400,000 bits—cont. (A) Label 1; (B) label 2.

(a)

C1	C2	C3	C4	C5	C6	C7	C8	C9	C10	P-VALUE	PROPORTION	STATISTICAL TEST
0	0	0	1	0	0	0	1	1	0	---	3/3	RandomExcursions
0	0	1	0	1	1	0	0	0	0	---	3/3	RandomExcursions
0	0	0	1	0	0	2	0	0	0	---	3/3	RandomExcursions
0	0	1	0	0	0	2	0	0	0	---	3/3	RandomExcursions
1	0	0	0	0	0	0	1	1	0	---	3/3	RandomExcursionsVariant
0	0	0	0	1	0	1	0	1	0	---	3/3	RandomExcursionsVariant
0	0	1	0	1	0	1	0	0	0	---	3/3	RandomExcursionsVariant
1	0	0	0	1	1	1	0	1	0	---	3/3	RandomExcursionsVariant
0	0	0	0	0	1	1	0	0	2	---	3/3	RandomExcursionsVariant
2	0	0	0	0	1	0	0	0	0	---	3/3	RandomExcursionsVariant
2	0	0	1	0	0	0	0	0	0	---	3/3	RandomExcursionsVariant
1	0	0	0	0	1	0	0	1	0	---	3/3	RandomExcursionsVariant
0	1	0	0	2	0	0	0	0	0	---	3/3	RandomExcursionsVariant
1	0	0	2	0	0	0	0	0	0	---	3/3	RandomExcursionsVariant
0	0	0	0	0	1	1	0	1	0	---	3/3	RandomExcursionsVariant
0	0	1	1	0	1	0	0	0	0	---	3/3	RandomExcursionsVariant
0	0	0	0	0	1	1	1	0	0	---	3/3	RandomExcursionsVariant
0	0	0	0	0	0	2	0	1	0	---	3/3	RandomExcursionsVariant
0	0	1	1	0	0	0	1	0	0	---	3/3	RandomExcursionsVariant
0	0	0	0	0	0	1	0	2	0	---	3/3	RandomExcursionsVariant
0	1	0	1	0	0	0	0	1	0	---	3/3	RandomExcursionsVariant
1	2	0	0	0	0	0	0	0	0	---	3/3	RandomExcursionsVariant
1	1	0	2	0	0	1	1	2	0	0.350485	10/10	Serial
1	2	1	0	1	0	2	0	2	1	0.534146	10/10	Serial
2	0	3	1	3	1	0	0	0	0	0.534146	10/10	LinearComplexity

The minimum pass rate for each statistical test with the exception of the random excursion (variant) test is approximately = 8 for a sample size = 10 binary sequences.

The minimum pass rate for the random excursion (variant) test is approximately = 2 for a sample size = 3 binary sequences.

For further guidelines construct a probability table using the MAPLE program provided in the addendum section of the documentation.

(b)

RESULTS FOR THE UNIFORMITY OF P-VALUES AND THE PROPORTION OF PASSING SEQUENCES

generator is <key_binary_data.txt>

C1	C2	C3	C4	C5	C6	C7	C8	C9	C10	P-VALUE	PROPORTION	STATISTICAL TEST
3	1	2	1	2	2	1	0	0	1	0.275709	15/15	Frequency
1	1	2	2	1	5	0	0	0	1	0.006196	15/15	BlockFrequency
3	2	1	2	0	1	1	0	0	0	0.275709	14/15	CumulativeSums
3	1	2	1	2	2	0	1	1	1	0.437274	15/15	CumulativeSums
2	3	1	1	1	0	2	1	1	0	0.006196	14/15	Runs
3	2	2	0	1	1	2	1	5	0	0.162606	15/15	LongestRun
2	2	3	0	1	1	2	0	0	2	0.162606	15/15	Rank
2	2	3	0	1	1	3	0	0	2	0.275709	15/15	FFT
1	0	0	1	0	2	0	0	2	1	0.025193	15/15	NonOverlappingTemplate
1	0	0	0	2	0	5	3	2	1	0.012650	15/15	NonOverlappingTemplate
0	2	0	3	0	2	1	3	1	3	0.048716	15/15	NonOverlappingTemplate
2	0	0	0	1	3	0	4	3	2	0.006196	15/15	NonOverlappingTemplate
1	3	3	0	3	1	0	3	0	0	0.090936	14/15	NonOverlappingTemplate
0	4	1	0	0	0	2	2	3	1	0.090936	14/15	NonOverlappingTemplate
2	1	3	1	1	1	0	0	2	3	0.025193	15/15	NonOverlappingTemplate
5	1	1	2	0	0	1	1	3	1	0.048716	13/15	NonOverlappingTemplate
0	3	2	2	0	1	1	0	3	3	0.001399	13/15	NonOverlappingTemplate
0	1	0	1	2	1	3	3	2	2	0.162606	15/15	NonOverlappingTemplate
2	1	2	4	0	1	0	2	0	4	0.637119	15/15	NonOverlappingTemplate
1	1	1	2	0	2	3	0	1	4	0.275709	15/15	NonOverlappingTemplate
2	1	0	4	0	2	0	4	1	1	0.048716	15/15	NonOverlappingTemplate
2	3	1	0	2	2	0	2	1	2	0.162606	15/15	NonOverlappingTemplate
1	3	1	2	1	0	1	4	1	1	0.162606	15/15	NonOverlappingTemplate
1	3	0	2	2	1	4	1	0	1	0.002971	15/15	NonOverlappingTemplate
2	1	3	2	1	2	0	2	1	1	0.437274	15/15	NonOverlappingTemplate
2	1	3	0	2	1	0	2	2	1	0.275709	15/15	NonOverlappingTemplate
2	1	0	1	1	0	1	2	3	3	0.437274	15/15	NonOverlappingTemplate
3	1	1	0	2	1	2	1	1	3	0.275709	15/15	NonOverlappingTemplate
1	1	0	4	2	1	2	1	2	4	0.162606	15/15	NonOverlappingTemplate
1	0	4	0	2	1	3	3	0	0	0.012650	15/15	NonOverlappingTemplate
1	1	0	1	1	1	4	1	1	0	0.162606	15/15	NonOverlappingTemplate
4	1	3	0	1	1	4	2	1	1	0.090936	14/15	NonOverlappingTemplate
1	3	0	2	3	2	0	2	1	1	0.090936	14/15	NonOverlappingTemplate
1	0	3	1	1	1	2	2	2	2	0.437274	14/15	NonOverlappingTemplate
2	1	1	2	3	1	2	1	0	1	0.437274	14/15	NonOverlappingTemplate

Fig. A.3 NIST statistical suite test result: (A) 10 blocks of binary data with block size of 400,000 bits and (B) 15 blocks of binary data with block size of 400,000 bits. (A) Label 1; (B) label 2.

(a)

(b)

Fig. A.4 NIST statistical suite test result: (A) and (B) 15 blocks of binary data with block size of 400,000 bits. (A) Label 1; (B) label 2.

(a)

C1	C2	C3	C4	C5	C6	C7	C8	C9	C10	P-value	Proportion	Test
1	2	1	3	2	2	1	0	2	2	0.437274	15/15	NonOverlappingTemplate
1	0	2	1	1	3	3	2	0	2	0.437274	15/15	NonOverlappingTemplate
3	2	4	2	2	0	1	1	0	0	0.275709	14/15	NonOverlappingTemplate
0	0	0	3	4	4	0	1	1	2	0.002971	15/15	NonOverlappingTemplate
0	3	1	1	0	2	1	5	1	1	0.012650	15/15	NonOverlappingTemplate
3	1	1	0	3	1	2	0	1	3	0.012650	15/15	NonOverlappingTemplate
1	0	3	3	1	0	2	2	3	0	0.025193	15/15	NonOverlappingTemplate
0	0	0	0	1	2	4	3	2	3	0.090936	15/15	NonOverlappingTemplate
4	1	0	0	3	0	3	0	2	2	0.048716	15/15	NonOverlappingTemplate
0	3	2	2	3	1	1	2	0	1	0.048716	15/15	NonOverlappingTemplate
3	1	2	0	1	4	3	0	1	0	0.025193	15/15	NonOverlappingTemplate
2	4	2	1	1	0	1	0	1	3	0.275709	15/15	NonOverlappingTemplate
2	2	1	0	0	1	3	3	0	3	0.275709	14/15	NonOverlappingTemplate
1	4	0	0	3	1	1	3	1	1	0.162606	15/15	NonOverlappingTemplate
2	2	2	2	0	0	1	1	3	2	0.090936	14/15	NonOverlappingTemplate
2	1	1	2	4	2	2	0	0	1	0.162606	14/15	NonOverlappingTemplate
1	1	0	3	3	0	1	2	3	1	0.090936	15/15	NonOverlappingTemplate
2	2	0	1	0	2	1	5	0	2	0.012650	15/15	NonOverlappingTemplate
2	0	3	1	2	2	1	0	3	1	0.002971	15/15	NonOverlappingTemplate
2	1	2	0	2	2	3	1	1	1	0.834308	15/15	NonOverlappingTemplate
3	2	4	1	1	0	2	1	0	1	0.048716	15/15	NonOverlappingTemplate
1	2	5	2	1	1	0	0	1	2	0.437274	15/15	NonOverlappingTemplate
2	2	1	3	2	0	1	2	0	2	0.012650	15/15	NonOverlappingTemplate
3	2	3	0	1	0	1	2	2	1	0.048716	15/15	NonOverlappingTemplate
1	1	1	3	2	0	1	1	1	4	0.048716	15/15	NonOverlappingTemplate
4	0	1	2	2	1	0	3	0	2	0.275709	15/15	NonOverlappingTemplate
1	1	3	2	1	1	2	1	0	3	0.275709	15/15	NonOverlappingTemplate
3	3	3	0	2	1	1	0	2	0	0.162606	15/15	NonOverlappingTemplate
2	0	2	4	2	1	0	1	0	3	0.162606	15/15	NonOverlappingTemplate
4	2	2	3	2	0	0	0	0	2	0.006196	15/15	NonOverlappingTemplate
2	1	0	3	3	0	1	3	2	0	0.162606	15/15	NonOverlappingTemplate
2	0	3	1	0	3	0	2	2	2	0.437274	15/15	NonOverlappingTemplate
4	4	1	2	2	0	1	0	0	1	0.090936	15/15	NonOverlappingTemplate
3	2	2	0	0	1	1	0	4	2	0.162606	15/15	NonOverlappingTemplate
0	1	2	0	4	1	3	2	1	1	0.006196	15/15	NonOverlappingTemplate
1	0	0	2	2	2	1	2	1	0	0.001399	14/15	Universal
0	1	1	1	0	1	0	0	0	1	0.637119	14/15	ApproximateEntropy
1	0	0	0	2	0	1	0	0	0	----	4/4	RandomExcursions
0	1	0	0	1	0	1	1	0	0	----	4/4	RandomExcursions
0	0	1	0	1	1	0	0	1	0	----	4/4	RandomExcursions

(a)

(b)

C1	C2	C3	C4	C5	C6	C7	C8	C9	C10	P-value	Proportion	Test
0	0	1	0	0	1	1	1	0	0	----	4/4	RandomExcursions
0	0	0	0	0	1	0	3	0	0	----	4/4	RandomExcursions
0	1	0	0	1	0	0	0	2	0	----	4/4	RandomExcursions
0	1	0	2	0	0	0	1	0	0	----	4/4	RandomExcursions
1	0	0	0	2	0	1	0	0	0	----	4/4	RandomExcursions
0	1	0	1	1	0	1	0	0	0	----	4/4	RandomExcursionsVariant
1	0	1	0	0	0	0	0	2	0	----	4/4	RandomExcursionsVariant
1	1	0	0	1	1	1	0	0	0	----	4/4	RandomExcursionsVariant
1	0	0	1	0	1	1	0	1	0	----	4/4	RandomExcursionsVariant
0	0	0	0	1	0	2	2	0	2	----	4/4	RandomExcursionsVariant
0	1	0	0	0	0	0	1	1	1	----	4/4	RandomExcursionsVariant
0	1	0	0	1	1	1	0	0	0	----	4/4	RandomExcursionsVariant
0	0	0	1	0	0	0	2	1	0	----	4/4	RandomExcursionsVariant
1	0	0	0	0	1	1	0	0	1	----	4/4	RandomExcursionsVariant
0	0	0	1	1	1	0	1	0	0	----	4/4	RandomExcursionsVariant
0	0	0	0	0	1	2	0	0	1	----	4/4	RandomExcursionsVariant
0	0	1	1	0	1	0	0	1	0	----	4/4	RandomExcursionsVariant
0	0	0	1	1	0	0	1	0	1	----	4/4	RandomExcursionsVariant
0	0	0	0	1	0	2	0	1	0	----	4/4	RandomExcursionsVariant
0	0	1	1	0	0	0	2	0	0	----	4/4	RandomExcursionsVariant
0	0	0	1	1	1	0	0	0	1	----	4/4	RandomExcursionsVariant
2	2	0	0	4	1	3	2	0	1	0.025193	15/15	Serial
2	2	2	4	1	1	2	2	0	2	0.048716	15/15	Serial
1	3	3	1	1	0	1	0	3	3	0.025193	15/15	LinearComplexity

The minimum pass rate for each statistical test with the exception of the random excursion (variant) test is approximately = 13 for a sample size = 15 binary sequences.

The minimum pass rate for the random excursion (variant) test is approximately = 3 for a sample size = 4 binary sequences.

For further guidelines construct a probability table using the MAPLE program provided in the addendum section of the documentation.

(b)

Fig. A.5 NIST statistical suite test result: (A) and (B) 15 blocks of binary data with block size of 400,000 bits. (A) Label 1; (B) label 2.

References

[1] K. Zhan, D. Wei, J. Shi, J. Yu, Cross-utilizing hyper chaotic and DNA sequences for image encryption, J. Electron. Imaging 26 (1) (2017) 013021 (1-11).

[2] T. Haq, T. Shah, Algebra-chaos amalgam and DNA transform based multiple digital image encryption, Journal of Information Security and Applications 54 (1) (2020) 102592 (1-17).

[3] A. Belazi, M. Talha, S. Kharbech, W. Xiang, Novel medical image encryption scheme based on chaos and DNA encoding, IEEE Access 7 (2019) 36667–36681, https://doi.org/10.1109/ACCESS.2019.2906292.

[4] R. Enayatifar, F.G. Guimarães, P. Siarry, Index-based permutation-diffusion in multiple-image encryption using DNA sequence, Opt. Lasers Eng. 115 (2019) 131–140, https://doi.org/10.1016/j.optlaseng.2018.11.017.

[5] M.A.B. Farah, R. Guesmi, A. Kachouri, M. Samet, A novel chaos based optical image encryption using fractional Fourier transform and DNA sequence operation, Opt. Laser Technol. 121 (2020) 105777, https://doi.org/10.1016/j.optlastec.2019.105777 (1-8).

[6] X. Chai, Z. Gan, K. Yuan, Y. Chen, X. Liu, A novel image encryption scheme based on DNA sequence operations and chaotic systems, Neural Comput Appl 31 (2019) 219–237, https://doi.org/10.1007/s00521-017-2993-9.

[7] X. Wang, Y. Hou, S. Wang, R. Li, A new image encryption algorithm based on CML and DNA sequence, IEEE Access 6 (2018) 62272–62285, https://doi.org/10.1109/ACCESS.2018.2875676.

[8] X. Wang, P. Li, Y. Zhang, L.Y. Liu, H. Zhang, X. Wang, A novel color image encryption scheme using DNA permutation based on the Lorenz system, Multimed. Tools Appl. 77 (2018) 6243–6265, https://doi.org/10.1007/s11042-017-4534-z.

[9] X. Fu, B. Liu, Y.Y. Xie, W. Li, Y. Liu, Image encryption-then-transmission using DNA encryption algorithm and the double chaos, IEEE Photonics J. 10 (03) (2018) 3900515 (1-16).

[10] A. Rehman, X. Liao, M. Abbas, R. Haider, An efficient mixed inter-intra pixels substitution at 2 bits-level for image encryption technique using DNA and chaos, Optik 153 (2018) 117–134, https://doi.org/10.1016/j.ijleo.2017.09.099.

[11] K.A. Patro, B. Acharya, V. Nath, Secure, Lossless, and noise-resistive image encryption using chaos, hyper-chaos, and DNA sequence operation, IETE Tech. Rev. 37 (3) (2020) 223–245.

[12] K.C. Jithin, S. Sankar, Colour image encryption algorithm combining Arnold map, DNA sequence operation, and a Mandelbrot set, J. Inf. Secur. Appl. 50 (2020) 102428, https://doi.org/10.1016/j.jisa.2019.102428 (1-22).

[13] T.T. Zhang, S.J. Yan, C.Y. Gu, R. Ren, K.X. Liao, Research on image encryption based on DNA sequence and chaos theory, J. Phys. Conf. Ser. 1004 (1) (2018) 012023 (1-7).

[14] M. Sokouti, B. Sokouti, A PRISMA-compliant systematic review and analysis on color image encryption using DNA properties, Comput. Sci. Rev. 29 (2018) 14–20, https://doi.org/10.1016/j.cosrev.2018.05.002.

[15] S.M. Hameed, H.A. Sa'adoon, M. Al-Ani, Image encryption using DNA encoding and RC4 algorithm, Iraqi J. Sci. 59 (1B) (2018) 434–446.

[16] X. Zhang, Z. Zhou, Y. Niu, An image encryption method based on the Feistel network and dynamic DNA encoding, IEEE Photonics J. 10 (4) (2018) 3901014 (1-14).

[17] J. Wu, X. Liao, B. Yang, Image encryption using 2D H'enon-Sine map and DNA approach, Signal Process. 153 (2018) 11–23, https://doi.org/10.1016/j.sigpro.2018.06.008.

[18] H. Nematzadeh, R. Enayatifar, M. Yadollahi, M. Lee, G. Jeong, Binary search tree image encryption with DNA, Optik 202 (2020) 163505 (1-10), https://doi.org/10.1016/j.ijleo.2019.163505.

[19] A. Babaei, H. Motameni, R. Enayatifar, A new permutation-diffusion-based image encryption technique using cellular automata and DNA sequence, Optik 203 (2020) 164000, https://doi.org/10.1016/j.ijleo.2019.164000 (1-16).

[20] Z. Azimi, S. Ahadpour, Color image encryption based on DNA encoding and pair coupled chaotic maps, Multimed. Tools Appl. 79 (2020) 1727–1744, https://doi.org/10.1007/s11042-019-08375-6.

[21] P.T. Akkasaligar, S. Biradar, Selective medical image encryption using DNA cryptography, Inf. Secur. J. A Global Perspect. 29 (2020) 91–101, https://doi.org/10.1080/19393555.2020.1718248.

[22] W. Feng, Y. He, H. Li, C. Li, A plain-image-related chaotic image encryption algorithm based on DNA sequence operation and discrete logarithm, IEEE Access 7 (2019) 181589–181609, https://doi.org/10.1109/ACCESS.2019.2959137.

[23] J. Zheng, L. Liu, Novel image encryption by combining dynamic DNA sequence encryption and the improved 2D logistic sine map, IET Image Process. 14 (2020) 2310–2320, https://doi.org/10.1049/ietipr.2019.1340.

[24] P. Liu, T. Zhang, X. Li, A new color image encryption algorithm based on DNA and spatial chaotic map, Multimed. Tools Appl. 78 (2019) 14823–14835, https://doi.org/10.1007/s11042-018-6758-y.

[25] M. Kumar, A. Iqbal, P. Kumar, A new RGB image encryption algorithm based on DNA encoding an elliptic curve Diffie-Hellman cryptography, Signal Process. 125 (2016) 187–202, https://doi.org/10.1016/j.sigpro.2016.01.017.

[26] P. Pavithran, S. Mathew, S. Namasudra, P. Lorenz, A novel cryptosystem based on DNA cryptography and randomly generated Mealy machine, Comput. Secur. 104 (1) (2020) 102–160.

[27] Monika, S. Upadhyaya, Secure communication using DNA cryptography with secure socket layer (SSL) protocol in wireless sensor networks", Procedia Comput. Sci. 70 (2015) 808–813.

[28] M. Poriye, S. Upadhyaya, DNA-based cryptography for security in wireless sensor networks, in: Cyber Security, Advances in Intelligent Systems and Computing, 729, 2018, pp. 111–118. vol.

[29] R. Dastres, M. Soori, Secure socket layer in the network and web security, Int. J. Comput. Sci. Inf. Technol. Res. 14 (10) (2020) 330–333.

[30] M. Poriye, S. Upadhyaya, Improved security using DNA cryptography in wireless sensor networks, Int. J. Comput. Appl. 155 (13) (2016) 32–35.

[31] S. Namasudra, Fast and secure data accessing by using DNA computing for the cloud environment, IEEE Trans. Services Comput. (2020) 2289–2300, https://doi.org/10.1109/TSC.2020.3046471.

[32] S. Namasudra, S. Sharma, G.C. Deka, P. Lorenz, DNA computing and table based data accessing in the cloud environment, J. Netw. Comput. Appl. 172 (2020), https://doi.org/10.1016/j.jnca.2020.102835.

[33] S. Namasudra, R. Chakraborty, S. Kadry, G. Manogaran, B.S. Rawal, FAST: fast accessing scheme for data transmission in cloud computing, Peer-to-Peer Netw. Appl. 14 (2021) 2430–2442, https://doi.org/10.1007/s12083-020-00959-6.

[34] S. Namasudra, Data access control in the cloud computing environment for bioinformatics, Int. J. Appl. Res. Bioinf. 11 (1) (2020) 40–50.

[35] S. Namasudra, G.C. Deka, Applications of Blockchain in Healthcare, Springer, 2021, https://doi.org/10.1007/978-981-15-9547-9.

[36] S. Namasudra, An improved attribute-based encryption technique towards the data security in cloud computing, Concurr. Comput. Pract. Exerc. 31 (9) (2017), https://doi.org/10.1002/cpe.4364.

[37] S. Namasudra, D. Devi, S. Choudhary, R. Patan, S. Kallam, Security, privacy, trust, and anonymity, in: Advances of DNA Computing in Cryptography, vol. 1, Taylor & Francis, 2018, pp. 138–150.

[38] S. Namasudra, R. Chakraborty, A. Majumder, N.R. Moparthi, Securing multimedia by using DNA based encryption in the cloud computing environment, ACM Trans. Multimed. Comput. Commun. Appl. 16 (3s) (2020) 1–19, https://doi.org/10.1145/3392665.

[39] S. Namasudra, G.C. Deka, R. Bali, Applications and future trends of DNA computing, in: Advances of DNA Computing in Cryptography, 2018, pp. 181–192, https://doi.org/10.1201/9781351011419-9x.

[40] J. Chen, L. Chen, Y. Zhou, Cryptanalysis of a DNA-based image encryption scheme, Inf. Sci. 520 (2020) 130–141, https://doi.org/10.1016/j.ins.2020.02.024.

[41] I.E. Hanouti, H.E. Fadili, K. Zenkouar, Cryptanalysis of an embedded systems' image encryption, Multimed. Tools Appl. 80 (2021) 13801–13820, https://doi.org/10.1007/s11042-020-10289-7.

[42] M. Kumar, K.L.S.J. Josyula, An interlaced secure algorithm for RGB image encryption in wavelet packet domain, Int. J. Wavelets Multiresolut. Inf. Process. 14 (03) (2016) 1650008 (1-14).

[43] A. Rukhin, J. Soto, J. Nechvatal, M. Smid, E. Barker, S. Leigh, M. Levenson, M. Vangel, D. Banks, A. Heckert, J. Dray, S. Vo, A Statistical Test Suite for Random and Pseudorandom Number Generators for Cryptographic Applications, NIST Special Publication, Report Number. 800-22 Rev1a, Available: http://csrc.nist.gov/groups/ST/toolkit/rng/documentation_software.html.

[44] M. Kumar, R.N. Mohapatra, S. Agarwal, G. Sathish, S.N. Raw, A new RGB image encryption using generalized Vigenère-types table over symmetric group associated with virtual planet domain, Multimed. Tools Appl. 78 (8) (2019) 10227–10263.

[45] Y. Wu, J.P. Noonan, S. Agaian, NPCR and UACI randomness tests for image encryption, IEEE J. Sel. Areas Commun. 1 (2) (2011) 31–38.

About the authors

Manish Kumar is presently working as an associate professor in the Department of Mathematics at the Birla Institute of Technology and Science, Pilani, at Hyderabad campus, Hyderabad, Telangana, India. He completed his Ph.D. at IIT (ISM) Dhanbad and his Master of Science at Banaras Hindu University (BHU). His research interests lie in pseudo-differential operators, distribution theory, wavelet theory and applications, digital image processing, and cryptography. He has collaborated actively with researchers in several other computer science disciplines, particularly encryption and cyber-security. Dr. Kumar guided several undergraduate/postgraduate students, including one Ph.D. candidate, and published various research papers in national and international journals of repute. Last year, he was awarded a Fellow of the Institute of Mathematics and its Application (FIMA), London, United Kingdom. Dr. Kumar chaired a session at the International Congress in Honor in the Faculty of Arts and Science, Department of Mathematics in Bursa, Turkey, in 2010 and organized a Symposium in ICNAAM 2013 at Rhodes in Greece. Kumar is a member of several national and international professional bodies and societies. Kumar has invited talks at several national and international conferences, including a talk on "Two-stage hyper-chaotic system-based image encryption in wavelet packet domain for wireless communication systems" at ICM 2018 in Rio de Janeiro, Brazil. Dr. Kumar actively reviews new projects in various funding schemes for the SERB (Science and Engineering Research Board) Government of India. He is also working as a reviewer in AMS Review, zbMath, etc., and an Editor in Mathematical Modelling of Engineering Problems, IIETA, Canada.

Sathish Gunasekaran has completed M. Tech. in Industrial Mathematics and Scientific Computing (IMSC) from IIT Madras, India and was JRF in BITS Pilani Hyderabad Campus, Hyderabad, Telangana, India. He also worked as a software developer in Auriss Technology, Coimbatore, Tamil Nadu, India. Currently, he is working in Verena Haptic & VR Systems Pvt Ltd, Chennai, Tamil Nadu, India. His current area of research interests includes Cryptography, Network Security, and Wireless Communication.

CHAPTER TEN

Research challenges and future work directions in DNA computing

Sachin Minocha[a] and Suyel Namasudra[b]

[a]School of Computing Science and Engineering, Galgotias University, Greater Noida, India
[b]Department of Computer Science and Engineering, National Institute of Technology Agartala, Tripura, India

Contents

Abstract

Deoxyribonucleic Acid (DNA) computing is an emerging field that can be responsible for the expansion of other promising technologies. The vital role of DNA computing is due to its high storage, energy efficiency, and large-scale parallelism capabilities. Such promising capabilities have gained the focus of numerous researchers to replace the traditional computers, i.e., silicon-based computers with biomolecular computers, i.e., DNA-based computers. This work discusses the various real-time challenges faced

Advances in Computers, Volume 129
ISSN 0065-2458
https://doi.org/10.1016/bs.adcom.2022.08.007

by different researchers to implement DNA computing in different application fields. This includes the challenges based on DNA chemistry, error probability, cost, combinatorial and finite-state problems. The future work directions of DNA computing have also been discussed in this chapter. The future work directions of DNA computing show its capabilities to solve the challenges faced by various new-age technologies like big data, cloud computing, blockchain, quantum computing, nanotechnology, and many more.

Abbreviations

A	adenine
AI	artificial intelligence
ANTLR	another tool for language recognition
C	cytosine
DNA	deoxyribonucleic acid
DSD	DNA strand displacement
G	guanine
IDE	integrated development environment
NGS	next-generation sequencing
NP	nondeterministic polynomial
RSA	Rivest–Shamir–Adleman
T	thymine
TB	terabyte
T-S	Takagi–Sugeno
XML	extensible markup language

1. Introduction

Modern computing technology concerns both hardware as well as software because of its exponential growth. Especially in the past seven decades, i.e., after the invention of the Von Neumann architecture. The progress of modern computing technologies is shown in Fig. 1. The current trend is towards minimization of the size of a silicon chip with improved

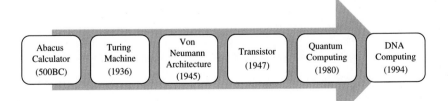

Fig. 1 Progress of modern computing technologies.

performance. As per Moore's law, the performance of silicon-chip doubles every 18 months [1]. However, one cannot expect to observe the same trend as we are getting close to the technical barriers.

The exponential growth of data and the emergence of new application domains require a new innovative and cost-effective alternative. DNA computing is an emerging technology that could overcome the limitation of silicon-based computing technology due to its high computation speed and parallelization capabilities [2]. The similarity between the working of DNA and the Turing machine, along with the high storage capacity of DNA sequences, motivates the researchers to explore this area [3]. A DNA molecule consists of two strands that wind around each other like a twisted ladder. Each DNA sequence has two ends with different polarities known as $3'$ and $5'$ ends. The hydrogen bond between the opposite polarity of DNA sequences gives the double-helix structure of the DNA molecule. A nucleotide is the basic building block of any DNA sequence that consists of Guanine (G), Thymine (T), Adenine (A), and Cytosine (C) as bases. A DNA sequence is generated by bonding different DNA bases. Here, A is always paired with T, and G is paired with C, which make (A, T) and (G, C) complementary base pairs. Different DNA sequences can be generated by utilizing various combinations of DNA bases [4]. The DNA sequences can be manipulated and retrieved by using enzymes and nucleic acid operations, respectively [3].

L.M. Adleman has first shown that DNA computing can be used to solve the Hamiltonian path problem, a renowned NP-complete problem [5,6]. An NP-complete problem cannot be solved by a deterministic algorithm in polynomial time. A directed graph G(V, E) having V and E as a set of vertices and edges, respectively, with V_i and V_o as designated vertices such that $V_i, V_o \in V$ is said to have a Hamiltonian path, if and only if there exists a path between V_i and V_o that traverse each vertex only once [7]. Adleman has solved the Hamiltonian path problem for a Graph G having seven vertices with $V_i = V_0$ and $V_o = V_6$, and used 50 pmol of each nucleotide for the encoding of DNA molecule.

Although the Hamiltonian path problem is an NP-complete problem, Adleman's experiment doesn't ensure that DNA computing can be used to solve an NP-complete problem. Lipton has extended Adleman's model in such a way that biological computers can change the way of computation [8,9]. At first, Lipton generalizes Adleman's solution and tries to solve the satisfiability problem, i.e., a basic NP-complete problem. The main contribution of Lipton's solution is the way to encode a binary string into

DNA strands [10]. The generalized operations of Adleman-Lipton were much successful that many authors have applied or extended the same for DNA addition, Gel-based computing, chess games, surface computing, etc. [11]. These generalized operations with further improvement lead DNA computing to apply it in many fields by using splicing systems and sticker models. The splicing system follows a set of rules, i.e., grammar, to produce a particular language. The sticker model of DNA computing is one of the most useful methods for parallel computing using DNA molecules. These models lead to the design of DNA computers that can replace traditional silicon-based computers due to their vast parallelism and high-density storage capacity.

Nowadays, DNA computing is widely used in various areas like quantum computing, big data storage, nanotechnology, cloud computing, cognitive machines, and security, including cryptography, steganography, and blockchain. DNA computing-based cryptography techniques are used to secure the huge amount of data transmitted over the internet. The authors of Refs. [12,13] have shown DNA-based cryptography by encrypting the data into nitrogen bases, i.e., A, C, T, and G, instead of binary bits, i.e., 0 and 1. DNA has high storage capacity, i.e., 1 g of DNA can store 700 TB of data. This capability of DNA motivates researchers to use DNA computing for storing data over the DNA-based cloud [14,15]. Quantum computing that focuses on speed can be combined with DNA computing. DNA computing-based quantum computers support the capability of both quantum computing and DNA computing [16]. In these computers, DNA molecules can be directly used either by nuclear magnetic resonance or by doping the DNA molecules.

However, DNA computing requires particular conditions to avoid the undesired reaction between enzymes. Moreover, the wet lab experiments based on DNA computing must consider the fidelity of biology protocols and the concentration of chemical components to avoid undesirable results. Such challenges and applications of DNA computing motivate the authors to study the research challenges faced by DNA computing for application in various fields, along with the possibilities for future research. The main contributions of this chapter are as follows:

- In this chapter, many research challenges are discussed, which must be considered to implement DNA computing in related fields.
- Here, many future work directions are presented, which can be beneficial for academicians, researchers, and other practitioners to work on DNA computing with advanced technologies.

The remaining chapter has been divided into three sections. The following section, i.e., Section 2, covers the research challenges faced by the community. Section 3 discusses the future work directions in DNA computing. Finally, Section 4 concludes the entire chapter.

2. Challenges

Various challenges are faced by the research community while implementing DNA computing. This section discusses all the key challenges, as well as solutions to solve these key challenges.

2.1 DNA chemistry

DNA computing requires chemical operations with optimization like a splicing system that uses the enzymatic. In a splicing system, a double-stranded DNA sequence is divided at a specific base sequence location, followed by the reassembling using ligation or hybridization. The required optimal conditions to control the enzyme reaction are difficult due to restrictions imposed by thermodynamics. The first factor that affects the feasibility is the reaction condition like temperature and time. The melting temperature at which 50% of the oligos are melted affects the reassembling of the base sequences. The melting temperature (M_T) is presented using Eq. (1).

$$M_T = \frac{\Delta E^\circ}{\Delta H^\circ + g \ln\left(\frac{M_t^s}{4}\right)} \tag{1}$$

where ΔH° and ΔE° are the entropy and enthalpy, respectively. M_t^s is the total concentration of strand in molar, and g is the gas constant. The melting curve is observed using the Ultraviolet (UV) absorbance technique because hybrid melt increases with the increase in melting temperature. The melting temperature increases due to the absorption of UV radiation and the corresponding width of the melting curve is given in Eq. (2).

$$\Delta M = \frac{6GM_T^2}{\Delta E^\circ} \tag{2}$$

The undesired hybridization occurs the reaction temperature (R_c) is higher than the sum of melting temperature and half of melting curve width, i.e., $M_T + \Delta M/2$. That is why melting temperature must be equal to or higher than the reaction temperature, i.e., $M_T \geq R_c$ for the proper hybridization.

Moreover, the thermodynamics of hybridization and reaction kinetics may result in unwanted reactions. The hybridization reaction in DNA computing is given in Eq. (3).

$$x + y \leftrightharpoons z \tag{3}$$

where z is the double-stranded hybridized DNA, while x, y are oligonucleotides. $\leftrightharpoons$ presents reversible reaction. As mentioned earlier, Eq. (3) is affected by the thermodynamics factor. Considering the volume as constant, Gibbs free energy is used as a parameter to express the equilibrium state. The change in Gibb's free energy dG is given in Eq. (4).

$$dG = \sum_{p=1}^{r} \left(\frac{\partial G}{\partial n_p} \right)_{n_q \neq p} dn_p \tag{4}$$

where n_p is the number of moles in p^{th} component out of r components. $\frac{\partial G}{\partial n_p}$ exhibits the changes in Gibb's free energy with a change in the number of moles. The chemical properties of the pth component is given in Eq. (5).

$$\rho_p \equiv \left(\frac{\partial G}{\partial n_p} \right)_{n_q \neq p} \tag{5}$$

If Eq. (3) is at an equilibrium state, then, dG is equal to zero. Using Eqs. (4) and (5) at equilibrium results, the change in Gibb's free energy is given in Eq. (6).

$$dG = \sum_{p=1}^{r} \rho_p dn_p \tag{6}$$

If the reaction is endothermic, it produces the heat, resulting $dG < 0$, while for the exothermic reaction, it consumes heat $dG > 0$. So, dn_p depends on the extent of reaction, which is given in Eq. (7).

$$dn_p = c_p d\varphi \tag{7}$$

where $d\varphi$ denotes the extent of reaction and c_p is the stoichiometric coefficient. Using Eq. (7), Eq. (8) can be calculated using Eqs. (5) and (4).

$$\frac{dG}{d\varphi} = \sum_{p=1}^{r} \rho_p c_p \tag{8}$$

where $\frac{dG}{d\varphi}$ is the change in Gibb's free energy with the extent of reaction. Therefore, Eq. (8) exhibits that the change in Gibb's free energy with reaction extent directly depends on the chemical properties of the component.

The change in Gibb's free energy is presented in Eq. (9).

$$\frac{dG}{d\varphi} = \Delta G^{\circ} + gM \log U \qquad (9)$$

where ΔG° is the amount of free energy available in ideal conditions. g and M are gas constant and melting temperature, respectively. U is the update factor for non-ideal conditions given in Eq. (10).

$$U = \prod_{p} \alpha_p{}^{c_p} \qquad (10)$$

where α_p is the activity for component p which is given in Eq. (11).

$$\alpha_p = e^{\left(\frac{\rho_p - \rho_p^{\circ}}{gM}\right)} \qquad (11)$$

where ρ_p° is the chemical capabilities of the component at an ideal state, while the activity coefficient is given in Eq. (12).

$$\sigma_p = \frac{\alpha_p}{\beta_p} \qquad (12)$$

where β_p is the mole fraction given in Eq. (13).

$$\beta_p = \frac{n_p}{\sum_{p} n_p} \qquad (13)$$

Eqs. (10)–(13) exhibit that change in free energy depends on the concentration of the reaction components. Thus, the free energy change in Eq. (3) determines the concentration of the generated product. It also determines the concentration of desired and undesired, as well as shifted hybridization. The proper encoding should exhibit the maximum free energy change for desired hybridizations and the minimum for undesired hybridizations.

The coupling among a set of hybridization reactions is another factor that affects the performance of DNA computing. Here, the relative concentration of reaction components plays an important role in extraction. Overall, the factors like reaction conditions, including the melting temperature and time, the undesired reaction between enzymes, the fidelity of biology protocols, and the concentration of chemical components must be considered while performing the wet lab experiment based on DNA computing. It is one of the critical challenges in DNA encoding [17,18].

2.2 Finite state problems

For the advancement of any technology, a basic need is the capability to perform basic operations, such as binary or boolean operations, in less time. The silicon-based computers are successful as they can solve such basic operations in a single step only. This requires the representation of non-negative numbers using the primers, i.e., short sequences of complementary bases. Different researchers have implemented various techniques to represent non-negative numbers in the form of DNA computing. Such works have been extended to perform arithmetic addition and other operations (subtraction, multiplication, etc.) using DNA computing. The main challenge faced by the researchers is the different DNA encoding of input and output numbers. This suppresses the chain operation and reduces the parallelism capabilities. Further expansion has resolved these issues, and advanced algorithms have been suggested by different authors to implement basic operations parallelly and also in a chained manner [3]. However, $O(\log_2 n)$, $O((\log n)^2)$ and $O(\log n)$ are achieved as bio-steps time complexities to perform the addition, multiplication, and subtraction operation reliably. Still, the key challenge is to perform the basic operations using the single bio step [19].

2.3 Reliable DNA codes

DNA codes are crucial for DNA nanotechnology, as well as for DNA computing. Different optimization techniques, including genetic algorithm, particle swarm optimization, simulated annealing, BAT algorithm, template matching, and exhaustive methods, have been developed to produce reliable DNA codes. However, deep learning has opened up a new hope for further optimization of DNA codes that has not been explored deeply [20].

2.4 High error probability

The massive data handling ability of DNA computing makes it prone to a high probability of error. This is due to the fact that error probability increases exponentially with the increase of data. The error probability in DNA computing is due to the mismatched pairs and less accurate enzyme operations. Moreover, an immature mathematical background of DNA computing makes it prone to errors in the security field like cryptography and steganography. The extraction of original data is challenging due to the availability of large permutations.

2.5 DNA scanning

Nowadays, physical records are replacing with digital records in this digital era. DNA scanning is a technique to digitalize biological records. This may result in discovering a piece of confidential information that can speed up research. A DNA scanner accepts nucleotide and amino acid sequences as the input and produces the digital DNA sequences. Hash tables and DNA motif finding are a few basic techniques used to perform DNA scanning in an optimized manner. This field is in the infancy stage, so a lot of possibilities can be explored [20].

2.6 DNA assembly

DNA assembly is the core of synthetic biology [21]. It is used to develop multiple DNA molecules. The work of DNA assembly in DNA computers is similar to the electronic circuit in traditional computers. Therefore, DNA assembly characteristics are often compared with the manufacturing of electronic circuits. The performance of the electronic circuit is measured using well-defined metrics, i.e., Q-metrics. However, establishing such metrics for DNA assembly is still in process. However, the metrics designed to measure the performance of DNA assembly methodologies should be sensitive to the constructs specified for a particular application. Various metrics designed by different authors have shown progress in this field. Still, the performance measures are not well-defined. Moreover, improvement in the DNA assembly methodology is also required.

3. Future work directions

Despite the above-mentioned challenges, DNA computing can be used in many fields. In this section, many future work directions in DNA computing are discussed in detail. Fig. 2 shows future work directions in DNA computing.

3.1 Smart DNA chips

DNA chips are still in their stage of infancy, so a lot of work can be done in the field.

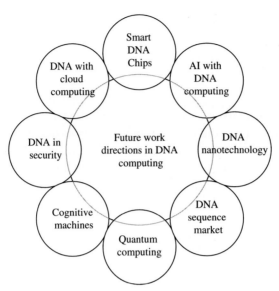

Fig. 2 Future work directions in DNA computing.

DNA chips are highly dense arrays of oligonucleotides that can be used for studying gene regulation and expression under different circumstances. This can be further used to study gene mutations, as well as mapping and sequencing the genes [22]. The steps to design a DNA chip is given in Fig. 3. The unknown agent, i.e., the full feature set is given as input to design a particular DNA chip for the classification of the agents. The automated DNA sequencing leads to generating a genome corresponding to the given agent. This genome is processed for oligonucleotide design to get the lead primer. The probe synthesis of lead primer through automated oligonucleotide gives the nucleotide probe, which generates the microarray, i.e., DNA prototype. This prototype is used for diagnostics assays. If the prototype fails, the probe is redesigned by using the same steps. Otherwise, DNA chips are manufactured on a large scale. DNA chips are then further used to identify species and inspect their phylogenies [15]. In Ref. [23], DNA chips are used for site-targeted mutagenesis, which further helps in developing variants of proteins that have acquired new characteristics. Moreover, such chips have a lot of pathogenic applications, including disease classification and drug design. The main challenge is to design intelligent techniques that explore the huge amount of information provided by the DNA chips. In Ref. [24], authors have shown improved classification of the tumor using the DNA chips. This is mainly achieved due to the high capabilities of DNA chips that

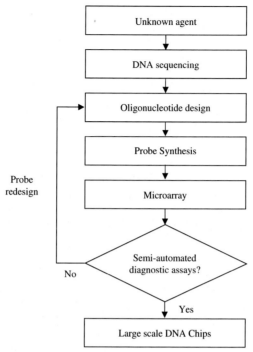

Fig. 3 Manufacturing of DNA chips.

can handle the imbalanced datasets and noise, as well as insignificant attributes. Thus, DNA chips can also be used in related fields to achieve high efficiency.

3.2 Artificial intelligence with DNA computing

With the increasing span of Artificial Intelligence (AI) in multiple domains, DNA computing can be used for enhancements in AI. The development of improving the predictability of artificial models is tremendous. However, further advancements can be made with DNA computing as it would be able to capture a more human-oriented approach to learning, which is much more complex than the existing state-of-the-art machine learning/deep learning models. DNA can be used to develop a more human-resembling learning architecture that can capture complex memory patterns dynamically on its own [25].

The present techniques in DNA computing reckon all the possible candidate solutions followed by the eradication of erroneous DNA. Due to this, the size of the candidate pool enhances exponentially with a linear increase in problem size. This leads to infeasible solution space due to limited computation capabilities. The problem can be handled by using the metaheuristic solutions that generate a random set of candidate solutions instead of the whole possible solution set. The metaheuristic algorithms perform the exploration and exploitation search to generate new solutions, as shown in Fig. 4. The new solutions are included in the existing solution set only if the fitness value of the new solution is better than the fitness value of the existing solution. The process is repeated until the termination criterion is not satisfied. This process is capable of generating the solution for massive size problems. The key issue with such solutions is the stochastic nature of the algorithm that gives a random solution and global optima are not guaranteed. Many authors have proposed strong metaheuristic techniques to produce reliable solutions using DNA computing [3,5]. Wood et al. [26] have implemented the DNA to learn poker by using the evolutionary algorithm. The main features of the implementation are waiting for the turn and learning

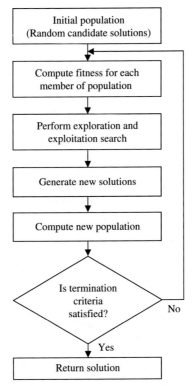

Fig. 4 Steps of Metaheuristic techniques to generate solution.

from payoffs. Ren et al. [27] have implemented a virus DNA–based evolutionary algorithm to self-learn the Takagi-Sugeno (T-S) fuzzy controller. They have demonstrated a new method to automatically construct a fuzzy controller by self-learning a T-S fuzzy controller in any non-linear system.

3.3 DNA nanotechnology

DNA nanotechnology is another budding field of study. It involves developing specific geometrical targets by the use of DNA computing [28]. Rather than focusing on DNA's biological aspect, this field relies on extracting information from a DNA strand to develop relevant structural motifs and connect them [29,30].

Although this concept was first coined in 1998, there are several challenges like the high cost of DNA synthesis, the high error rate of self-assembly, etc. [31]. Further developing a gigadalton of DNA nanostructure, around 1 Mb long scaffold molecule is required, which becomes so long that it becomes difficult to synthesize and also very fragile. But, with the recent advancements in the field of DNA computing, the nanotechnology field has also picked up the pace due to numerous applications as shown in Fig. 5. Moreover, there are huge scopes to use DNA computing in biomimetic systems in which artificial systems can mimic most of the functional behaviors of human DNA [32].

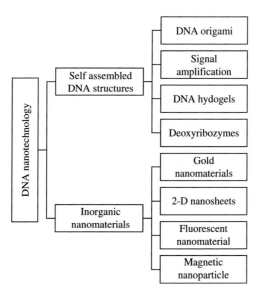

Fig. 5 Applications of DNA nanotechnology.

3.4 DNA sequence market

DNA sequencing is the process of extracting the order of nucleotides from the DNA molecule. One of the milestones of sequencing is the Human Genome Project, a 13-year long project cost of $3 billion that was completed in 2003 [33]. Since then, there has been a huge demand for cheap DNA sequencing. This led to the development of the Next-Generation Sequencing (NGS), which promises slashed sequencing prices and higher throughput than its predecessor techniques [34]. The basic workflow of the NGS experiment completes in four phases, namely sample collection, template generation, sequencing reaction and detection, and data analysis as given in Fig. 6.

The first phase of sample collection may lead to whole-genome sequencing or epigenomic sequencing. The DNA sequencing generated in phase 1 is used for the template generation by undergoing fragmentation and size selection, adapter ligation, amplification, template immobilization, and spatial separation. The short DNA sequence template undergoes complete coverage using the single or pair-end sequencing, while the large DNA sequence template undergoes partial coverage. The detection of sequencing is followed by data analysis as a huge amount of data is generated. Initial data analysis is performed by the proprietary software, which is followed by the alignment of the sequence as a reference sequence. If a reference sequence is

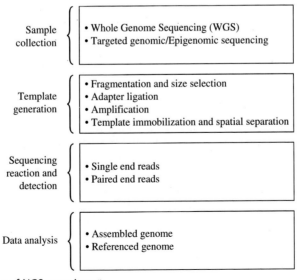

Fig. 6 Phases of NGS experiment.

available, de novo assembly is conducted. This market is also expanding due to its application with cloud computing for improved performance. In 2016, the cost of the global DNA market was $5000 M and it is likely to become three times by 2023 [33].

3.5 Quantum computing

Quantum computing is a concept based on the quantum theory of physics, which deals with nature at the quantum level. A quantum is the minimum amount of a physical entity that can be considered for computing. The idea behind quantum computing is to operate the computers on extremely short pulses of light. This is practical, can speed up the operation of computers by 100,000 times. The data in such computers are not stored directly rather it is converted to quantum bits. However, storing this kind of data is difficult. This is where DNA computing can come into the phase, given the recent advancements made in the field. Also, quantum computing in combination with sentiment analysis and DNA computing can be applied to different fields, such as healthcare, medicine, etc.

The immense computation power of DNA computing and quantum computing has motivated authors to propose DNA computing-based quantum computers [16]. Quantum computing focuses on speed, while DNA computing focuses on storage. Therefore, a combination of both fields can change the world. The DNA molecules can be directly used in quantum computers either by using nuclear magnetic resonance or by dopping the DNA molecules for implementing the quantum gate. Such possibilities are already explored by many authors and a lot of work can be done in the field [16].

3.6 Cognitive machines

Cognitive computing refers to mimicking the human thinking ability as close as possible. In other words, it can be termed as simulating the human brain to develop a computerized model for thinking as close as possible to it. The long-term goal of cognitive computing is to develop a system that will be able to solve complex problems without human assistance.

Researchers from South Korea have been developing biologically-inspired machines based on physical intelligence and DNA computing [35]. This machine encodes information based on DNA computing. It is also expected that these systems will be able to provide state-of-the-art data mining techniques, including dynamic programming, rule-based methods, genetic programming, etc. However, in order to develop such applications, a perfect DNA sequence is required without any prior assumptions. If the idea gets successful, the system will be able to analyze patterns in images and other multimedia systems and finally infer distinguishable characteristics from them.

3.7 DNA in security

DNA computing can be used to enhance security. It can be applied in cryptography, steganography, and blockchain to enhance security as discussed in the following subsections.

3.7.1 Cryptography

With the increasing use of DNA computing since 1994 [36], new applications are being developed. One of the applications includes using DNA computing in cryptography. DNA can store approx. 1000 billion times more information than the traditional storage disks, which makes it an ideal candidate to be replaced with or optimize the one-time encryption techniques that require a large overhead to produce a unique and random cipher each time [37]. Some works include using DNA computing with the RSA algorithm to make it more robust [38]. Here, at first, the plaintext is encoded by using the nucleotides of DNA (Adenine, Guanine, Thymine, and Cytosine). Each of the nucleotides is given a random number, and then the encrypted plaintext is further encrypted by mapping the nucleotides to their corresponding numerical value.

The general steps used in cryptography, i.e., converting plaintext to ciphertext and decryption, i.e., converting ciphertext to plaintext using DNA computing, are given in Figs. 7 and 8, respectively. DNA computing-based encryption process encodes the plaintext to the DNA sequence followed by the implication of complementary rules and primers, as shown in Fig. 7. The complementary rule and addition of primer improve the randomness of DNA bases in ciphertext that improve security. As shown in Fig. 8, the decryption process initiates with the removal of primers and codons from the ciphertext. Then, the complementary rule is applied to get the original DNA sequence. At last, the original DNA sequence is decoded to retrieve the plaintext [39]. The above-mentioned steps discussed the basic DNA cryptography technique. However, few authors recently encrypted the images using DNA computing using chaotic maps, etc. Still, there is a huge scope in the DNA cryptography field for developing novel models.

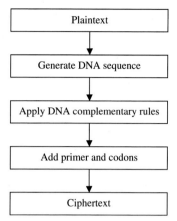

Fig. 7 Encryption process using DNA computing.

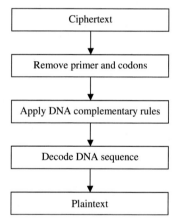

Fig. 8 DNA computing-based decryption process.

3.7.2 Steganography

Steganography is the field where secret data is hidden into non-secret data to avoid detection by attackers and malicious users. This is also known as data hiding as this process hides original data inside a DNA strand. With the immense storage capacity of DNA, it is also possible to improve or replace the existing data hiding algorithms. The first use of DNA in steganography was proposed in Ref. [40] in 2001. In this technique, the authors encode the data into a DNA sequence and then hide this DNA sequence using a key and dummy generated DNA sequence. The dummy DNA sequence is circulated after hiding the original DNA sequence, i.e., encoded text. The extraction of original data is done using the key and dummy DNA sequence.

Here, at first, the plaintext is encrypted using the DNA sequence with a random one-time substitution key, and then, the encoded ciphertext is mixed with a large DNA sequence to improve security. One of the obvious benefits of DNA steganography is that several random DNA sequence chains can be made to accompany the encrypted data to hide it [41]. The authors of Ref. [42] combined DNA steganography with the Hyperelliptic Curve Cryptography to encrypt image files. This scheme has achieved high security, lower noise ratio, and mean squared error than the traditional steganography techniques.

3.7.3 Blockchain

Blockchain has gained pace since 2008 by Satoshi Nakamoto by proposing a peer-to-peer transaction system called Bitcoin [43]. It has found applications in many domains like e-commerce, healthcare, logistics, and many more. Despite providing robust security protocols and tremendous transparency, blockchain has its issues [44]. One of the significant issues faced by the current blockchain architecture is scalability. Due to the increasing use of data for analytics, it has become essential for businesses and researchers to store data safely and transparently. With this increasing data, blockchain technology faces issues regarding time like latency and time lag. In Ref. [45], Edkawy et al. have proposed a solution to overcome this scalability issue and time issue using DNA computing. The similarities between the blockchain architecture and DNA molecule indicate that the two fields can be combined to optimize the overall problem efficiently. Since DNA can handle much more data than a usual storage disk as discussed above, it can be leveraged along with the blockchain to provide a more scalable yet equally secure and transparent as a traditional blockchain solution.

3.8 Cloud computing

As different kinds of data are generated every day, users are using cloud computing environments to store the data. Rather than storing data locally, cloud computing enables users to store data in a virtual space called "cloud." Cloud services allow users to pay only for what they use, thus also cutting the overall cost and not worrying about the space overhead. DNA computing with cloud computing can be used to store a huge amount of data due to its high storage capability. For example, the authors of Ref. [46] used a DNA-based encryption system for multimedia data in the cloud environment to protect the data against hackers and malicious users. The high searching and accessing time of data in the cloud environment create further problems by charging users more for using the cloud services. So, there is a need for a fast and secure data access control model in the cloud, which can be achieved by using DNA computing [46]. Therefore, developing a data access control model in a cloud environment is one such research area where ample research can be done [15].

3.9 Big data

Big Data can be defined as a group of data that is huge in volume and grows exponentially with respect to time [44]. Any big data is much large in size and high complexity, so it cannot be processed or stored by using traditional data processing techniques. Big data consist of several characteristics, such as high volume, value, visualization, veracity, viscosity, virality, velocity, and variety. Due to these features, big data requires advanced technologies to store and process the data. DNA has a huge storage capacity as 1 g of DNA can store 700 TB of data. This means few grams of DNA can store all the available data of the world [47].

DNACloud is a tool that is designed to store the data in the form of DNA sequences. This tool mainly performs three tasks, as shown in Fig. 9, including encoding data into DNA sequences, estimating the requirement, and decoding DNA sequencing to the original data. DNA encoding task uses techniques like Huffman encoding to encode the plaintext into DNA sequences. Many researchers are working on DNA encoding techniques to perform the task fast and efficiently [48]. The estimator phase is used to analyze the memory requirement and the biochemical properties. The memory requirement estimates the amount of DNA required along with the memory requirement and file size. Researchers are working on this phase also to estimate the requirements accurately [4]. The decoding task uses the decoding technique corresponding to the encoding technique to produce the original data accurately. The high storage capabilities of DNA and the success of the DNACloud tool have motivated many researchers to explore this field in depth.

3.10 DNA computing-based compilers

The design of DNA computers needs a DNA-based compiler to encode the data and retrieve the data automatically. The DNA-based compiler must be able to execute any code written in any programming language by using DNA molecules. This is possible only if the compiler has the capability of

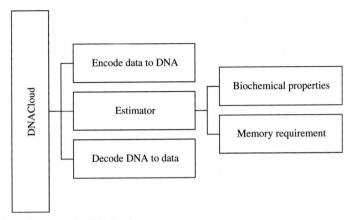

Fig. 9 Functionality of DNACloud.

encoding and decoding the data from DNA efficiently, along with the capability to perform basic operations.

As shown in Fig. 10, the architecture of such a DNA-based compiler is proposed by Yan and Wong [49]. This compiler takes the source code written in any programming language, i.e., C, java, python, etc. This source code is given as input to the parser as well as the ANTLR IDE, i.e., Another Tool for Language Recognition Integrated Development Environment. The ANTLR IDE is a framework that checks the grammar followed by the source code and recognizes the language used in the source code. In such a way, the parser gets the source code and parses the source code to form a parse tree. The parse tree is converted to the syntax tree by using the syntax tree generation phase of the compiler. Then, the target code, i.e., object code or binary code is generated using this syntax tree. This binary code is simulated using the DNA Strand Displacement (DSD) simulation technique to generate the XML file, which can be visualized using the DNA strand displacement visualizer. The DSD technique is used to transform the data by using specified DNA reactions. This architecture is promising and seems to handle the issue of limited computation power. However, this architecture is not capable of performing a general-purpose calculation. Such systems should be maintained by the experts of bio-techniques, which may increase the maintenance cost. Thus, another issue (i.e., cost) is to be addressed. Moreover, communication with the same or different types of systems is still not explored. The deployment of a DNA computing-based system is challenging due to the required environment for DNA reactions. Thus, there is a wide scope of research to replace silicon-based devices with DNA computing-based devices.

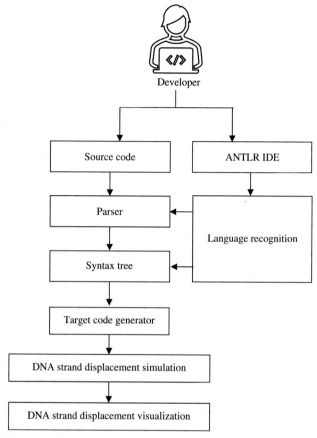

Fig. 10 Architecture of DNA computing-based compiler.

4. Conclusions

DNA computing can be referred to as an emerging branch of computing that utilizes DNA, molecular biology hardware, and biochemistry instead of traditional electronic computing. It uses molecular reaction techniques that are executed on DNA molecules. DNA computing is mainly based on four nitrogen bases, i.e., guanine, thymine, adenine, and cytosine. DNA computing provides high computation power and storage capacity due to its structure. The key challenge for implementing DNA computing is the wet lab experiments in a controlled manner. In this chapter, many research challenges are discussed faced by the research community while implementing DNA computing. Moreover, many future work directions in DNA computing are also presented in this chapter. The ability of

DNA computing to perform with quantum computing and nanotechnology has widened the scope of research in the DNA computing field. In the future, this work can be expanded in any of the domains mentioned in the future work directions, including DNACloud, quantum computing, DNA nanotechnology, security, and DNA computing-based compilers.

References

[1] S. Varghese, J.A.A.W. Elemans, A.E. Rowan, R.J.M. Nolte, Molecular computing: paths to chemical Turing machines, Chem. Sci. 6 (11) (2015) 6050–6058, https://doi.org/10.1039/c5sc02317c.

[2] C. Zhang, et al., DNA computing for combinational logic, Inf. Sci. (NY) 2 (2018) 1–17, https://doi.org/10.1360/n112019-00007.

[3] Z. Ezziane, DNA computing: applications and challenges, Nanotechnology 17 (2) (2006) R27–R39, https://doi.org/10.1088/0957-4484/17/2/R01.

[4] S. Namasudra, G.C. Deka, R. Bali, Applications and future trends of DNA computing, in: S. Namasudra, G.C. Deka (Eds.), Advances of DNA Computing in Cryptography, CRC Press, 2018, pp. 181–192.

[5] L. Adleman, Molecular solution to computational problems.pdf, Science (80-.) 266 (5187) (1994) 1021–1024.

[6] S.A. El-Seoud, R. Mohamed, S. Ghoneimy, DNA computing: challenges and application, Int. J. Interact. Mob. Technol. 11 (2) (2017) 74–87, https://doi.org/10.3991/ijim.v11i2.6564.

[7] J. Watada, R.B.A. Bakar, DNA computing and its applications, in: Proc.—8th Int. Conf. Intell. Syst. Des. Appl. ISDA, vol. 2, 2008, pp. 288–294, https://doi.org/10.1109/ISDA.2008.362.

[8] R.J. Lipton, DNA solution of hard computational problem, Science (80-.) 268 (5210) (1995) 542–545.

[9] R. Lipton, Using DNA to solve NP-complete problems, Science (80-.) 268 (i) (1995) 542–545.

[10] H. Eghdami, M. Darehmiraki, Application of DNA computing in graph theory, Artif. Intell. Rev. 38 (3) (2012) 223–235, https://doi.org/10.1007/s10462-011-9247-5.

[11] R. A. M, M. Amos (Eds.), DNA computing. In computational complexity: theory, techniques and applications, in: Encyclopedia of Complexity and Systems Science, Springer, 2008, pp. 882–896.

[12] Y. Zhang, The image encryption algorithm based on chaos and DNA computing, Multimed. Tools Appl. 77 (16) (2018) 21589–21615, https://doi.org/10.1007/s11042-017-5585-x.

[13] F.E. Ibrahim, M.I. Moussa, H.M. Abdalkader, A symmetric encryption algorithm based on DNA computing, Int. J. Comput. Appl. 97 (16) (2014) 41–45, https://doi.org/10.5120/17094-7634.

[14] S. Namasudra, S. Sharma, G.C. Deka, P. Lorenz, DNA computing and table based data accessing in the cloud environment, J. Netw. Comput. Appl. 172 (April) (2020) 1–13, https://doi.org/10.1016/j.jnca.2020.102835.

[15] S. Namasudra, Fast and secure data accessing by using DNA computing for the cloud environment, IEEE Trans. Serv. Comput. 15 (4) (2022) 2289–2300, https://doi.org/10.1109/TSC.2020.3046471.

[16] R. Deaton, DNA and quantum computers, in: Proceedings of the 3rd Annual Conference on Genetic and Evolutionary Computation, 2001, pp. 989–996).

[17] R. Deaton, M. Garzon, Thermodynamic constraints on DNA-based computing 1 introduction 2 thermodynamic and chemical constraints, in: Computing With Biomolecules, 1998, pp. 138–152. http://aleteya.cs.buap.mx/~jlavalle/papers/dnacomputing/dna_thermo.pdf.

[18] R. Deaton, A. Suyama, DNA Computing and Molecular Programming, in: Proceedings of the 5th International Conference on DNA, DNA, 2009, pp. 8–11.

[19] R. Barua, J. Misra, Binary arithmetic for DNA computers, in: Lect. Notes Comput. Sci. (including Subser. Lect. Notes Artif. Intell. Lect. Notes Bioinformatics), vol. 2568 (13) 2003, pp. 124–132, https://doi.org/10.1007/3-540-36440-4_11.

[20] M.S. Ullah, W. Aslam, H. Nazir, DNA computing: a review of promises and potential, J. Inf. Commun. Technol. Robot. Appl. 2019 (2019) 93–101.

[21] D.I. Walsh, et al., Standardizing automated DNA assembly: best practices, metrics, and protocols using robots, SLAS Technol. 24 (3) (2019) 282–290, https://doi.org/10.1177/2472630318825335.

[22] M. Gabig, G. Węgrzyn, An introduction to DNA chips: principles, technology, applications and analysis, Acta Biochim. Pol. 48 (3) (2001) 615–622, https://doi.org/10.18388/abp.2001_3896.

[23] D. Saboulard, et al., High-throughput site-directed mutagenesis using oligonucleotides synthesized on DNA chips, Biotechniques 39 (3) (2005) 363–367, https://doi.org/10.2144/05393ST04.

[24] S.L. Pomeroy, et al., Prediction of central nervous system embryonal tumour outcome based on gene expression, Nature 415 (6870) (2002) 436–442, https://doi.org/10.1038/415436a.

[25] C. Baek, Molecular AI with DNA Computing-Molecular Machine Learning for in vitro Pattern Classification (Doctoral dissertation, Graduate School of Seoul National University), 2020.

[26] D.H. Wood, H. Bi, S.O. Kimbrough, D.J. Wu, J. Chen, DNA starts to learn poker, Lect. Notes Comput. Sci. (including Subser. Lect. Notes Artif. Intell. Lect. Notes Bioinformatics) 2340 (June) (2002) 92–103, https://doi.org/10.1007/3-540-48017-x_9.

[27] L. Ren, Y. Ding, H. Ying, S. Shao, Emergence of self-learning fuzzy systems by a new virus DNA-based evolutionary algorithm, Int. J. Intell. Syst. 18 (3) (2003) 339–354.

[28] N.C. Seeman, DNA nanotechnology: novel DNA constructions, Annu. Rev. Biophys. Biomol. Struct. 27 (1998) 225–248, https://doi.org/10.1146/annurev.biophys.27.1.225.

[29] N.C. Seeman, H.F. Sleiman, DNA nanotechnology, Nat. Rev. Mater. 3 (2017), https://doi.org/10.1038/natrevmats.2017.68.

[30] S. Kogikoski, W.J. Paschoalino, L. Cantelli, W. Silva, L.T. Kubota, Electrochemical sensing based on DNA nanotechnology, Trends Anal. Chem. 118 (July) (2019) 597–605, https://doi.org/10.1016/j.trac.2019.06.021.

[31] A.V. Pinheiro, D. Han, W.M. Shih, H. Yan, Challenges and opportunities for structural DNA nanotechnology, Nat. Nanotechnol. 6 (12) (2011) 763–772, https://doi.org/10.1038/nnano.2011.187.

[32] T. Chen, et al., DNA nanotechnology for cancer diagnosis and therapy, Int. J. Mol. Sci. 19 (6) (2018), https://doi.org/10.3390/ijms19061671.

[33] J.M. Rizzo, M.J. Buck, Key principles and clinical applications of 'next-generation' DNA sequencing, Cancer Prev. Res. 5 (7) (2012) 887–900, https://doi.org/10.1158/1940-6207.CAPR-11-0432.

[34] A. Grada, K. Weinbrecht, Next-generation sequencing: methodology and application, J. Invest. Dermatol. 133 (8) (2013) 1–4, https://doi.org/10.1038/jid.2013.248.

[35] R. Chandra, Towards an affective computational model for machine consciousness, Lect. Notes Comput. Sci. (including Subser. Lect. Notes Artif. Intell. Lect. Notes Bioinformatics) 10638 LNCS (2017) 897–907, https://doi.org/10.1007/978-3-319-70139-4_91.

[36] L.M. Adleman, Adleman1994, Science (80-.) 266 (1994) 1021–1024.

[37] C. Guangzhao, Q. Limin, W. Yanfeng, Z. Xuncai, Information security technology based on DNA computing, in: 2007 IEEE Int. Work. Anti-counterfeiting, Secur. Identification, ASID, vol. 1, 2007, pp. 288–291, https://doi.org/10.1109/IWASID.2007.373746.

[38] X. Wang, Q. Zhang, DNA computing-based cryptography, in: *BIC-TA 2009 - Proceedings, 2009 4th Int. Conf. Bio-Inspired Comput. Theor. Appl*, 2009, pp. 67–69, https://doi.org/10.1109/BICTA.2009.5338153.

[39] S. Namasudra, An improved attribute-based encryption technique towards the data security in cloud computing, Concurr. Comput. Pract. Exp. (2017) 1–15, https://doi.org/10.1002/cpe.4364.

[40] V.I. Risca, Dna-based steganography, Cryptologia 25 (1) (2001) 37–49, https://doi.org/10.1080/0161-110191889761.

[41] G. Cui, C. Li, H. Li, X. Li, DNA computing and its application to information security field, in: 5th Int. Conf. Nat. Comput. ICNC 2009, vol. 6(C), 2009, pp. 148–152, https://doi.org/10.1109/ICNC.2009.27.

[42] P. Vijayakumar, V. Vijayalakshmi, G. Zayaraz, An improved level of security for DNA steganography using Hyperelliptic curve cryptography, Wirel. Pers. Commun. 89 (4) (2016) 1221–1242, https://doi.org/10.1007/s11277-016-3313-x.

[43] S. Nakamoto, Bitcoin: A Peer-to-Peer Electronic Cash System, 2009. https://scholar.google.co.in/scholar_url?url=https://www.debr.io/article/21260.pdf&hl=en&sa=X&ei=6WiHY5elEeqK6rQP7ZGx8A4&scisig=AAGBfm1a970bM4TlB3AM3jhGG-EWZpGFyg&oi=scholarr.

[44] S. Namasudra, G.C. Deka, Applications of Blockchain in Healthcare, Springer, 2020.

[45] A.M. El-Edkawy, M.A. El-Dosuky, T. Hamza, DNA based network model and blockchain, arXiv (2019). https://arxiv.org/abs/1908.07829.

[46] S. Namasudra, R. Chakraborty, A. Majumder, N.R. Moparthi, Securing multimedia by using DNA-based encryption in the cloud computing environment, ACM Trans. Multimed. Comput. Commun. Appl. 16 (3s) (2020) 1–19, https://doi.org/10.1145/3392665.

[47] L. Ceze, J. Nivala, K. Strauss, Molecular digital data storage using DNA, Nat. Rev. Genet. 20 (8) (2019) 456–466, https://doi.org/10.1038/s41576-019-0125-3.

[48] S. Shah, D. Limbachiya, M.K. Gupta, DNACloud: A Potential Tool for storing Big Data on DNA, 2013. [Online]. Available: http://arxiv.org/abs/1310.6992.

[49] S. Yan, K.C. Wong, Future DNA computing device and accompanied tool stack: towards high-throughput computation, Future Gener. Comput. Syst. 117 (2021) 111–124, https://doi.org/10.1016/j.future.2020.10.038.

About the authors

Dr. Sachin Minocha is working as an Assistant professor in School of Computing Science and Engineering, Galgotias University. He has received his Ph.D. degree from Department of Computer Science and Engineering, Sant Longowal Institute of Engineering and Technology. Dr. Minocha has 5 patents, 10 publications in conference proceedings, book chapters, and refereed journals like CAEE, JARS, Expert Systems, CIBM. His research area is Machine Learning, Nature Inspired Techniques, Hyperspectral Images, and DNA computing.

Dr. Suyel Namasudra is an assistant professor in the Department of Computer Science and Engineering at the National Institute of Technology Agartala, Tripura, India. Before joining the National Institute of Technology Agartala, Dr. Namasudra was an assistant professor in the Department of Computer Science and Engineering at the National Institute of Technology Patna, Bihar, India, and a post-doctorate fellow at the International University of La Rioja (UNIR), Spain. He has received Ph.D. degree in Computer Science and Engineering from the National Institute of Technology Silchar, Assam, India. His research interests include blockchain technology, cloud computing, IoT, and DNA computing. Dr. Namasudra has edited 4 books, 5 patents, and 60 publications in conference proceedings, book chapters, and refereed journals like IEEE TII, IEEE T-ITS, IEEE TSC, IEEE TCSS, ACM TOMM, ACM TALLIP, FGCS, CAEE, and many more. He has served as a Lead Guest Editor/ Guest Editor in many reputed journals like ACM TOMM (ACM, IF: 3.144), CAEE (Elsevier, IF: 3.818), CAIS (Springer, IF: 4.927), CMC (Tech Science Press, IF: 3.772), Sensors (MDPI, IF: 3.576), and many more. Dr. Namasudra has participated in many international conferences as an Organizer and Session Chair. He is a member of IEEE, ACM, and IEI. Dr. Namasudra has been featured in the list of the top 2% scientists in the world in 2021 and 2022, and his h-index is 25.

Printed in the United States
by Baker & Taylor Publisher Services